BASIC CHEMISTRY

BASIC CHEMISTRY

Glenn H. Miller
with the assistance of

Frederick B. Augustine
Chabot College

Canfield Press
San Francisco
A Department of Harper & Row, Publishers, Inc.
New York Evanston London

Basic Chemistry

 For information address Harper & Row, Publishers, Inc., 10 East 53rd St., New York, N.Y. 10022.

75 76 77 10 9 8 7 6 5 4 3 2 1

Library of Congress Cataloging in Publication Data

Miller, Glenn H. 1922–
Basic chemistry.

Includes index.
1. Chemistry. I. Augustine, Frederick B. II. Title.
QD31.2.M53 540 75-1408
ISBN 0-06-383625-4

Cover design by Robert Conover

To

M. A. M.

L. E. M.

M. A. M.

J. A. M.

preface

Today, students with a minimal background in mathematics and little or no previous exposure to chemistry have an opportunity to begin a study of chemistry at the college level. Since a large group of students are becoming interested in the health sciences, environmental studies, and other professional fields but find that they did not receive adequate preparation in science during their high school career, it is primarily for this group of students that this text was written. For those students who plan to major in the physical or biological sciences, this test will adequately prepare them for the major, full-year college chemistry sequence.

This text makes few assumptions about the mathematical skills of the student. Numerous example problems and their step-by-step solutions are given throughout every chapter. The factor (or conversion analysis) method of problem solving is emphasized and used throughout the text. An average of thirty-five graded problems are given at the end of each chapter that are similar to the worked examples. The Answers to Problem Sets A are given in the Appendix; the answers to Problem Sets B are reserved for use by the instructor and are given in the Instructor's Manual.

There are several study aids that will help students to evaluate their progress through the material in the text. Each chapter is introduced with a list of objectives. With these in mind, the first reading of the chapter should be for the purpose of obtaining an overall picture of the material. Before studying the chapter in detail, students should first check to make sure they understand the important new terms; these are emphasized by the use of boldface type and are summarized in the glossary at the end of each chapter. After a second reading of the chapter, students should again refer to the list of objectives to see if they now fully comprehend the topics discussed. Finally, the problem sets provide a comprehensive review of the subject.

One unique feature of the text is the programmed sequences contained in the Appendix for self-testing. These are designed so that a student can check his or her progress and understanding of many of the basic, more important, and often more difficult concepts. The first sequence, for example, consists of a review of basic conversion manipulations. These sequences are referred to in the body of the text, and students are directed to them at the proper time in their study. Some of the topics covered are: significant figures, negative numbers, exponential notation, percent composition, the mole concept, stoichiometry, and equation balancing.

Most instructors will find that the text contains more material than can be covered in a one-semester course. However, the book is arranged so that portions can be omitted at the instructor's discretion without loss of continuity. For example, chapters 17 and 18—kinetic theory, equilibrium, and organic chemistry—are commonly omitted if the class is primarily being taught as a preparatory course.

Only portions of the SI system of units have been adopted. For example, in the interest of clarity for beginning students, the older system of using a degree sign

(°K rather than K) and one-digit before the decimal point in exponential notation (2.00×10^{-4} rather than the recommended 20.0×10^{-3}) has been retained.

The mole concept is emphasized throughout the text. It is introduced early in the discussion of atoms and elements, and is then reintroduced in the next chapter when molecules are discussed. This double exposure gives students a chance to understand the significance of the mole as it relates to atoms before they encounter formulas and molecules.

The periodic table is also introduced early and is used throughout the text. The first discussion occurs in Chapter 3 where it is used as a method for classifying the elements. It is again used as an aid to formula writing in Chapter 6. The main presentation, however, occurs in Chapter 12 following the treatment of the electronic structure of atoms.

Students are taught how to predict chemical formulas from oxidation numbers in Chapter 6 and this is followed by the rules for basic inorganic nomenclature. Grouping all the inorganic nomenclature together provides the student with a convenient reference section.

The kinetic–molecular theory and the gas laws are introduced relatively early in Chapter 7 so that the students can be prepared for their laboratory work. The kinetic-molecular concepts are then utilized in Chapter 8 for the discussion of the liquid and solid states of matter.

The chemistry of sulfuric acid is used to introduce stoichiometry in Chapter 9. This format gives the instructor the opportunity to present some important descriptive chemistry and at the same time to minimize the burden of the mathematical manipulations which usually hinder an understanding of stoichiometry.

Two chapters are devoted to atomic structure. Beginning with a discussion of radioactivity and proof for the existence of subatomic particles, the important concepts of nuclear structure are presented. Electronic structure is introduced by means of atomic spectra and quantization. Quantum numbers are discussed, but only in just enough detail to enable students to accomplish the main goal of making the proper electron assignments to atoms.

The chapter on chemical bonding contains, in addition to the basic presentation of covalent and electrovalent bonding, a section on the four principal types of bonding that exist in solids. Also, in order to include some elementary orbital theory, the structure of the tetrahedral carbon atom and sp^3 hybridization is presented. Here, as in many other sections of the text, organic molecules are introduced so that the student gets some appreciation of organic chemistry and not the impression that this topic is to be left to the last chapter and that it is therefore of lesser importance.

Only molar concentrations are emphasized in Chapter 14 where solutions are discussed. For reference purposes, other methods for expressing concentration such as parts per million, molal, and volume and mass percent are briefly defined, but the problems at the end of the chapter are concerned only with molarity.

Two methods for balancing oxidation-reduction equations are given in Chapter 16. The instructor may elect to teach both, one, or even to omit the topic according to the nature of the course.

The concluding chapter on organic chemistry is a brief survey. Proceeding from hydrocarbons and functional groups the discussion moves to polymers—nylon—and concludes with molecules of biological importance, such as proteins.

An Instructor's Manual is available from the publisher on request by instructors.

The author welcomes comments, criticisms, and suggestions from instructors and students who use this text. I personally wish to thank those who have been so helpful in the preparation of the manuscript: my wife Mary; Frederick B. Augustine, Chabot College, for his assistance in writing and assembling the book; Jack Healey, Chabot College, for his reviews of several stages of manuscript; Leo Kallan; El Camino College, for the many valuable suggestions in his reviews, and for his contribution of most of the programmed sequences; and Peter Scott, Linn-Benton Community College, for his reviews. Naturally, any errors or flaws in the book must be considered the responsibility of the author only.

Glenn H. Miller
Monrovia, Liberia
February 1975

contents

preface vii

1 **introduction to chemistry** 1
HOW TO USE THIS BOOK
WHAT IS CHEMISTRY?
THE SCIENTIFIC METHOD
THE BRANCHES OF CHEMISTRY

2 **the metric system and scientific measurements** 5
MASS AND WEIGHT
THE METRIC SYSTEM
METRIC CONVERSIONS
THE ENGLISH SYSTEM
CONVERTING FROM METRIC TO ENGLISH
DENSITY
ACCURACY OF MEASUREMENTS AND SIGNIFICANT FIGURES
EXPONENTIAL NOTATION
HEAT AND TEMPERATURE
MEASURING MASS

3 **elements, atoms, and moles** 26
STATES OF MATTER: SOLID, LIQUID, GAS
PROPERTIES OF MATTER: PHYSICAL AND CHEMICAL
CHANGES OF MATTER: PHYSICAL AND CHEMICAL
FORMS OF MATTER: HETEROGENEOUS AND HOMOGENEOUS
ATOMS, MOLECULES, AND COMPOUNDS
ELEMENTS
NAMES AND SYMBOLS FOR ELEMENTS
THE RELATIVE MASSES OF ATOMS
THE MASS SPECTROGRAPH
ISOTOPES
THE MOLE
AVOGADRO'S NUMBER
THE SIZE OF ATOMS
THE CLASSIFICATION OF ATOMS: THE PERIODIC TABLE

4 **molecules and moles** 44
LAW OF DEFINITE COMPOSITION
LAW OF MULTIPLE PROPORTIONS
LAW OF CONSERVATION OF MASS-ENERGY
THE MOLE CONCEPT APPLIED TO MOLECULES AND COMPOUNDS
THE SIZE OF MOLECULES

5 **chemical formulas and equations** 57
DETERMINING EMPIRICAL FORMULAS
DETERMINING MOLECULAR FORMULAS
CALCULATING PERCENT COMPOSITION FROM FORMULAS
WRITING AND BALANCING CHEMICAL EQUATIONS
THE TYPES OF CHEMICAL EQUATIONS

6 **how to write formulas and name compounds** 67
PREDICTING FORMULAS FOR COMPOUNDS
OXIDATION NUMBERS
NAMING COMPOUNDS
CONCLUSIONS

7 **gases and the kinetic-molecular theory** 82
THE KINETIC-MOLECULAR THEORY
TEMPERATURE
MOLECULAR MOTION
PRESSURE
STANDARD CONDITIONS OF TEMPERATURE AND PRESSURE
BOYLE'S LAW
DALTON'S LAW
CHARLES'S LAW
THE ABSOLUTE TEMPERATURE SCALE
BOYLE'S AND CHARLES'S LAWS COMBINED
AVOGADRO'S LAW
IDEAL GASES
THE EQUATION OF STATE FOR AN IDEAL GAS
MOLAR GAS VOLUME

8 **liquids, solids, and the kinetic-molecular theory** 103
CRYSTALLINE SOLIDS
DIFFUSION
SURFACE TENSION
EVAPORATION AND SUBLIMATION
VAPOR PRESSURE
BOILING POINT

ENERGY AND CHANGES OF STATE
COOLING BY EVAPORATION

9 **stoichiometry: calculations from chemical equations** 118
SULFURIC ACID
WHAT CHEMICAL EQUATIONS TELL US
THE MOLE-FACTOR CALCULATION METHOD
THE CONTACT PROCESS FOR SULFURIC ACID
THE PROPERTIES OF SULFURIC ACID
ADDITIONAL EXAMPLES OF STOICHIOMETRY PROBLEMS
LIMITING-REAGENT PROBLEMS
PERCENT YIELD
PROBLEMS INVOLVING GAS VOLUMES
THERMOCHEMICAL EQUATIONS

10 **subatomic particles and the nucleus** 137
THE DISCOVERY OF RADIOACTIVITY AND SUBATOMIC PARTICLES
ELECTRICAL PHENOMENA
ELECTRONS
THE PROTON
THE NATURE OF RADIOACTIVITY
THE NUCLEUS: RUTHERFORD'S EXPERIMENT
ATOMIC NUMBER
NUCLEAR STRUCTURE

11 **electronic structure of atoms** 149
ELECTROMAGNETIC RADIATION
SPECTROSCOPY
THE BOHR THEORY OF THE ATOM
HOW BOHR'S THEORY EXPLAINS ATOMIC SPECTRA
ORIGIN OF THE FOUR QUANTUM NUMBERS
THE WAVE-THEORY APPROACH TO THE ELECTRONIC STRUCTURE OF ATOMS
WAVE-MECHANICAL MODEL OF THE ATOM
A DESCRIPTION OF THE QUANTUM NUMBERS
THE EXCLUSION PRINCIPLE
ELECTRON ASSIGNMENTS IN ATOMS
A PHYSICAL PICTURE OF THE ATOM

12 **the periodic table** 167
THE PERIODIC LAW
THE BASIS FOR THE PERIODIC ARRANGEMENT
TRANSITION ELEMENTS

TRANSURANIUM ELEMENTS
IONIZATION POTENTIALS
PERIODIC RELATIONSHIPS IN OTHER PHYSICAL PROPERTIES
PREDICTING THE PROPERTIES OF NEW ELEMENTS
NOBLE GASES
ALKALI METALS
HALOGENS

13 chemical bonds **182**
THE OCTET RULE
ELECTROVALENT BONDS
COVALENT BONDS
ELECTRON SPIN AND BONDING
LEWIS STRUCTURES
THE CARBON ATOM AND THE STRUCTURE OF METHANE
MULTIPLE BONDS
RESONANCE
ELECTRONEGATIVITY
POLAR MOLECULES
THE MOLECULAR STRUCTURE OF H_2O
HYDROGEN BONDING
TYPES OF BONDING IN SOLIDS

14 solutions **199**
TYPES OF SOLUTIONS
ELECTROLYTES AND NONELECTROLYTES
WHY DO COMPOUNDS DISSOLVE?
SOLUBILITY AND THE RATE OF SOLUTION
FACTORS AFFECTING SOLUBILITY
THE CONCENTRATION OF SOLUTIONS
MOLARITY
DIFFERENT UNITS OF CONCENTRATION
THE COLLIGATIVE PROPERTIES OF SOLUTIONS
IDEAL SOLUTIONS

15 solutions of electrolytes, acids, and bases **220**
ELECTROLYTIC SOLUTIONS
THE ARRHENIUS THEORY
THE PROPERTIES OF SOLUTIONS EXPLAINED BY THE ARRHENIUS THEORY
WRITING IONIC EQUATIONS
ACIDS AND BASES
THE BRØNSTED-LOWRY DEFINITION OF ACIDS AND BASES

HYDROLYSIS
THE STRENGTHS OF ACIDS AND BASES
ACID-BASE REACTIONS
TITRATION

16 oxidation-reduction reactions and electrochemistry 243
WHAT ARE OXIDATION-REDUCTION REACTIONS?
THE REDOX TERMINOLOGY
BALANCING REDOX EQUATIONS
VOLTAIC CELLS
ELECTRICAL UNITS
OXIDATION POTENTIALS
DISPLACEMENT REACTIONS: WHICH DIRECTION WILL THEY PROCEED?
THE LEAD STORAGE BATTERY

17 reaction rates and chemical equilibrium 263
RATES OF CHEMICAL REACTIONS
THE NATURE OF REACTANTS
THE LAW OF MASS ACTION: THE EFFECT OF CONCENTRATION ON REACTION RATES
TEMPERATURE AND REACTION RATE
CATALYSIS
CHEMICAL EQUILIBRIA
EQUILIBRIUM CONSTANTS
THE EFFECT OF CONCENTRATION ON CHEMICAL EQUILIBRIUM
LE CHÂTELIER'S PRINCIPLE
THE EFFECT OF TEMPERATURE AND CATALYSTS ON CHEMICAL EQUILIBRIUM
AMMONIA SYNTHESIS: AN EXAMPLE OF RATE AND EQUILIBRIUM CONSIDERATIONS
IONIC EQUILIBRIA
COMMON ION EFFECT
SOLUBILITY PRODUCT

18 introduction to organic chemistry 281
HOW ORGANIC CHEMISTRY IS ORGANIZED
HYDROCARBONS
ALKANES
STRUCTURAL ISOMERISM
THE NOMENCLATURE FOR ALKANES
ALKANES AND ALKYNES
BENZENE
SOME CHEMICAL PROPERTIES OF HYDROCARBONS
GIANT MOLECULES

POLYETHENE (POLYETHYLENE), AN IMPORTANT ADDITION POLYMER
CELLULOSE
DACRON, A POLYESTER FIBER
PROTEINS

appendix A: programmed sequences 301

INSTRUCTIONS FOR USING PROGRAMMED SEQUENCES
1 CONVERSION OPERATIONS
2 UNITS AND DIMENSIONS
3 SCIENTIFIC MEASUREMENT AND SIGNIFICANT FIGURES
4 THE MOLE CONCEPT
5 EMPIRICAL FORMULAS AND PERCENT COMPOSITION
6 NEGATIVE NUMBERS
7 STOICHIOMETRY
8 MOLARITY
9 NET IONIC EQUATIONS

appendix B: answers to problems 335

index 351

1

introduction to chemistry

THIS WILL SAVE YOU TIME:

Refer to this list as you read through the chapter. In answering the questions, you will understand the important concepts you are most likely to be asked about later on.

After reading this chapter you should be able to:

1. *Describe the subject matter of chemistry and its relationship to the other sciences.*
2. *Explain the terms* law, hypothesis, *and* theory.
3. *List six branches of chemistry.*

In the minds of many students, the word *chemistry* calls up visions of a difficult and boring science conducted in laboratories filled with chemicals ready to burst into flames or explode in your face. Such a picture is hardly accurate. Like all sciences, chemistry can be complicated, but we have purposely limited this book to the important and basic concepts of the field. The mathematics has been kept simple, and all examples are worked out step by step to show you precisely what is happening. Nor is chemistry boring. We live in the golden age of chemistry, surrounded by its products—fabrics, dyes, paints, plastics, paper, glass, metals. Likewise, chemistry can tell us a great deal about such modern troubles as disposing of indestructible waste, limiting pollution, and searching for new energy resources. As for the dangers of the chemistry laboratory, they are greatly overrated. Improper handling of household cleaning products is far more dangerous than working in a lab. In short, chemistry can be understandable, intriguing, and safe.

HOW TO USE THIS BOOK

There are five ways to use this book that will make learning chemistry easier:

1. Use the list of concepts and objectives at the beginning of each chapter—the ones listed under THIS WILL SAVE YOU TIME. As you go through the chapter, check off the list. This procedure will save you time by pointing out what is important.
2. Read the chapter. In fact, read it twice. People who teach speed reading say that the best way to approach new material is to "preread"—read the chapter quickly to grasp the main ideas, topic sentences, headings, illustrations, and so on—then reread more slowly to absorb the details. If the chapter still seems difficult, go back over it a third time. To aid your reading, important concepts are indicated in **boldface type,** and their definitions are given in *italics.* The terms and definitions are repeated in the NEW TERMS list at the end of the chapter.
3. Look at the NEW TERMS list. It will reinforce in your mind what you have just finished reading. Many students are apt to think of the NEW TERMS list as optional, something to skip over in their hurry to get on to the next chapter. That is a mistake; use the list to help yourself. You can also refer back to the NEW TERMS later if you forget the meaning of a word.
4. Use the PROGRAMMED SEQUENCES found in Appendix A. These are self-teaching devices designed to help you over troublesome areas. All answers are provided, there are no trick questions, and you can work at your own speed. If you have had difficulty with mathematics in the past or have forgotten parts of it, you will find the programmed sequences particularly helpful.
5. Finally, work the PROBLEM SETS at the end of each chapter. Doing problems is the most difficult task, but it solidifies learning as nothing else can. There are two sets of problems, A and B. Answers for Problem Set A are provided in the back of the book so that you can check yourself. Your instructor has the answers for Problem Set B in his teacher's manual.

Before getting into the details of chemistry, we should define the field and consider the meaning of science and the scientific method.

WHAT IS CHEMISTRY?

Chemistry is the science that describes how matter—that is, the observable materials that make up the physical universe—is structured, what it is composed of, how it is changed, and the reasons for the change.

Science, in turn, can be defined as systematized knowledge. Chemistry, then, is systematized knowledge of things in our universe.

Chemistry overlaps other branches of science, such as physics and biology. Physics deals with matter and motion, the phenomena of matter that involve no change in chemical composition. Biology's interest in the nature of life involves studying the chemical processes of living organisms. Other areas also overlap chemistry: medicine, agriculture, oceanography, engineering, and mathematics, for example.

THE SCIENTIFIC METHOD

All sciences follow the **scientific method**—the accepted process for organizing and using knowledge to predict and explain nature. Ideally, the scientific method entails a series of steps leading from observation to theory.

Observation: Observing water in an uncovered dish could prompt you to ask, Why does water evaporate? Do other liquids evaporate? Why does water evaporate more rapidly when it is heated?

Experimentation: You might then experiment in order to get more information. It is experimentation that distinguishes the natural sciences from the social sciences.

Law: After many experiments, you conclude that all liquids behave the same way: they all evaporate, they evaporate more rapidly when heated, and with enough heat they reach an unchanging boiling temperature. You have thus discovered some **laws** of nature—*statements about how this particular part of nature behaves*. A law does not attempt to explain why liquids evaporate or boil; it simply says that they always do these things.

Hypothesis: The next step is to explain tentatively why nature behaves as it does. The explanation is called an hypothesis, an unproved assumption based on the facts that you gathered in your experiments. To account for the evaporation of liquids, you might hypothesize that small particles of the liquid are in constant motion and that some of these particles have more energy than the others and so are able to escape from the liquid to become a gas.

Theory: If further experiments show this hypothesis to be consistent with the behavior of all known liquids, it is then elevated to the level of a theory, the most important status in the scientific method.

A theory must hold true for all known cases; it can allow no exceptions. If an experiment produces results that are not consistent with the theory, then the theory must be discarded or amended. A theory is often only a model that describes submicroscopic phenomena in terms of the more familiar larger world. It may not claim to be the whole truth, but even so, it is useful to scientists. Not only does a theory organize our knowledge about things in nature, but it also permits us to speculate and predict how nature should behave. Our hypothesis about evaporation of liquids, now elevated to a theory, enables us to explain boiling, freezing, temperature, and a lot of other important observations and phenomena.

THE BRANCHES OF CHEMISTRY

Traditionally, chemistry is divided into several subject areas, many of which overlap with each other a great deal:

1. *General chemistry* covers practically the whole field, but it is particularly concerned with the basic theories underlying chemistry.
2. *Analytical chemistry* has two parts. The first, *qualitative analysis,* is concerned with the chemical components (elements or compounds) of a given piece of material. The second, *quantitative analysis,* is concerned with the exact amount of each component in a piece of material.
3. *Organic chemistry* is the study of chemical components (compounds) containing carbon. In particular, it is concerned with compounds in which carbon is combined with hydrogen, nitrogen, sulfur, and a family of elements called the halogens.
4. *Inorganic chemistry* is concerned with all chemical components (elements and compounds) except most of those containing carbon.
5. *Physical chemistry* is the science of the basic laws of chemistry and of the hypotheses and theories used to explain them.
6. *Biochemistry* is concerned with the chemistry of living things.

Actually, a better scheme for dividing chemistry into its branches is to use important subject areas such as photography, glass, fuels, wood, explosives, paints, detergents, sugars, cement, plastics, and so on. The journal, *Chemical Abstracts,* the key to the world's chemical literature, uses 80 such groupings.

This completes the introduction. In the next chapter, we will describe the measurements of chemistry. At this point, however, we urge you to take a look at the NEW TERMS, to read the instructions to the PROGRAMMED SEQUENCES on p. 301, and to familiarize yourself with the sequences themselves.

NEW TERMS

Chemistry: science concerned with the structure and composition of matter, the changes it undergoes, and the reasons for those changes.

Hypothesis: an unproved assumption, a tentative explanation of what causes a certain behavior.

Law: a statement describing how a particular part of nature behaves.

Science: systematized knowledge.

Scientific method: a process for organizing and using knowledge to predict and explain nature. Its steps are: observation, experiment, law, hypothesis, theory.

Theory: a verified hypothesis; a statement of general principles that explains certain phenomena.

2

the metric system and scientific measurements

THIS WILL SAVE YOU TIME:

Refer to this list as you read through the chapter. In answering the questions, you will understand the important concepts you are most likely to be asked about later on.

After reading this chapter you should be able to:

1. *Name the basic metric units for mass, volume, and length.*
2. *Explain the meanings of the prefixes used for the metric units.*
3. *Convert a measurement from one metric unit to another.*
4. *Explain the relationships between length, volume, and mass units in the metric system.*
5. *Convert measurements from the English system of units to the metric system and vice versa.*
6. *Define the term density and be able to work problems using densities.*
7. *Tell what is meant by accuracy of measurements and be able to state numbers with the proper number of significant figures.*
8. *Write numbers using exponential notation.*
9. *Multiply, divide, add, and subtract numbers written in exponential notation.*
10. *Distinguish between heat and temperature.*
11. *Explain the construction and calibration of a Celsius thermometer.*
12. *Convert degrees Celsius to degrees Fahrenheit and vice versa.*
13. *Distinguish between weight and mass, and tell how a simple balance is used for mass measurements.*

MASS AND WEIGHT

When a spaceship leaves the earth's gravity behind, it does not need power to keep moving. Once in motion, it will move at the same speed until something interferes with it. This phenomenon is due to **inertia,** *the tendency of an object to maintain a constant state of motion.*

Mass is a measure of inertia. More strictly, mass is *a measure of the quantity of matter in a body.* It does not vary from one place to another in the universe. The mass of an astronaut, for example, is the same in outer space, on the moon, and on earth.

Weight, on the other hand, does differ from place to place because it varies with gravity or the lack of it. An astronaut who weighs 180 pounds on earth would weigh only 30 pounds on the moon—where gravity is one-sixth as strong as on earth—and nothing at all in outer space—where there is no gravity to speak of. Thus weight is *a measure of the force of gravitational attraction between objects.*

Since chemists are usually restricted to earthbound laboratories, the distinction between mass and weight is generally overlooked, and the weight is commonly used to denote mass. However, in this book, we will use the proper term mass.

THE METRIC SYSTEM

The international standard for measurements of length is the meter. The **meter** is a little longer than a yard, 39.37 inches.

For volume, the international standard is the volume of a perfectly square box, or a perfect cube, that has inside edges exactly 0.1 of a meter long. The unit is called the **liter,** and it equals 1.06 quarts.

For mass measurements, the international standard is a metal cylinder made of platinum and iridium to which the value of 1 **kilogram** has been assigned. The mass of this standard equals the mass of 1 liter of water at 3.98° Celsius (the temperature at which water has its greatest density). A kilogram equals 2.20 pounds.

The metric system is a decimal system in which prefixes are used to designate fractions or multiples of the standard unit. These prefixes are italicized in Table 2–1, which also lists the commonly used metric units and their abbreviations.

TABLE 2–1 METRIC UNITS

Length	*milli*meter	mm =	0.001	m
	*centi*meter	cm =	0.01	m
	*deci*meter	dm =	0.1	m
	meter	m =	1.0	m
	*kilo*meter	km =	1000.	m
Volume	*milli*liter	ml =	0.001	liter
	liter	l =	1.0	liter
Mass	*micro*gram	μg =	0.000001	g
	*milli*gram	mg =	0.001	g
	gram	g =	1.0	g
	*kilo*gram	kg =	1000.	g

One of the greatest virtues of the metric system, in addition to the ease of doing calculations in a decimal system, is that the units of mass, volume, and length are related. A box 0.1 m square holds 1 liter of water that has a mass of 1 kg (when temperature = 3.98°C). Because 0.1 m equals 10 cm, the volume of the perfect cube, which holds 1 liter, can also be expressed as 1000 cubic centimeters. The commonly used volume unit, the cubic centimeter, cm^3 or cc, equals 0.001 liter or 1 ml.

PROGRAMMED SEQUENCE:

Turn now to p. 302, and work through Set 1 of the Programmed Sequences. These problems review basic conversion operations. If you have had difficulty with mathematics in the past, these sequences will give you needed practice. Even if you feel confident about your mathematical ability, the sequences will serve as a self-test to point out anything you may have forgotten.

METRIC CONVERSIONS

It is very easy to convert from one unit in the metric system to another. The following group of examples will illustrate the steps involved.

Example 2–1: Converting length units *How many cm are in 25.0 m?*

Two methods can be used to solve this problem. We will consider the conversion-factor method first. We know that 100 cm = 1.00 m. (Most all of the problems in this text will employ numbers with three significant figures; see page 13 for a discussion of significant figures. Therefore, rather than writing 100 cm = 1 m, the notation 100 cm = 1.00 m is usually used.) If we divide both sides of this equation, 100 cm = 1.00 m, by 1.00 m, the value of the equality remains unchanged:

$$\frac{100\text{ cm}}{1.00\text{ m}} = \frac{1.00\text{ m}}{1.00\text{ m}} = 1.00 \quad \left(\text{Note that } \frac{100\text{ cm}}{1.00\text{ m}} \text{ means } 100\text{ cm} \div 1.00\text{ m.}\right)$$

The relationship 100 cm/1.00 m is called a **conversion factor.** Since it is equal to one, we can multiply any number by it and change only the units, not the value. For example, when the conversion factor is applied to the 25.0 m of our example, the distance is not changed, just the means of expressing it:

$$25.0\ \cancel{\text{m}} \times \frac{100\text{ cm}}{1.00\ \cancel{\text{m}}} = 25.0 \times 100\text{ cm} = 2500\text{ cm}$$

Note the cancellation of the units. Note also that we might have divided both sides of our original equation, 100 cm = 1.00 m, by 100 cm to obtain a second conversion factor:

$$\frac{100\text{ cm}}{100\text{ cm}} = \frac{1.00\text{ m}}{100\text{ cm}} = 1.00$$

Using this conversion factor, however, would not allow the units to be canceled, and a meaningless answer would result:

$$25.0\text{ m} \times \frac{1.00\text{ m}}{100\text{ cm}} = \frac{25.0\text{ m}^2}{100\text{ cm}} \quad (\text{m}^2 = \text{m} \times \text{m})$$

We choose the conversion factor that allows cancellation of old units and conversion to the desired new units.

The second method for solving the problem is simply to move the decimal point. There are 100 cm in 1.00 m; thus to change from meters to centimeters or from centimeters to meters, all you have to do is move the decimal point two places. You must, however, decide which direction the decimal point should be moved. This problem concerns a certain fixed distance. Measured in terms of a larger size unit (the meter), it will have a certain numerical value (25.0 m). Measured in terms of a smaller size unit (the centimeter), it will have a larger value (greater than 25.0 m). Or, put another way, it takes more of the smaller units to cover the same distance. Therefore, the decimal point must be moved to the right.

$$25.00\ \text{m} = 2500\ \text{cm}$$

This method, which might be called the intuitive approach, is most useful as a way of checking the reasonableness of the answer gotten by the conversion-factor method.

Example 2–2: Converting length units *How many cm are in 5.00 km?*

First, you must have memorized from Table 2–1 the equalities 1.000 km = 1000 m and 1.00 m = 100 cm. Four conversion factors can be written from these two equations, each of which equals one. They are:

$$\frac{1.000\ \text{km}}{1000\ \text{m}} \qquad \frac{1000\ \text{m}}{1.000\ \text{km}} \qquad \frac{1.00\ \text{m}}{100\ \text{cm}} \qquad \frac{100\ \text{cm}}{1.00\ \text{m}}$$

To solve the problem, we choose the factors that will allow the conversion of km to m and then m to cm.

$$5.00\ \cancel{\text{km}} \times \frac{1000\ \cancel{\text{m}}}{1.000\ \cancel{\text{km}}} \times \frac{100\ \text{cm}}{1.00\ \cancel{\text{m}}} = 500{,}000\ \text{cm}$$

In multistep problems such as this, first set up all the necessary conversion factors, one after the other, and then cancel units as you progress. When you arrive at the desired units, then—and only then—perform the mathematical manipulations on the numbers.

Does the answer 5.00 km = 500,000 cm seem reasonable? It does, because the centimeter is a much smaller unit than the kilometer and it takes more centimeters than kilometers to describe the same distance. Carrying the intuitive approach through indicates a total change of five decimal places (three for km to m, and two for m to cm), and the decimal point is moved to the right to give a larger number:

$$5.000\ 000\ \text{km} = 500{,}000\ \text{cm}$$

Example 2–3: Converting volume units *How many cubic centimeters are contained in three liters?*

We know that 1000 cc = 1.000 liter. Applying this conversion factor gives

$$3.000\ \cancel{\text{liters}} \times \frac{1000\ \text{cc}}{1.000\ \cancel{\text{liter}}} = 3000\ \text{cc}$$

Check your answer using the decimal-point method. The cc, or cubic centimeter, unit is equal to a milliliter; there are 1000 of them per liter. Thus the decimal point is shifted three places to the right:

3,000 liters = 3000. ml

Example 2–4: Converting mass units *How many kg in 30.0 g?*

Answer:

$$30.0\ \cancel{g} \times \frac{1.000\ \text{kg}}{1000\ \cancel{g}} = \frac{30.0\ \text{kg}}{1000} = 0.0300\ \text{kg}$$

Example 2–5: Relationship between length and volume *A square box has inside dimensions of length, width, and height all equal to 1.000 m. What is the volume of the box in liters?*

Since the volume unit cc, or cm^3, equals ml, one convenient method for solving this problem is to find the volume of the box in cm^3 or ml and then convert this value to liters. Since 100.0 cm = 1.000 m, the volume of the box is

$$(100.0\ \text{cm})(100.0\ \text{cm})(100.0\ \text{cm}) = 1{,}000{,}000\ \text{cm}^3 = 1{,}000{,}000\ \text{ml}$$

To convert to liters

$$1{,}000{,}000\ \cancel{\text{ml}} \left[\frac{1.000\ \text{liter}}{1000\ \cancel{\text{ml}}}\right] = 1000\ \text{liters}$$

THE ENGLISH SYSTEM

The English system of mass, length, and volume now used in the United States has no legal material standards like those for the metric system. Instead, its units are defined in terms of metric units. Thus an inch is 0.0254 m, and a foot 12×0.0254 m. The metric system was authorized for use in the United States in 1864 and was made a legal system by the Metric Act of 1886. All scientific professions, except engineering, now use the metric system. Numerous bills proposing the abolition of the English system have been put before the Congress, and it is just a matter of time until the metric system becomes the American way of measurement.

CONVERTING FROM METRIC TO ENGLISH

Table 2–2 lists three conversion factors that must be memorized.

TABLE 2–2 METRIC-ENGLISH CONVERSIONS

1 in. =	2.54 cm (exactly)
1.06 qt =	1.00 liter
1.000 lb =	453.6 g (when pounds are used to signify mass units)

With these factors, you can make any desired change from the English to the metric system or vice versa, provided, of course, that you still remember the relationships between units in the English system. A few of these are listed in Table 2–3 to refresh your memory.

TABLE 2–3 ENGLISH UNITS

Length	12 in = 1 ft
	3 ft = 1 yd
	5280 ft = 1 mile
Volume	2 pt = 1 qt
	4 qt = 1 gal
Mass	16 oz = 1 lb
	2000 lb = 1 ton

Example 2–6: Converting length units, English to metric *How many cm are in 12.0 inches?*

The correct equality is the one relating cm to in.

2.54 cm = 1 inch

The conversion factors are

$$\frac{2.54\text{ cm}}{1.00\text{ in}} \quad\text{or}\quad \frac{1.00\text{ in}}{2.54\text{ cm}}$$

Selecting the proper conversion factor and multiplying gives

$$12.0\,\cancel{\text{in}} \times \frac{2.54\text{ cm}}{1.00\,\cancel{\text{in}}} = 30.5\text{ cm}$$

Example 2–7: Converting length units, metric to English *An olympic-size swimming pool is 50.0 m long. How long is the pool in feet?*

$$50.0\,\cancel{\text{m}} \times \frac{100\,\cancel{\text{cm}}}{1.00\,\cancel{\text{m}}} \times \frac{1.00\,\cancel{\text{in.}}}{2.54\,\cancel{\text{cm}}} \times \frac{1.00\text{ ft}}{12.0\,\cancel{\text{in.}}} = 164\text{ ft}$$

Example 2–8: Converting volume units, English to metric *How many milliliters are contained in 4.00 pt?*

Because this problem is about volume, the correct conversion factor is 1.06 qt = 1.00 liter.

$$4.00\,\cancel{\text{pt}} \times \frac{1.00\,\cancel{\text{qt}}}{2.00\,\cancel{\text{pt}}} \times \frac{1.00\,\cancel{\text{liter}}}{1.06\,\cancel{\text{qt}}} \times \frac{1000\text{ ml}}{1.00\,\cancel{\text{liter}}} = 1887\text{ ml}$$

Always retain the units along with the numbers in equations. Only the desired units should remain after canceling; this test serves as a check on the procedure.

Example 2–9: Converting mass units, metric to English *How many pounds are in 1.00 kg?*

$$1.00\ \cancel{kg} \times \frac{1000\ \cancel{g}}{1.000\ \cancel{kg}} \times \frac{1.00\ lb}{454\ \cancel{g}} = 2.20\ lb$$

Example 2–10: Converting mass units, metric to English *An expensive chemical costs $20.00 for 50.0 g. What is the cost per pound?*

First find the mass in pounds.

$$50.0\ \cancel{g} \times \frac{1.00\ lb}{454\ \cancel{g}} = 0.110\ lb$$

The necessary conversion factor is either

$$\frac{0.110\ lb}{\$20.00} \quad \text{or} \quad \frac{\$20.00}{0.110\ lb}$$

Selecting the correct one and solving, we get

$$\frac{\$20.00}{0.110\ \cancel{lb}} \times 1.00\ \cancel{lb} = \$182$$

A shorter route is

$$\frac{\$20.00}{50.0\ \cancel{g}} \times \frac{454\ \cancel{g}}{1.00\ lb} = \$182/lb$$

The first factor is taken directly from the statement of the problem. $20.00/50.0 g means $20.00 per 50.0 g. The word *per* designates the division sign in a conversion factor.

DENSITY

Density is defined as *the mass of a given volume of a substance.* It is the ratio of mass to volume, that is, mass divided by volume.

$$\text{density} = \frac{\text{mass}}{\text{volume}}$$

Any suitable units may be used, but for solids or liquids, the density is usually expressed in terms of grams per milliliter or cubic centimeter:

$$d = g/ml \quad \text{or} \quad d = g/cc$$

For gases, the customary units are grams per liter:

$$d_{gases} = g/liter$$

The density of water at 3.98°C is 1.000 g/ml. This fact, of course, is a consequence of the standards used for relating length, volume, and mass units in the metric system. The densities of some selected substances are given in Table 2–4.

TABLE 2–4 DENSITIES OF SOME SUBSTANCES

Solids and liquids g/ml at 20°C		Gases g/liter at 0°C 1 atmosphere pressure	
platinum	21.45	chlorine	3.17
gold	19.3	carbon dioxide	1.96
mercury	13.55	oxygen	1.43
lead	11.34	air	1.29
iron	7.86	nitrogen	1.25
aluminum	2.70	neon	0.90
magnesium	1.74		
water	1.00		
ethyl alcohol	0.79		
ether	0.71		

The density of water and all other substances varies with the temperature. Thus, when writing densities it is customary to show the temperature at which the measurements are made. For example,

$$d_{\text{water}}^{3.98^\circ\text{C}} = 1.000 \text{ g/ml}$$

Example 2–11: Finding density *What is the density of a metal alloy, 680 g of which has a volume of 128 cc at 25°C?*

Answer:

$$d_{\text{alloy}}^{25^\circ\text{C}} = \frac{680 \text{ g}}{128 \text{ cc}} = \frac{5.31 \text{ g}}{1.00 \text{ cc}} \quad \text{or} \quad 5.31 \text{ g/cc}$$

Example 2–12: Finding mass, given volume and density *What mass of water will occupy a rectangular container with the inside dimensions 2.00 cm × 3.00 cm × 8.00 cm?*

The volume of the box is given by the product of length times width times height.

$$(2.00 \text{ cm})(3.00 \text{ cm})(8.00 \text{ cm}) = 48.0 \text{ cm}^3 = 48.0 \text{ ml}$$

Because the density of water is 1.00 g/ml, the box would hold

$$48.0 \cancel{\text{ml}} \times 1.00 \frac{\text{g}}{\cancel{\text{ml}}} = 48.0 \text{ g}$$

Example 2–13: Finding mass, given volume and density *Silver has a density of 10.5 g/cc. What is the mass of 200 ml of silver?*

Density = mass/volume   or   mass = density × volume

Substituting,

$$\text{mass} = (10.5\ \text{g}/\cancel{\text{ml}})\ (200\ \cancel{\text{ml}})$$
$$\text{mass} = 2100\ \text{g}$$

Example 2–14: Finding volume, given mass and density *What volume of the alloy of Example 2–11 is needed to give a 100 g sample?*

There are three variables in the equation $d = \text{g/cc}$; given any two, you can solve for the third. In this problem the density and mass are known; hence, by substituting the given values, you can solve for the volume.

$$d = \frac{\text{g}}{\text{cc}}$$

$$5.31\ \frac{\cancel{\text{g}}}{\text{cc}} = \frac{100\ \cancel{\text{g}}}{V}$$

To solve for V, first clear the fractions by multiplying both sides by V cc

$$V\ \cancel{\text{cc}} \times \frac{5.31}{\cancel{\text{cc}}} = \frac{100}{\cancel{V}} \times \cancel{V}\ \text{cc}$$

to give

$$5.31\ V = 100\ \text{cc}$$

Then, dividing both sides by 5.31 gives the answer

$$\frac{\cancel{5.31}\ V}{\cancel{5.31}} = \frac{100\ \text{cc}}{5.31} = 18.8\ \text{cc}$$

PROGRAMMED SEQUENCE:

Turn now to p. 304 and work through Set 2 of the Programmed Sequences. This section tests you on the units of measurement and on conversions from one system to another. The problems will solidify what you have learned and point up any difficulties. Be sure that you understand these operations well before reading further in the text.

ACCURACY OF MEASUREMENTS AND SIGNIFICANT FIGURES

Consider the following example:

Example 2–15: Significant figures *What volume of water is needed to fill a rectangular box with the inside dimensions 122 mm by 34 mm by 11 cm?*

Converting the given data to cm units and multiplying gives

$(12.2 \text{ cm})(3.4 \text{ cm})(11 \text{ cm}) = 456.28 \text{ cm}^3$

Several things are wrong with Example 2–15 and its answer. First, look at the statement of the problem. The box was measured with an instrument, probably a ruler, calibrated in units as small as a mm (two of the values are reported to the nearest mm). This ruler is capable of giving accurate measurements to the nearest cm, but when the measurements are made to the nearest mm, there is probably some uncertainty in the accuracy of the last digit. Scientific measurements are always reported with numbers whose last digits are somewhat in doubt. The correct value of the 12.2 cm or 122 mm reading, for example, is not necessarily 122 mm, but may be somewhere between 121.5 and 122.5 mm. The number 122 contains two digits that are certain and one that has some degree of uncertainty; we say that the number has three significant figures. *The digits of a number that are certain, plus one more digit,* are called **significant figures.** The number 122 mm can also be written to three significant figures as 0.122 m. Zeros used to locate the decimal point are not themselves significant figures.

The number of digits used to record data should always indicate the precision with which the measured quantity is known. One of the measurements of Example 2–15 is recorded to two significant figures as 11 cm. This is not correct. The measuring instrument is capable of accurate measurements to cm and permits a reliable estimate of the length to mm, the smallest unit on the scale; hence, this number should have been recorded as 11.0 cm or 110 mm. The added zero is significant. The actual value is probably between 109.5 and 110.5 mm and not between 10 and 12 cm, as the 11 cm reading would indicate.

The answer given for Example 2–15, 456.28 cm^3, incorrectly indicates an accuracy of five significant figures. The accuracy of the result can never be greater than the accuracy of the least precise measurement in the problem. In this case, it is the measurement of the 34-mm side where the measurement has been taken with a precision of one part in 34—an accuracy of two significant figures. The answer must therefore be rounded off to give a value with the same degree of precision. One correct way to do this is to write 4.6×100 (or 4.6×10^2). Simply rounding the number to 460 is not necessarily correct, for the zero in this case may or may not be a significant figure, and 460 would imply an accuracy of three significant figures. Another correct method is to express the answer in liters, as 0.46 liter.

Example 2–16: Significant figures *Express the following numbers with an accuracy of three significant figures: 395.8, 620.5, 621.5, and 622.5.*

The number 395.8 would round off to 396. The number 620.5 expressed to three significant figures is 620. When a number is halfway between two equally correct answers, in this case 620 and 621, convention dictates choice of the even number. The numbers 621.5 and 622.5 would both be rounded to 622.

When numbers are added or subtracted, the number of significant digits in the answer depends upon the position of the decimal point. The answer can have no more significant digits to the right of the decimal point than are contained in the number with the fewest digits to the right of the decimal point. For example, add the following numbers:

$$
\begin{array}{rl}
39.32 & \text{(two digits to the right of the decimal point)} \\
\underline{+\ \ 0.5432} & \text{(four digits to the right of the decimal point)} \\
39.8632 = 39.86 &
\end{array}
$$

The number must be rounded off to 39.86 so as to have only two digits to the right of the decimal point.

Example 2–17: Significant figures, addition *Add the following numbers and express the answer in the correct number of significant figures: 21.06, 0.00976, and 112.3.*

$$
\begin{array}{r}
21.06 \\
0.00976 \\
\underline{112.3} \\
133.36976
\end{array}
$$

This must be rounded off to 133.4.

Example 2–18: Significant figures, addition *What is the calculated total mass of a beaker containing a glass bead if the beaker is reported to have a mass of 55.35 g, and the bead (weighed on a more sensitive balance), 0.155 g?*

Simple addition yields

$$
\begin{array}{r}
55.35\ \text{g} \\
\underline{0.155\ \text{g}} \\
55.505\ \text{g}
\end{array}
$$

However, uncertainty in the mass of the beaker occurs in the hundredths column—the second decimal place—requiring that the sum also have uncertainty in the same column. Hence the original answer of 55.505 g must be rounded off to 55.50 g.

A multiplication or division calculation that also involves addition or subtraction within one of the factors requires special care. The addition or subtraction must be performed before the number of significant figures for that factor can be established. An example will illustrate this situation.

Example 2–19: Significant figures, subtraction *How many significant figures should the answer to this calculation contain:* $\frac{(12{,}079.0 - 12{,}076.0)}{1.97}$

When the subtraction is performed, it is evident that only two significant figures are permitted:

$$\frac{(12{,}079.0 - 12{,}076.0)}{1.97} = \frac{3.0}{1.97} = 1.52 = 1.5$$

EXPONENTIAL NOTATION

The exponential form of notation, illustrated in Table 2–5, is a convenient way of expressing very large or very small numbers and a good system for expressing the accuracy of any measurement.

TABLE 2–5 EXPONENTIAL NOTATION

1,000,000,000,000	$= 1 \times 10^{12}$
1,000,000,000	$= 1 \times 10^{9}$
1,000,000	$= 1 \times 10^{6}$
1,000	$= 1 \times 10^{3}$
100	$= 1 \times 10^{2}$
10	$= 1 \times 10^{1}$
1	$= 1 \times 10^{0}$
0.1	$= 1 \times 10^{-1}$
0.01	$= 1 \times 10^{-2}$
0.001	$= 1 \times 10^{-3}$
0.000 001	$= 1 \times 10^{-6}$
0.000 000 001	$= 1 \times 10^{-9}$
0.000 000 000 001	$= 1 \times 10^{-12}$

Note that with exponential notation, it is customary to use only one digit before the decimal point. For example, the numbers 4.56×10^{7} and 45.6×10^{6} express the same value, but the former is the more commonly used notation.

Example 2–20: Exponential notation *Write the number 33.0×10^{-5} in different forms using various powers of 10.*

The number can be written in the following forms

330×10^{-6}	330 micro(units)
33.0×10^{-5}	
3.30×10^{-4}	This is the commonly used form, one digit before the decimal point.
0.330×10^{-3}	0.330 milli(units)

Example 2–21: Exponential notation *Write the number 478,321 using exponential notation, and express the answer with an accuracy of three significant figures.*

Shift the decimal point to the left and count the number of moves—that is, going from

478,321 to 4.78321 entails a shift of five places. Hence the answer is 4.78×10^5. The exponent always equals the number of places the decimal point is moved.

Example 2–22: Exponential notation *Write the number 0.000 000 000 564 using exponential notation.*

A shift of the decimal point 10 places to the right gives 5.64×10^{-10}.

Note that moving the decimal point to the left results in a positive exponent, and moving it to the right results in a negative exponent. This probably seems intuitively correct to you, but it helps to rationalize the process in the following way:

We can multiply the 478,321 of Example 2–21 by $\frac{100{,}000}{100{,}000}$ without changing its value because $\frac{100{,}000}{100{,}000}$ is equal to one. Thus

$$478{,}321 \times \frac{100{,}000}{100{,}000} = \frac{478{,}321}{100{,}000} \times 100{,}000 = 4.78 \times 10^5$$

Similarly, in Example 2–22, we got the answer by multiplying 0.000 000 000 564 by $10^{10}/10^{10}$.

To add or subtract terms involving exponentials, all the terms must first be converted to the same power of 10, as illustrated in the next Example.

Example 2–23: Exponential notation, addition *Add 2.07×10^3 to 3.03×10^2*

The numbers 2.07 and 3.03 cannot be added directly because they are associated with different exponential factors. In this case, it is more convenient to convert the second term to the same power of 10 as the first.

$3.03 \times 10^2 = 0.303 \times 10^3$

Then,

$$\begin{array}{l} 0.303 \times 10^3 \\ \underline{2.07\ \ \times 10^3} \\ 2.373 \times 10^3 \end{array}$$

The answer, to the correct number of significant figures, is 2.37×10^3.

The reason for dwelling on exponential notation is that it is the easiest method of handling problems involving very large or very small numbers. In the following example, notice how much exponentials ease the handling of many-digit numbers.

Example 2–24: Exponential notation, division and multiplication *Solve the following arithmetical problem:*

$$\frac{(8{,}800{,}000{,}000)(210{,}000{,}000)}{0.000\ 000\ 000\ 000\ 000\ 33}$$

First write the numbers, using the exponential notation.

$$\frac{(8.8 \times 10^{9})(2.1 \times 10^{8})}{3.3 \times 10^{-16}}$$

Then, since multiplication and division can proceed in any order,

$$\frac{(8.8)(2.1)}{3.3} \times \frac{(10^{9})(10^{8})}{10^{-16}} = 5.6 \times \frac{10^{(9+8)}}{10^{-16}} = 5.6 \times 10^{(17+16)} = 5.6 \times 10^{33}$$

To multiply numbers written as powers of 10, you add the exponents; to divide, you subtract the denominator exponent from the numerator exponent.

PROGRAMMED SEQUENCE:

Turn now to p. 309 and work through Set 3 of the Programmed Sequences. These problems deal with scientific measurements, significant figures, and exponentials. Since all these topics are crucial to what follows in the rest of the text, it will pay to learn them well now. Work the problems and review the text until you feel comfortable with these ideas.

HEAT AND TEMPERATURE

Heat is a form of energy. Some other forms of energy are light, chemical, mechanical, and electrical. Energy can be converted from one form into others. When coal is burned, for example, the chemical energy of coal is converted to heat and light energy. This heat energy can be used to boil water and generate steam which, when harnessed, can drive a turbine—converting heat energy to mechanical energy. The turbine can be used to turn a generator to produce electrical energy which everyone uses in his home to produce heat and light.

The unit of heat energy that we will use is the calorie. A **calorie** *is the amount of heat necessary to raise the temperature of one gram of water 1°C* (specifically, the degree from 14.5 to 15.5°C).

Example 2–25: Heat, calorie units *How many calories of heat are required to raise the temperature of 10.0 g of water from 30° to 40°C?*

One calorie is needed to raise the temperature of 1 g of water 1°C; hence

$$\frac{1.00\ \text{cal}}{(1.00\ \cancel{g})(1\cancel{^\circ C})} \times 10.0\ \cancel{g} \times 10\cancel{^\circ C} = 100\ \text{calories}$$

When heat is added to an object the temperature usually increases. **Temperature** is a *measure of the intensity of heat.* Temperature does not tell you how much heat energy a body possesses—a large piece of metal, for example, contains more heat energy than a smaller piece at the same temperature. At this point in our discussion, it is only necessary to know the units used to express temperature measurements; temperature will be defined more precisely later in the book.

Temperature measurement

Devices used to measure temperature are called **thermometers.** Thermometers are all built around some material with a physical property that varies predictably with changes in temperature. Thermometers are calibrated at certain fixed points, and the interval between these points is divided into a fixed number of spaces called **degrees.**

The most familiar thermometers contain a fixed volume of mercury that expands and contracts as the temperature rises and falls. A typical mercury-in-glass thermometer is pictured in Figure 2–1. Pure mercury is contained in a reservoir bulb connected to a capillary tube of uniform bore. A mark is usually made on the thermometer to indicate the extent of immersion during calibration. To obtain accurate measurements, the thermometer must then be immersed to the same mark in the material whose temperature is being measured. At least two fixed calibration points are needed. The 0° ice-water and the 100° water-steam points are used for two temperature marks on the thermometer illustrated, and the interval between them is divided into 100 equal degrees. If the expansion of the mercury is uniform and if the bore of the capillary does not vary, then the height of the mercury will give an accurate reading of the temperature in degrees Celsius.

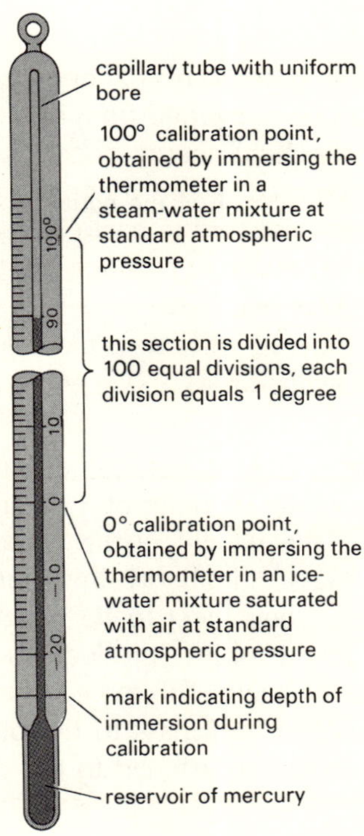

Figure 2–1 Mercury thermometer.

Temperature-scale conversions

The Fahrenheit temperature scale, like the English system of weights and measures, is not convenient for scientific work. The universally adopted temperature scale, the **Celsius* scale,** uses the freezing point of water as the 0° calibration point and the boiling point of water as the 100° calibration point. The scale between these two points is divided into 100 equal degrees.

Example 2–26: Temperature scale conversion *Convert 30°C into °F.*

The algebraic relationship between the Fahrenheit scale and the Celsius scale is

$$\frac{°C}{°F - 32} = \frac{5}{9}$$

Therefore,

$$\frac{30}{°F - 32} = \frac{5}{9}$$

$$(9)(30) = 5(°F - 32)$$

$$\frac{(9)(30)}{5} = °F - 32$$

$$\frac{(9)(30)}{5} + 32 = °F$$

$$°F = 86$$

To convert °F to °C, simply replace the F in the equation with the given value and solve for C.

MEASURING MASS

In the laboratory, a **balance** is used to determine mass. Figure 2–2 schematically illustrates the principle behind the working of a balance. The key to all measurements of mass is comparing the weight of the object under investigation to the weight of objects of known mass. Admittedly, this sounds like so much double talk, until we look at an example. The object to be weighed is placed on one of the balance pans. The weight of the object produces a downward force (weight = force) on this side of the balance beam. This force is then counterbalanced by adding metal cylinders of known weight and mass to the other balance pan. When one force equals the other, the indicator fastened to the center of the beam comes to

* Formerly, the Celsius scale was known as the Centigrade scale.

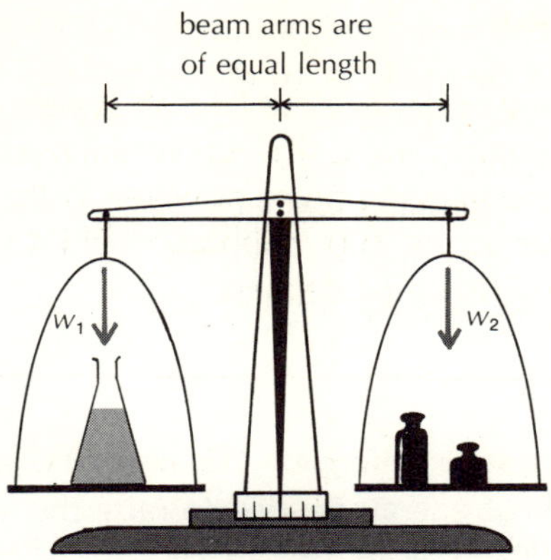

Figure 2–2 Determination of mass by use of a balance. When the pointer is at the zero mark, $w_1 = w_2$, and $m_1 = m_2$.

rest at the zero mark. At this point, the downward force (weight) on the left pan equals the downward force on the right pan. When the weights are equal, the masses are equal.

Two of the many types of balances commonly used in the laboratory are pictured in Figure 2–3. Balance **A** is a triple-beam balance; the beams contain known

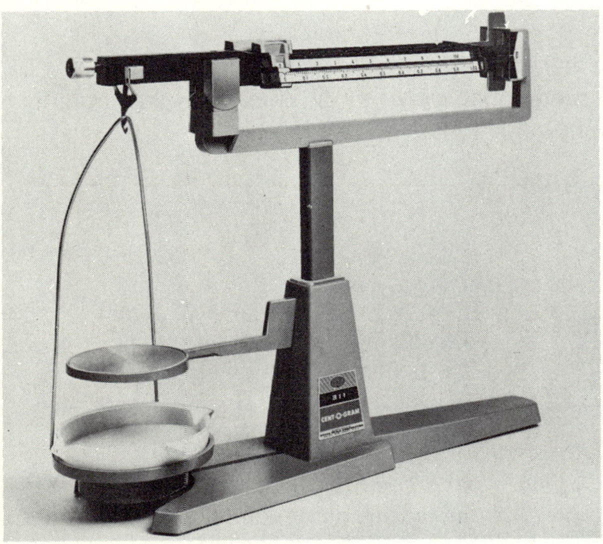

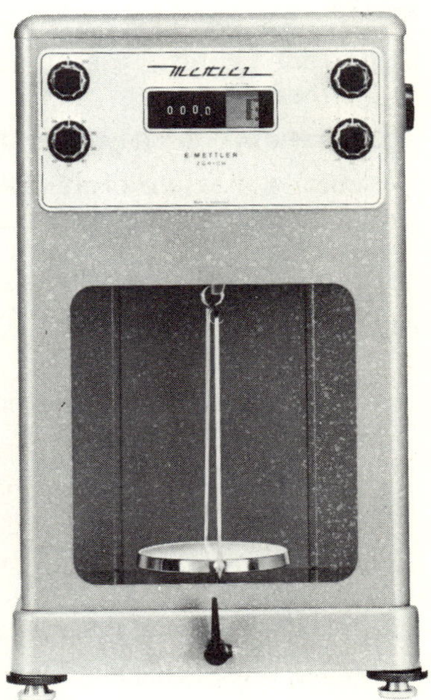

Figure 2–3 Two types of laboratory balances. **A** is a triple-beam balance for measurements to 0.01 g or 10 mg (courtesy of Ohaus Scale Corporation); **B** is an analytical balance for measurements to 0.1 mg (courtesy of Mettler Instrument Corporation, Princeton, New Jersey.)

masses that can be moved to calibrated positions. This balance has a total capacity of 2.6 kg, and readings can be made to 10 mg with a sensitivity of 0.05 g.

Balance **B** is an analytical balance used for more exact weighings. It has a sensitivity of 0.1 mg and a capacity of 300 g. Only a single pan is used. A mechanical arrangement joined to the calibrated weight knobs on the front of the instrument adds the weight. To make a weighing, you only have to load the pan, dial the weights, and read the results on the projected scale.

NEW TERMS

calorie: the amount of heat energy needed to raise the temperature of 1 gram of water 1°C (precisely, the degree from 14.5 to 15.5°C).

Celsius temperature: the temperature scale that uses the freezing and boiling points of water as the 0° and 100° calibration points.

Density: mass divided by volume.

Energy: the ability to do work.

Exponential notation: a method for expressing numbers as powers of 10.

Heat: a form of energy.

Kilogram: the standard for mass measurements in the metric system. It equals 1000 grams.

Liter: the standard for volume measurements in the metric system. It equals 1000 ml, or 1000 cc.

Mass: a measure of the quantity of matter in a body. It is equal to the weight of an object divided by the acceleration due to gravity.

Meter: the standard for length measurements in the metric system.

Significant figures: the number of digits in a number that indicate the accuracy of the measurement.

Temperature: a measure of the intensity of heat.

Weight: a measure of the force of attraction between an object and the earth. It is equal to mass times the acceleration due to gravity.

PROBLEM SET A

1. Make the following conversions within the metric system:
   a. 50 cm = ____ m = ____ km
   b. 87 liters = ____ ml = ____ cc
   c. 90 g = ____ kg = ____ mg
   d. 0.763 m = ____ km = ____ mm
   e. 349 ml = ____ liters = ____ cc
   f. 39.0 mg = ____ g = ____ kg
2. Add the following lengths, expressing your answer in meters: 56.0 cm, 340 mm, 0.0100 km.
3. Add the following volumes, expressing your answer in liters: 870 ml, 0.066 liter, 54 cc.
4. What is the volume in liters of a box with the internal dimensions of length 300 cm, height 500 cm, and width 40 cm? Assume all digits given are significant.
5. What is the mass of water needed to fill the box in Problem 4? Take the density of water to be 1.00 g/ml.
6. The following table gives the masses of five different objects. Make the metric system conversions to complete all the boxes in the table.

μg	mg	g	kg	Mg
830				
	1.50			
		6.32		
			9.33	
				387

7. Express the following lengths of time in seconds. Show the cancellation of units, using only conversion factors you have memorized, and report your answers to three significant figures.
   a. 300 hours
   b. 5 days
   c. 100 centuries (use 365 days/year)
8. Calculate the mass in kilograms and the height in centimeters of a person who weighs 150 lb and has a height of 5 ft 10 in.
9. Make the following metric-English conversions. Use conversion factors derived from Tables 2–1 and 2–2; show cancellation of units.
   a. 350 ft = ____ m
   b. 24.0 yd = ____ cm
   c. 1000 km = ____ mile
   d. 86.0 cm = ____ inch
   e. 4.00 qt = ____ liter
   f. 6.00 gal = ____ liter
   g. 250 cc = ____ qt
   h. 90.0 liter = ____ qt
   i. 7.00 lb = ____ g
   j. 5000 g = ____ lb
10. A **fifth** is a unit of measure of liquor equal to one-fifth of a gallon. What is the mass of water (in grams) required to fill a fifth?
11. When gasoline cost an American tourist in Mexico 9 cents/liter, what was the cost per U.S. gallon?
12. A liquid chemical sells for $0.80 per ml. What is the cost per fluid ounce? (The **fluid ounce,** an unfortunately named unit of the English system, is a measure of **volume,** not mass; 1 quart = 32 fluid ounces.)
13. If 300 g of a liquid has a volume of 200 ml, what is the liquid's density?
14. What is the volume of a 250-g sample of the liquid described in Problem 13?
15. What is the mass of a 250-ml sample of the liquid described in Problem 13?
16. Aluminum has a density of 2.70 g/ml. What is the mass of a rectangular block of aluminum that has the dimensions 12.0 cm by 3.00 cm by 4.00 cm?
17. If a block of lead has the same dimensions as the block of aluminum in Problem 16, how much heavier is the lead than the aluminum? Lead has a density of 11.3 g/ml. ("How much heavier?" means "What is the ratio of the mass of the lead to the mass of the aluminum?" Do not leave the ratio as a fraction, but carry out the division to three significant figures.)
18. Calculate the density of ether given the following data:
   mass of empty container.........................140.0 g
   mass of container plus 30.0 ml ether.........163.7 g
19. The density of a gas is often expressed in units of g/liter, and always at some stated temperature and pressure. If at 0°C and 1 atmosphere pressure the mass of air in a 2.50-liter container is 3.21 g, what is the density of air under these conditions?

20. Express the following numbers in exponential notation with an accuracy of three significant figures:
    a. 245,400
    b. 0.000 000 065 36
    c. 9 million
    d. 788,649

21. Solve the following statements, and express the answers to the correct number of significant figures:
    a. $2.86 + 0.039 + 11.70$
    b. $\dfrac{36\,483}{301}$
    c. 52×52
    d. $\dfrac{(3.14)(72.70 - 72.58)}{0.555}$

22. Carry out the following manipulations using exponential notation.
    a. Add: 3.50×10^5, 4.1×10^6, and 9.20×10^7
    b. Multiply: $(5.0 \times 10^{-16})(6.0 \times 10^8)(3.0 \times 10^{18})$
    c. Divide: 9.3×10^{12} by 3.1×10^{22}
    d. $\dfrac{(0.000\,000\,000\,08)(0.000\,000\,000\,000\,000\,007)}{400{,}000{,}000{,}000}$

23. How much heat energy (calories) is required to increase the temperature of 50 g of water 10°C?

24. Make the following temperature scale conversions:
    a. 72°F = ____ °C
    b. 300°C = ____ °F
    c. 20°F = ____ °C
    d. −100°C = ____ °F
    e. −60°F = ____ °C
    f. −40°C = ____ °F

PROBLEM SET B

Make the following conversions within the metric system:

1. a. 5.00×10^2 cm = ____ km
   b. 0.500 ml = ____ liter
   c. 0.500 mm = ____ m
   d. 1.00 mg = ____ g

2. Add the following lengths, expressing your answer in meters: 2.70×10^3 cm, 3.00×10^4 mm, 1.82×10^{-2} km,

3. Add the following volumes, expressing your answer in ml: 40.0 ml, 2.00×10^{-2} liter, 2.5 cc.

4. What is the volume in liters of a cube with an edge length of 1.00 m?

5. What is the mass, in kg, of a cubic meter (1 m^3) of sea water? Take the density of sea water to be 1.03 g/ml.

6. Make the following metric-English conversions. Use conversion factors derived from Tables 2–1 and 2–2; show the cancellation of units.
   a. 5280 ft = ____ m
   b. 100 cm = ____ yd
   c. 2.54 in = ____ cm
   d. 2.00 kg = ____ lb
   e. 4.00 qt = ____ liter
   f. 1.00 oz = ____ g

7. Given the following data, calculate the density of mercury in g/ml:
   mass of empty container.............................. 75.35 g
   mass of container plus 13.0 ml mercury.........252.55 g

8. Magnesium has a density of 1.74 g/ml. What is the mass of a cylindrical rod of magnesium 1.00 m long and 2.0 cm in diameter? (Volume of a cylinder = $\pi r^2 l$, where r = radius and l = length.)

9. If a cylinder of aluminum has the same dimensions as the one in Problem 8, how much heavier is it than the magnesium cylinder? (See comments in Problem 17, Set A.)

10. If 500 ml of hydrogen at 0°C and 1 atmosphere pressure weighs 45 mg, what is the density of the gas in g/liter under these conditions of temperature and pressure?

11. Carry out the following manipulations using exponential notation:
   a. Add: 2.70×10^5, 2.25×10^4, 1.25×10^3
   b. Multiply: $(5.0 \times 10^{-3})(1.5 \times 10^{-3})(2.0 \times 10^3)$
   c. Divide: 7.2×10^{-8} by 8.0×10^{-11}
   d. Divide: 1.00 by 6.02×10^{23}

12. Calculate the kilocalories required to raise the temperature of 1.00 kg of water from 10°C to 20°C.

13. Make the following temperature scale conversions:
   a. 98.6°F = ____°C
   b. 1000°C = ____°F
   c. 0°F = ____°C
   d. −10°F = ____°C
   e. −10°C = ____°F

3

elements, atoms, and moles

THIS WILL SAVE YOU TIME:

Refer to this list as you read through the chapter. In answering the questions, you will understand the important concepts you are most likely to be asked about later on.

After reading this chapter, you should be able to:

1. *Explain what is meant by physical and chemical properties and be able to give examples.*
2. *Distinguish among the three states of matter.*
3. *Give examples of physical and chemical changes.*
4. *Know the meaning of the terms element, compound, atom, and molecule.*
5. *List the most abundant elements in the earth's crust, the whole earth, and the solar system.*
6. *Name the most common elements, and give their symbols.*
7. *Explain what an atomic mass unit is.*
8. *Tell what is meant by isotopes and explain why the recorded atomic masses for elements are often not close to whole numbers.*
9. *Explain what a mole is.*
10. *Using Avogadro's number, calculate the number of atoms in a given mass of an element or the mass of a given number of atoms of an element.*
11. *Calculate the number of moles in a given mass of an element or the mass of a given number of moles.*
12. *Sketch a periodic table and show the location of the noble gases, the metals, and the nonmetals.*

This chapter begins our discussion of the heart of chemistry with a look at matter, its forms in nature, how it is described, how it can change, and how it is classified. We will find that all matter is composed of building blocks called elements and that each element has its own unique set of properties. In examining the elements, we will learn their names, their symbols, their classification, and their masses. The end of the chapter shows how the mole concept is applied to the elements.

STATES OF MATTER: SOLID, LIQUID, GAS

Matter refers to *anything that has mass and occupies space.* Matter exists in three states: **solid, liquid,** or **gas.** Consider water as an example. Water as a solid, which is ice, has a definite shape. In a solid, the particles of matter are fixed in a definite position. Water as a liquid must take the shape of the container it is placed in. Liquid particles are free to move and to take the shape—but not necessarily fill the total volume—of their container.

PROPERTIES OF MATTER: PHYSICAL AND CHEMICAL

A **property** is a characteristic by which a substance can be identified and described. Properties can be either physical or chemical. **Physical properties** *can be determined without changing the composition of the material.* Common physical properties of a substance are its state of matter (solid, liquid, or gas), color, odor, taste, density, melting temperature, and boiling temperature. **Chemical properties** *describe how a substance behaves when its composition changes.*

An inventory of the physical properties of water shows that it is tasteless, odorless, and colorless, that it freezes at 0°C and boils at 100°C. We know that these are physical properties because all of them can be determined without changing the composition of the water. The water remains water.

Not so with the chemical properties of water. Put a drop of water on a piece of sodium metal, and a chemical change results. The matter remaining after the interaction of the sodium and the water is neither sodium nor water. Both substances have changed.

CHANGES OF MATTER: PHYSICAL AND CHEMICAL

Physical changes *occur with no alteration of the chemical composition of matter.* For example, ice can be changed to water and then to steam (that is, from solid to liquid to gas). In these physical changes, no new substance is formed, and what remains at the end has the same composition as the original substance.

Chemical changes *always alter the composition of matter.* If you burn the paper in this book, you will no longer have paper when the fire is out. The original materials will be completely altered, and different substances will have been produced. Thus a chemical change forms new substances. The burning of paper in the presence of air is a chemical property of paper.

Both chemical and physical changes occur constantly in nature. Physical changes alter the landscape: Water evaporates, condenses, and then falls as rain. Streams and winds continuously erode the land. Rocks crumble from the expansion of freezing water. The human body is the site of continuous chemical change: Oxygen is consumed, food is digested, and tissues are built.

FORMS OF MATTER: HETEROGENEOUS AND HOMOGENEOUS

Table 3–1 illustrates the basic classification of matter as heterogeneous and homogeneous. In **heterogeneous matter,** more than one part can be detected, and each part has different properties. Granite, for example, is a heterogeneous solid consisting of three distinct parts visible to the naked eye: quartz, feldspar, and mica. Each of these parts, though, is itself **homogeneous**—that is, only one part can be detected, and it has the same properties throughout. Each of the parts of the granite is also known as a **pure substance,** because all known specimens of each substance have the same composition and identical properties. One piece of quartz is chemically and physically the same as any other specimen of quartz.

Mixtures can be either homogeneous or heterogeneous. The naked eye can tell that a mixture of iron and sulfur is heterogeneous. The iron can be easily separated from the sulfur by passing a magnet over the mixture; the iron will cling to the

TABLE 3–1 THE CLASSIFICATION OF MATTER

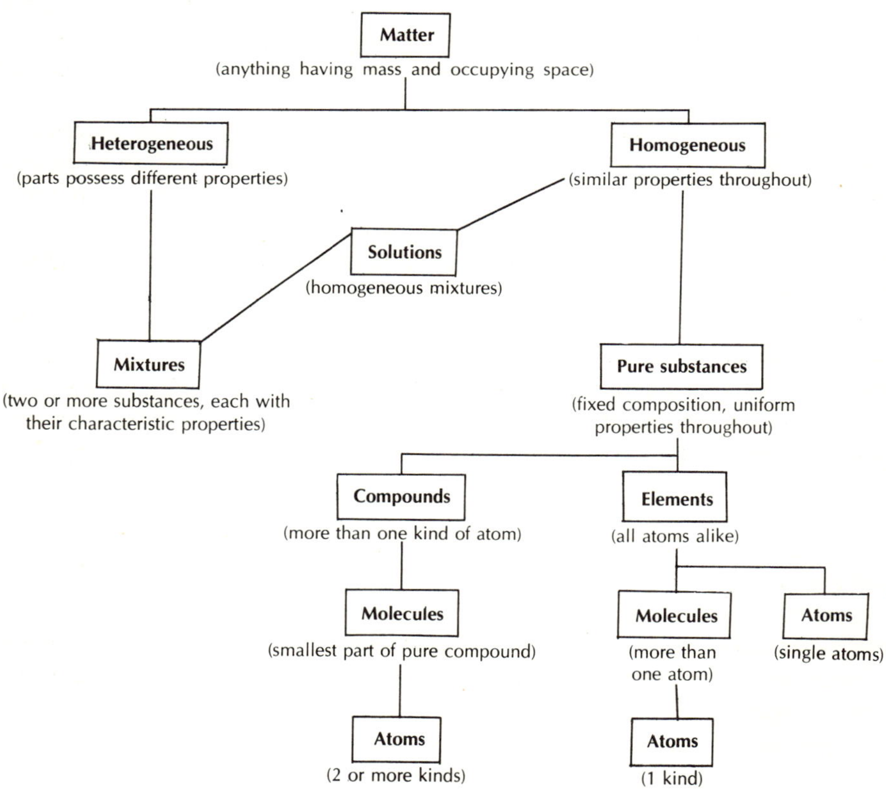

magnet, while the sulfur remains behind. Sugar and water, though, give a homogeneous mixture; the resulting liquid has the same properties throughout. We know, however, that this liquid is a mixture because the components can be separated by allowing the water to evaporate, leaving only the sugar. Such homogeneous mixtures are called **solutions.** Air is another common example of a solution; it is a homogeneous mixture of several pure substances, mainly nitrogen and oxygen.

ATOMS, MOLECULES, AND COMPOUNDS

What would happen if we took a piece of a pure substance—say, a lump of table sugar—and divided it into smaller and smaller pieces? We can assume that we have instruments fine enough to see what we come up with, even though the actual particles are far too small to appear under even the most powerful microscope. As we divide the sugar into smaller and smaller pieces, we can identify each of the pieces as sugar by its chemical and physical properties. Eventually, however, we get to a point where the substance cannot be subdivided further without destroying its characteristic properties. We have arrived at *the smallest bit of matter that retains certain properties of the pure substance;* this smallest part of sugar is called a sugar **molecule.** Actually, each sugar molecule is made up of three different kinds of building blocks: carbon, hydrogen, and oxygen. These building blocks are called **elements.** *The smallest particles of each of these three substances—or elements—that retain certain properties of the substance are called* **atoms.** Sugar is made up of molecules, each of which contains 22 atoms of the element hydrogen, 12 atoms of the element carbon, and 11 atoms of the element oxygen, a total of 45 atoms per molecule. Sugar is a **compound,** *a pure substance whose molecules are composed of more than one kind of atom.*

Molecules are clusters of atoms held together by certain forces of attraction. Some molecules, like sugar molecules, contain atoms of different kinds of elements. Water is a compound whose molecules consist of two atoms of hydrogen and one of oxygen. Molecules can also be composed of atoms of the same element. Oxygen and nitrogen gases are examples; each nitrogen molecule consists of two atoms of nitrogen, and each oxygen molecule contains two atoms of oxygen.

There are about 100 different kinds of elements in nature, the building blocks from which all matter in the universe is composed. Actually, 106 different kinds of elements are known, but a number exist only in the laboratory. The atoms of these 106 elements can unite with one another or with unlike atoms to form an almost infinite number of different molecules; such combinations are the very stuff of ourselves and our world.

ELEMENTS

Elements are pure substances composed of only one kind of atom. All the elements are listed inside the back cover of this book. Most elements are solids, two are liquids at room temperature, and twelve exist in nature as gases. Some elements are radioactive, some are extremely rare, and others can only be made in the laboratory.

Some elements are more plentiful than others. For example, eight elements make up 98.5 percent of the mass of the earth's crust, which is about 10.6 miles deep. In order of abundance, these elements are: oxygen, silicon, aluminum, iron, calcium, sodium, potassium, and magnesium. The next most abundant elements are titanium, phosphorus, hydrogen, and manganese. The remaining elements account for only 0.6 percent of the earth's crust.

Relative abundance changes when one considers the entire planet rather than just the crust. Iron, for example, composes 34.6 percent of the whole earth, but only 5 percent of the crust. If the whole solar system is taken into account, hydrogen and helium are the most abundant elements, with oxygen, neon, carbon, and nitrogen next.

A number of the elements occur very rarely, if at all, on earth. These include the four elements technetium, promethium, astatine, and francium, along with all the elements heavier than uranium: lawrencium, nobelium, einsteinium, and so on.

NAMES AND SYMBOLS FOR ELEMENTS

To save the time, trouble, and long-windedness of writing out the name in full, each element has been assigned a **symbol,** a conventional and accepted form of abbreviation. Some of the symbols are obvious, but others cannot be predicted from the name of the element. Carbon, calcium, and cadmium, for instance, all have the same first and second letter. Thus the symbol C has been assigned to carbon, Ca to calcium, and Cd to cadmium. The symbols for many elements are derived from their original Latin names, not the modern English names. Examples are Fe for iron (*ferrum,* in Latin), Au for gold (*aurum*), Ag for silver (*argentum*), and Pb for lead (*plumbum*). The symbol Na for sodium comes from the eighteenth-century word *natrium*. Table 3–2 lists many of these symbols derived from old names.

When a new element is discovered, the person who discovered it has the privilege of naming it. As a result, elements have been named after geographical

TABLE 3–2 SYMBOLS FOR ELEMENTS DERIVED FROM OLDER NAMES

Present name	Symbol	Old name
Antimony	Sb	Stibium
Copper	Cu	Cuprum
Gold	Au	Aurum
Iron	Fe	Ferrum
Lead	Pb	Plumbum
Mercury	Hg	Hydrargyrum
Potassium	K	Kalium
Silver	Ag	Argentum
Sodium	Na	Natrium
Tin	Sn	Stannum
Tungsten	W	Wolfram

locations, people, planets, or anything else that happened to capture the fancy of the discoverer. The town of Ytterby in Sweden has been honored in the naming of four elements—yttrium, ytterbium, erbium, and terbium—because the scientists who discovered these elements were located there. Americans will not find it difficult to guess where the elements berkelium and californium were discovered.

Table 3–3 shows the names and symbols of the elements you will encounter most commonly. Note, by the way, that symbols are always composed of one or two letters, that the first letter is always capitalized and the second one is not (for example, O for oxygen, and Ni for nickel). Since the symbols will be used often, you should commit them to memory. A list of all the elements appears inside the back cover.

TABLE 3–3 NAMES AND SYMBOLS OF THE MOST COMMON ELEMENTS

Element	Symbol	Element	Symbol	Element	Symbol
Aluminum	Al	Fluorine	F	Phosphorus	P
Antimony	Sb	Gold	Au	Platinum	Pt
Argon	Ar	Helium	He	Potassium	K
Arsenic	As	Hydrogen	H	Radium	Ra
Barium	Ba	Iodine	I	Silicon	Si
Bismuth	Bi	Iron	Fe	Silver	Ag
Boron	B	Lead	Pb	Sodium	Na
Bromine	Br	Lithium	Li	Strontium	Sr
Cadmium	Cd	Magnesium	Mg	Sulfur	S
Calcium	Ca	Manganese	Mn	Tin	Sn
Carbon	C	Mercury	Hg	Titanium	Ti
Chlorine	Cl	Neon	Ne	Tungsten	W
Chromium	Cr	Nickel	Ni	Uranium	U
Cobalt	Co	Nitrogen	N	Zinc	Zn
Copper	Cu	Oxygen	O		

THE RELATIVE MASSES OF ATOMS

It is important to realize that atoms of different elements have different masses, but these different masses are relative to an arbitrary standard devised by scientists. We can best illustrate the relativity of atomic masses with a little history.

In 1826, the Swedish chemist Jöns Jakob Berzelius used careful analytical techniques to compute the relative masses of most of the elements known to scientists of the time. To compile this list, he had to assign a given value to the mass of one of the elements and then record the masses of the other elements in relation to that assigned standard. Berzelius chose oxygen as his standard, assigned its atom an arbitrary mass value of 100, and then computed the mass values of the single atoms of the other elements in relation to the oxygen atom. His values are given in Table 3–4.

TABLE 3–4 RELATIVE ATOMIC MASSES TABULATED BY BERZELIUS IN 1826 (OXYGEN ATOM ASSIGNED A STANDARD OF 100.000)

Element	Symbol	Atomic mass
Aluminum	Al	171.161
Calcium	Ca	256.019
Carbon	C	76.437
Chlorine	Cl	221.325
Gold	Au	1243.013
Hydrogen	H	6.2398
Iron	Fe	339.213
Lead	Pb	1265.822
Nitrogen	N	88.518
Oxygen	O	100.000
Potassium	K	489.916
Silicon	Si	277.478
Silver	Ag	1351.607
Sodium	Na	290.897
Sulfur	S	201.165
Tin	Sn	735.294

Example 3–1: Relative masses *According to the data of Table 3–4, Berzelius found the mass of a carbon atom to have a value of 76.437 if the mass of an oxygen atom has a value of 100. What would the mass of an oxygen atom be if the mass of the carbon atom were assigned a standard value of 12.00?*

From the table we get the ratio

$$\frac{\text{mass of C atom}}{\text{mass of O atom}} = \frac{76.437}{100}$$

If we let x equal the mass of an oxygen atom when a carbon atom has a mass of 12.00, then

$$\frac{\text{mass of C atom}}{\text{mass of O atom}} = \frac{12.0}{x} = \frac{76.4}{100}$$

Solving for x,

$$x(76.4) = 12.0(100)$$

$$x = \frac{12.0(100)}{76.4}$$

$x = 15.7$, the relative mass of an oxygen atom if the mass of carbon is 12.0.

THE MASS SPECTROGRAPH

In 1920, Francis William Aston (1877–1945, Nobel prize in chemistry 1922) invented the **mass spectrograph,** a powerful tool used to determine atomic mass ratios with great accuracy. The principle behind the instrument is quite simple. If two balls of equal size, one a ping-pong ball and the other a ball of metal, are dropped from a height when a stiff wind is blowing, the balls will not hit the ground at the same spot. The lighter ball will be blown much farther off course than the heavy ball. This principle can be applied to atoms and molecules to determine their relative masses. Atoms or molecules are caused to move in a straight line at a constant speed, and some force analogous to the breeze is created. If you observe the place where the particles strike a target and measure the extent of deflection from a straight path, you can calculate the relative masses of the particles.

In the mass spectrograph, the atoms and molecules are given a positive charge and set into motion by their attraction to a negatively charged electrode. As the particles stream along, they pass through a magnetic field that acts as a deflecting force analogous to the wind in the example above. Lighter atoms and molecules are deflected more than heavier ones. Figure 3–1 shows how a mass spectrograph is used to separate helium and neon.

Mass spectrographs are now used for routine chemical determinations. Some instruments are sensitive enough to separate atoms that have mass differences of only 1 part in 30,000.

With a mass spectrograph, the masses of all the elements can be tabulated in comparison to some arbitrary standard. Instead of using oxygen with a value of 100, as Berzelius did, the modern standard is the mass of the commonest carbon atom to which a value of 12. (exact) has been assigned. This list of relative masses is known as the **atomic mass scale** (atomic weight scale). Because of the manner in which they are determined, the atomic mass values have no units. It is convenient, however, to speak of these relative masses in terms of **atomic mass units,** abbreviated amu. One amu is equal to one twelfth the mass of the most common kind of carbon atom; or, in other words, the most common kind of carbon atom (mass 12 isotope, see next paragraph) has a mass of 12 amu. The atomic masses of the most

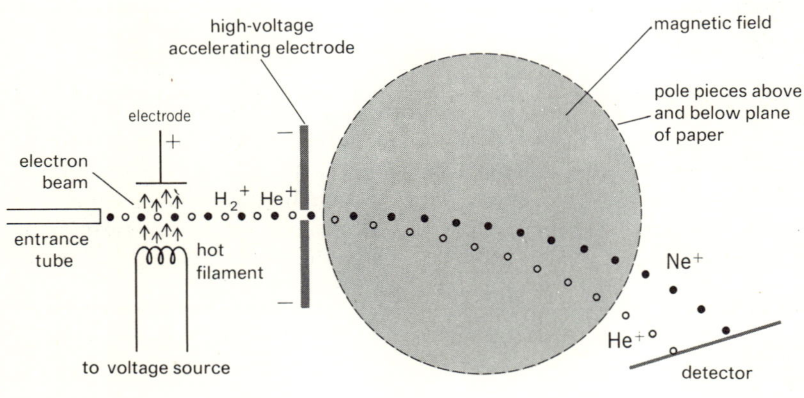

Figure 3–1 Mass spectrograph.

common elements are given in Table 3–5. The atomic masses for all the elements are given inside the back cover of this book. On this scale, hydrogen, the lightest element, has a relative mass of 1.008; helium, 4.0026; oxygen, 15.9994; and so forth.

TABLE 3–5 ATOMIC MASSES FOR THE MOST COMMON ELEMENTS

Name	Symbol	Atomic mass	Name	Symbol	Atomic mass
Aluminum	Al	26.98154	Lithium	Li	6.939
Antimony	Sb	121.75	Magnesium	Mg	24.312
Argon	Ar	39.948	Manganese	Mn	54.9380
Arsenic	As	74.9216	Mercury	Hg	200.59
Barium	Ba	137.34	Neon	Ne	20.183
Bismuth	Bi	208.9804	Nickel	Ni	58.71
Boron	B	10.811	Nitrogen	N	14.0067
Bromine	Br	79.909	Oxygen	O	15.9994
Cadmium	Cd	112.40	Phosphorus	P	30.97376
Calcium	Ca	40.08	Platinum	Pt	195.09
Carbon	C	12.01115	Potassium	K	39.098
Chlorine	Cl	35.453	Radium	Ra	(226)
Chromium	Cr	51.996	Silicon	Si	28.086
Cobalt	Co	58.933	Silver	Ag	107.870
Copper	Cu	63.542	Sodium	Na	22.98977
Fluorine	F	18.99840	Strontium	Sr	87.62
Gold	Au	196.967	Sulfur	S	32.064
Helium	He	4.0026	Tin	Sn	118.69
Hydrogen	H	1.0079	Titanium	Ti	47.90
Iodine	I	126.9044	Tungsten	W	183.85
Iron	Fe	55.847	Uranium	U	238.03
Lead	Pb	207.19	Zinc	Zn	65.38

Example 3–2: Atomic masses *What is the mass of a chromium atom?*

From Table 3–5, we find that the atomic mass of chromium is 52.00 amu. This figure means that a chromium atom has a mass of 52 units if a carbon atom is assigned an arbitrary mass of 12 units or that the mass of a chromium atom is 52/12 times the mass of a carbon atom. We can calculate the actual mass of a chromium atom in grams if we know the mass of an atom of some other element. Given the actual mass of any one element, we can calculate the masses of the atoms of all elements using the relationships of the atomic mass scale.

ISOTOPES

During the period from 1907 to 1913, a number of scientists found values for the atomic mass of lead, Pb, that varied from 206.4 to 208.4. We now know that all elements can have atoms of different masses. Each of these different atoms is called an **isotope.** For example, the assignment of 12. (exact) amu to carbon as the atomic mass standard refers to only one particular isotope of carbon, since not all carbon atoms have the same mass. In the conventions of chemical notation, this isotope is called carbon-12.

The atomic masses of individual isotopes of most elements are very close to whole numbers. The tabulated atomic mass for the element chlorine, however, is 35.453. This value is not close to a whole number, but its oddity is easily explained. The tabulated atomic mass for each element is the average of the masses of the element's isotopes in the proportion found in nature. The mass spectrogram peaks for pure chlorine gas are shown in Figure 3–2. Two prominent isotopes are detected, one with a mass of 35.0 and the other with a mass of 37.0. Because the height of these mass spectrogram peaks is proportional to the quantity of each species present, it is easy to calculate that ordinary chlorine is made up of 75.8 percent of the 35 isotope and 24.2 percent of the 37 isotope. The average atomic mass reported in the table is thus 24.2 percent of 37.0 plus 75.8 percent of 35.0, a value equal to the recorded 35.45.

Fluorine, F, is an element with only one naturally occurring isotope; most other elements have more than one. Tin, Sn, has a total of ten naturally occurring isotopes, with masses 120, 118, 116, 119, 117, 124, 122, 112, 114, and 115, in the order of their abundance.

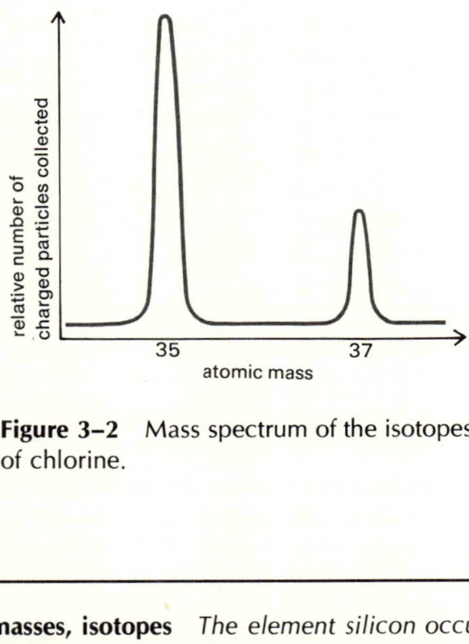

Figure 3–2 Mass spectrum of the isotopes of chlorine.

Example 3–3: Atomic masses, isotopes *The element silicon occurs in nature as a mixture containing 92.2 percent of mass 28.0, 4.70 percent of mass 29.0, and 3.09 percent of the mass 30.0 isotopes. What should be the recorded value for the atomic mass of silicon?*

Assume that we have a total of 1000 silicon atoms. Then we would have

92.2 percent of 1000 = 922 atoms with a mass of 28.0 amu
4.70 percent of 1000 = 47 atoms with a mass of 29.0 amu
3.09 percent of 1000 = 31 atoms with a mass of 30.0 amu

The total mass of the 1000 atoms is

$$922 \cancel{\text{atoms}} \times \frac{28.0 \text{ amu}}{\cancel{\text{atom}}} = 25{,}816 = 25.8 \times 10^3 \text{ amu}$$

$$47 \cancel{\text{atoms}} \times \frac{29.0 \text{ amu}}{\cancel{\text{atom}}} = 1{,}363 = 1.36 \times 10^3 \text{ amu}$$

$$31 \cancel{\text{atoms}} \times \frac{30.0 \text{ amu}}{\cancel{\text{atom}}} = 930 = \underline{0.930 \times 10^3 \text{ amu}}$$

$$\text{total mass of 1000 atoms} = 28.090 \times 10^3 = 28.1 \times 10^3 \text{ amu}$$

The average mass per atom, the atomic mass, is therefore

$$\frac{28.1 \times 10^3 \text{ amu}}{1000} = 28.1 \text{ amu}$$

A more direct route is:

92.2 percent of 28.0 = 25.8 amu
4.70 percent of 29.0 = 1.36 amu
3.09 percent of 30.0 = $\underline{0.93}$ amu
28.09 = 28.1 amu, the atomic mass of Si

THE MOLE

Suppose you are given a sample of known mass of oxygen and asked to determine how large a sample of carbon must be for both samples to contain the same number of atoms. Given atomic masses, this is a relatively easy calculation. 12.01 g of carbon must contain the same number of atoms as 16.00 g of oxygen. The reason is that an oxygen atom has a mass 16.00/12.01 times the mass of a carbon atom. Similarly, 12.01 lb of carbon contains the same number of atoms as 16.00 lb of oxygen.

Example 3–4: Atomic masses and numbers of atoms *What mass of the element silver contains the same number of atoms as 10 g of nitrogen?*

The atomic mass of silver is 107.9 amu and that of nitrogen is 14.01 amu; thus, silver atoms are 107.9/14.01 times as heavy as nitrogen atoms. It would take, therefore,

$$\frac{107.9}{14.01} \times 10 \text{ g} = 77.0 \text{ g}$$

of silver to have the same number of atoms as 10 g of nitrogen.

Because the atomic mass scale is a relative scale, the quantity represented by the atomic mass of one element expressed in grams will contain exactly the same number of atoms as the atomic mass of any other element expressed in grams. For convenience, chemists have called this unit the mole; a **mole** is *the mass of an element in grams numerically equal to the atomic mass,* that is, a collection of atoms of an element whose total mass in grams is numerically equal to the atomic mass.

Example 3–5: Moles *What is the mass of one mole of iron?*

Because the atomic mass of iron is 55.847 amu, the mass of 1 mole is 55.847 g.

If we know the number of atoms in 12. (exact) g of carbon, we also know the number of atoms in 15.9994 g oxygen, 1.008 g hydrogen, 55.847 g iron, and so on. By substituting gram units for atomic mass units, we have not changed the relative values on the atomic mass scale. The ratio of the mass of an oxygen atom to that of a carbon atom is 15.9994/12.0000, whether we assign amu or grams as the units in the ratio. Thus it follows that the number of atoms in one mole of one element should be the same as the number of atoms in one mole of any other element.

Example 3–6: Number of atoms in a given mass of an element *Given the fact that there are 6.02×10^{23} atoms in 12.000 g of carbon, how many atoms are contained in 55.847 g of iron?*

Because the actual masses of the carbon and iron atoms are in the ratio 55.847/12.000, 55.847 g of iron will contain the same number of atoms as 12.000 g of carbon. Hence, 1 mole, 55.847 g, of iron also contains 6.02×10^{23} atoms.

AVOGADRO'S NUMBER

The actual number of atoms in one mole of an element has been determined by more than a dozen different methods, and it can now be expressed to an accuracy of six significant figures. This number, 6.02252×10^{23}, is called **Avogadro's number,** after the famous Italian physicist Amedeo Avogadro (1776–1856), who formulated a hypothesis that led to the determination of the number.

Example 3–7: Number of atoms in a given mass of an element *How many atoms are in 8.00 g of sulfur?*

One mole of sulfur is 32.1 g; therefore, this number of grams is the mass of sulfur that contains Avogadro's number of atoms. The number of atoms in 8.00 g is

$$8.00\ \cancel{g} \times \frac{1\ \cancel{\text{mole}}}{32.1\ \cancel{g}} \times \frac{6.02 \times 10^{23}\ \text{atoms}}{1\ \cancel{\text{mole}}} = 1.50 \times 10^{23}\ \text{atoms}$$

Example 3–8: Mass of a single atom *What is the actual mass of a single sulfur atom?*

Using Avogadro's number,

$$1\ \cancel{\text{atom}} \times \frac{32.1\ \text{g}}{6.02 \times 10^{23}\ \cancel{\text{atoms}}} = 5.33 \times 10^{-23}\ \text{g}$$

Example 3–9: Amu *What is the mass in grams of 1 amu?*

Because 6.02×10^{23} atoms of carbon have a mass of exactly 12.0 g, we can solve for the mass of a single atom:

$$1\ \cancel{\text{atom}} \times \frac{12.0\ \text{g}}{6.02 \times 10^{23}\ \cancel{\text{atoms}}} = 1.99 \times 10^{-23}\ \text{g}$$

Because one carbon atom also has a mass of 12 amu, 1 amu must equal

$$\frac{1.99 \times 10^{-23}\ \text{g}}{12\ \text{amu}} = 1.66 \times 10^{-24}\ \text{g/amu}$$

THE SIZE OF ATOMS

With the information we now have, we can predict the size of an atom. From Table 2–4, we know that the density of gold is 19.3 g/ml. In addition, we know that the atomic mass expressed in grams, one mole or 197 g, of gold contains 6.02×10^{23} atoms. From these data we can find a value for the average volume available to an atom of gold in the solid state. First calculate the number of atoms in 19.3 g.

$$19.3\ \cancel{\text{g}} \left[\frac{6.02 \times 10^{23}\ \text{atoms}}{197\ \cancel{\text{g}}}\right] = 5.90 \times 10^{22}\ \text{atoms in 19.3 g of gold}$$

Then, since 19.3 g of gold has a volume of one ml, the volume available per atom is

$$\frac{1.00\ \text{ml}}{5.90 \times 10^{22}\ \text{atoms}} = 1.69 \times 10^{-23}\ \text{ml available per atom of gold}$$

This result, of course, is not an accurate determination of the size of a gold atom, for we do not know whether the atoms in the solid state are packed close together or far apart. The result does, however, represent a maximum value and shows how very, very small atoms are.

The actual size of an atom is a difficult thing to measure; however, the values of atomic radii have been obtained from spectral studies. These radii (covalent radii, half the distance between two like atoms joined by a covalent bond) vary from 3.7×10^{-11} m for hydrogen to 23.5×10^{-11} m for the element cesium, Cs. The relative sizes of atoms are shown in Figure 3–3.

The dimensions of atoms can be illustrated in many ways. It is actually easier

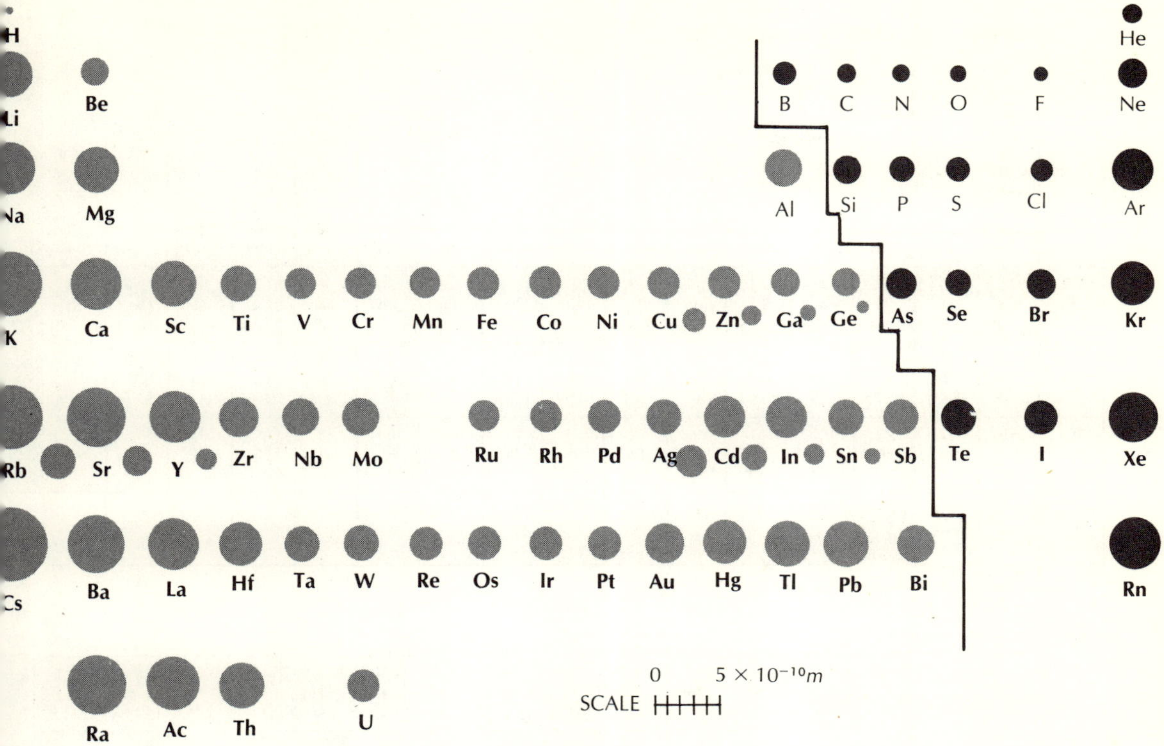

Figure 3–3 Relative sizes of neutral atoms.

to count the national debt (the largest number one can normally think of) dollar by dollar than it is to count the atoms in a grain of sand. It would take all the people in the United States, including the newborn, a total of about 40,000 years to count the atoms in a drop of water, provided, of course, that they did not sleep on the job. The ink on this page contains over 1,000,000,000,000, or 10^{12}, atoms of carbon.

THE CLASSIFICATION OF ATOMS: THE PERIODIC TABLE

Several scientists working in the 1860s found that a chart of the elements arranged in order of increasing atomic mass revealed a repetition in both the chemical and physical properties of the elements. Two scientists, Julius Lothar Meyer (1830–1895) and Dmitri Ivanovich Mendeleev (1834–1907), completed such an arrangement of the elements almost simultaneously in 1869. However, because Mendeleev contributed more to the understanding of the table, the **periodic chart** is usually attributed to him. He formulated a **periodic law,** which stated: *The properties of the elements are a periodic function of their atomic weights* (*atomic masses*). As we will discover, this law had to be changed after the discovery of isotopes. A modern periodic table, giving the symbols and atomic masses of the elements, is shown in Figure 3–4. Note that several of the elements are not listed in the order of increasing atomic masses.

Period	IA	IIA	IIIB	IVB	VB	VIB	VIIB	VIII	VIII	VIII	IB	IIB	IIIA	IVA	VA	VIA	VIIA	Noble gases
1	1 **H** 1.008																	2 **He** 4.0[illegible]
2	3 **Li** 6.939	4 **Be** 9.012											5 **B** 10.81	6 **C** 12.011	7 **N** 14.007	8 **O** 15.999	9 **F** 19.00	[illegible]
3	11 **Na** 22.990	12 **Mg** 24.31											13 **Al** 26.98	14 **Si** 28.086	15 **P** 30.974	16 **S** 32.064	17 **Cl** 35.453	[illegible]
4	19 **K** 39.102	20 **Ca** 40.08	21 **Sc** 44.956	22 **Ti** 47.90	23 **V** 50.94	24 **Cr** 51.996	25 **Mn** 54.94	26 **Fe** 55.85	27 **Co** 58.93	28 **Ni** 58.71	29 **Cu** 63.54	30 **Zn** 65.37	31 **Ga** 69.72	32 **Ge** 72.59	33 **As** 74.92	34 **Se** 78.96	35 **Br** 79.909	[illegible]
5	37 **Rb** 85.47	38 **Sr** 87.62	39 **Y** 88.91	40 **Zr** 91.22	41 **Nb** 92.91	42 **Mo** 95.94	43 **Tc** 99	44 **Ru** 101.07	45 **Rh** 102.91	46 **Pd** 106.4	47 **Ag** 107.870	48 **Cd** 112.40	49 **In** 114.82	50 **Sn** 118.69	51 **Sb** 121.75	52 **Te** 127.60	53 **I** 126.90	[illegible]
6	55 **Cs** 132.91	56 **Ba** 137.34	57 **La** 138.91	72 **Hf** 178.49	73 **Ta** 180.95	74 **W** 183.85	75 **Re** 186.2	76 **Os** 190.2	77 **Ir** 192.2	78 **Pt** 195.09	79 **Au** 196.97	80 **Hg** 200.59	81 **Tl** 204.37	82 **Pb** 207.19	83 **Bi** 208.98	84 **Po** 210	85 **At** 210	[illegible]
7	87 **Fr** 223	88 **Ra** 226.05	89 **Ac** 227	104 **(Rf)*** **(Ku or Kh)**	105 **(Ha)***													
	119																	

Group numbers as printed above the table; "transition elements" spans IIIB–IIB.

Key: atomic number / symbol for element / atomic mass

lanthanide series	58 **Ce** 140.12	59 **Pr** 140.91	60 **Nd** 144.24	61 **Pm** 145	62 **Sm** 150.35	63 **Eu** 151.96	64 **Gd** 157.25	65 **Tb** 158.92	66 **Dy** 162.50	67 **Ho** 164.93	68 **Er** 167.26	69 **Tm** 168.93	70 **Yb** 173.04	71 **Lu** 174.97
actinide series	90 **Th** 232.04	91 **Pa** 231	92 **U** 238.03	93 **Np** 237	94 **Pu** 242	95 **Am** 243	96 **Cm** 247	97 **Bk** 249	98 **Cf** 251	99 **Es** 254	100 **Fm** 253	101 **Md** 256	102 **No** 253	103 **Lw** 257
	122													

*Symbols not officially assigned.

Figure 3–4 Periodic table of the elements.

The group of elements at the extreme right of the table, labeled noble gases, all have similar properties. They are all monatomic gases at room temperature; that is, they do not form molecules, but exist as individual atoms. Also they are very unreactive chemically.

The heavy diagonal (stairstep) line on the right side of the table divides metals from nonmetals; all elements to the left of and below the line are metals. **Metals** are substances that typically have a particular luster, conduct heat and electricity well, are opaque, are flexible enough to be deformed, and can be melted. Some of the elements touching the diagonal line have both metal and nonmetal properties. The position of hydrogen places it in the metal group, although some authors prefer to put it above fluorine and next to helium. It actually does not fit conveniently into either position. Note that the metallic elements greatly out number the nonmetallic elements.

Only two elements are liquid at room temperature, mercury and bromine. In addition to the noble gases, the gaseous elements are oxygen, nitrogen, fluorine, chlorine, and hydrogen. All of the non-noble gases exist in nature as diatomic molecules—that is, two atoms are combined to form a molecule. They do not exist as individual atoms.

The periodic table will be discussed in greater detail in Chapter 12.

NEW TERMS

amu: abbreviation for atomic mass units. The most common isotope of carbon has been assigned a mass of exactly 12 amu; the masses of all other atoms are determined in reference to this value.

Atomic mass: the mass of an atom expressed in amu units. The tabulated atomic mass of an element is the average mass in amu of the mixture of isotopes as they exist in nature.

Avogadro's number: 6.02×10^{23}, the number of particles in one mole—the number of atoms in 12. (exact) g of the carbon-12 isotope.

Chemical change: a change in which new compounds are formed.

Chemical property: a description that relates to a change in the chemical composition of the substance.

Compound: a pure substance composed of more than one kind of atom.

Diatomic molecule: a molecule composed of two atoms.

Element: one of the building blocks of matter. There are 106 known elements.

Gas: one of the physical states of matter, in which the particles are free to move so that they occupy the total space into which they are put.

Heterogeneous: parts have different properties.

Homogeneous: similar properties throughout.

Isotopes: atoms of the same elements that have different masses.

Liquid: one of the three physical states of matter. The particles are free to move so that they assume the shape (but not necessarily the total volume) of their container.

Mass spectrograph: an instrument for determining relative masses of atoms and molecules.

Matter: anything that has mass and occupies space.

Metal: a substance that typically has a particular luster, is a good conductor of heat and electricity, is flexible enough to be deformed, and can be melted.

Mixture: a combination of two or more substances, each with its own characteristic properties.

Mole: the mass of an element in grams numerically equal to the atomic mass. The amount of material that contains Avogadro's number of particles; one mole of carbon, 12.0000 g, contains 6.02×10^{23} atoms.

Molecule: the smallest unit of a compound.

Monatomic gas: a gas that exists in nature as uncombined, single atoms.

Noble gases: a group of elements that are monatomic gases and have little chemical reactivity.

Periodic law: the properties of the elements are a periodic function of their atomic numbers. (The original law stated: The properties of the elements are a periodic function of their atomic masses.)

Physical change: changes that occur with no change in the chemical composition of matter.

Physical property: properties of a substance that can be determined with no change in the substance.

Properties: the characteristics by which a substance can be identified and described.

Solid: one of the states of matter, in which the particles are in definite positions and the object has a definite shape.

Solution: a homogeneous mixture.

PROBLEM SET A

1. The compound called ammonia has molecules composed of three atoms of hydrogen and one of nitrogen. The percent of nitrogen in the compound is 82.3. What is the ratio of the mass of a nitrogen atom to that of a hydrogen atom?
2. Compare the atomic masses of the elements iron, calcium, and sodium as tabulated by Berzelius with the present-day values. Do this by recalculating the Berzelius values to the carbon-12 standard.
3. Calculate the relative-mass ratio for an atom of silver to an atom of copper.
4. Is it possible for two different elements to have the same atomic mass?
5. If the standard value of 12. (exact) has been assigned to carbon for the atomic mass scale, then why is the reported mass value for carbon in the table 12.01?
6. Silver has two isotopes, masses 107 and 109. The naturally occurring metal has an atomic mass of 107.870. What percentage of ordinary silver is the mass 107 isotope?
7. How many atoms are in 100 g of copper?
8. Calculate the mass of a single zinc atom in grams.
9. What mass of nitrogen has the same number of atoms as 100 g of mercury?
10. Calculate the mass of one atom of the lightest element, hydrogen, and compare it to the mass of lawrencium, one of the heaviest elements.
11. How many moles of the element iron are contained in 300 g?
12. How many grams of carbon are contained in 2.50 moles?
13. How many amu are equal to one gram?
14. The density of solid platinum is 21.4 g/ml. Calculate the volume available per atom.

PROBLEM SET B

1. The compound titanium tetrachloride has molecules composed of four atoms of chlorine and one of titanium. The percent of titanium in the compound is 25.2. What is the ratio of the mass of a titanium atom to that of a chlorine atom?
2. A chemist during the time of Berzelius could have come up with his own atomic-mass scale. Assuming that some unknown chemist did just that and that he assigned a standard value of 5.00 to elemental hydrogen, compare the relative atomic masses for elemental iron, mercury, and phosphorus to those based on the carbon-12 standard.
3. Calculate the relative mass ratio of an atom of mercury to an atom of potassium.
4. Referring to Figure 3–4, explain why elements number 100 and 102 have the same reported atomic masses.
5. Lithium has two isotopes, masses 6.02 and 7.02. The naturally occurring metal has an atomic mass of 6.941. What percentage of ordinary lithium is the mass 6.02 isotope?

6. Calculate the number of atoms contained in a 300 g sample of calcium.
7. Calculate the mass of an isolated atom of arsenic in grams.
8. Calculate the mass of gold that has the same number of atoms as 50.0 g of aluminum.
9. Calculate the mass of one atom of each of the following atoms: carbon, phosphorus, and mercury.
10. How many moles of nickel are contained in a 500 g sample?
11. How many grams of uranium are contained in 5.00 moles?
12. Calculate the number of amu in 24.3 g (1.00 mole) of magnesium atoms.
13. The density of solid titanium is 4.50 g/ml. Calculate the volume available per atom.

4

molecules and moles

THIS WILL SAVE YOU TIME:

Refer to this list as you read through the chapter. In answering the questions, you will understand the important concepts you are most likely to be asked about later on.

After reading this chapter, you should be able to:

1. *Explain the law of definite composition.*
2. *Work problems involving percentages of elements in compounds.*
3. *Know the meaning of chemical equations and formulas.*
4. *State the law of multiple proportions and give examples illustrating it.*
5. *Distinguish between empirical and molecular formulas.*
6. *Tell what is meant by the law of conservation of mass-energy.*
7. *Apply the mole concept to molecules.*
8. *Calculate the molecular mass of a compound from its formula.*
9. *Calculate the number of moles in a given mass of a compound.*
10. *Calculate the mass of a given number of moles of a compound.*
11. *Calculate the number of atoms and molecules in a given amount of a compound.*
12. *Calculate the mass of a single molecule of a compound.*
13. *Discuss the sizes of molecules and know what is meant by polymers and macromolecular chemistry.*

In Chapter 3, we saw that all matter is composed of atoms and that these atoms typically join together to form molecules. Working from these elemental but significant observations, scientists formulated three laws of nature: the law of definite composition, the law of multiple proportions, and the law of conservation of mass-energy. In this chapter, we will look at each of these laws in turn, describe how the mole concept is applied to molecules, and discuss the size of molecules.

LAW OF DEFINITE COMPOSITION

The **law of definite composition** states that *a given compound always contains the same elements in the same proportion by mass.* This means that, no matter where you obtain a sample of water, for example, it always contains 88.81 percent oxygen and 11.19 percent hydrogen by mass, and this composition never varies. In other words, this law says that water consists of a combination of oxygen and hydrogen atoms and that these atoms have united to form molecules.

Hydrogen and oxygen atoms are discrete particles with definite masses, and they can combine to form molecules only as whole units. There is simply no way that one-half a hydrogen atom may combine with one-third an oxygen. The water molecule consists of two hydrogens and one oxygen, and the mass of the molecule is the added mass of the component atoms. Thus the law of definite composition reflects the fact that the countless molecules in any sample of water all have the same components in the same proportions all the time.

The formation of a water molecule from atoms can be illustrated by writing **a chemical equation** employing the symbols for the elements:

$$2H + 1O \rightarrow 1HOH$$

The arrow means yields, produces, or gives. The number in front of the symbol or formula notation tells the number of atoms or molecules; if the number is 1, it is omitted.

$$2H + O \rightarrow HOH$$

The equation can also be written

$$2H + O \rightarrow H_2O$$

Here the formula H_2O is equivalent to HOH; the subscript refers to the number of hydrogen atoms in the molecule. However, if this equation is supposed to describe the reaction of oxygen gas with hydrogen gas, it is not correct. Both elements exist in nature as *diatomic molecules* rather than as individual atoms; that is, oxygen gas is O_2 and hydrogen gas is H_2. Taking this fact into account, the correct equation is

$$2H_2 + O_2 \rightarrow 2H_2O$$

Note that the number before a formula tells how many of that species (molecule, atom, etc.) participate as reactants or products. The subscripts show how many atoms are in a given molecule. This is the basic form for all chemical equations. We will consider them in more detail later.

Example 4–1: Formulas *The formula for a pure compound, chloroform, is $HCCl_3$. Explain what this formula means and how it is related to the law of definite composition.*

The formula says that each basic unit of this pure compound is composed of one atom of hydrogen, one atom of carbon, and three atoms of chlorine. These atoms are joined to form one molecule. The same formula can be written for chloroform regardless of the source of the substance because the compositions of all pure compounds are constant. Once the identity and number of each of the atoms in a given compound are known, a formula can be written that describes its composition.

Example 4–2: Chemical equations *Explain the meaning of the equation:*
$CH_4 + 2O_2 \rightarrow CO_2 + 2H_2O$

CH_4 is the compound methane, CO_2 is carbon dioxide, and O_2 is oxygen. The equation says: One molecule of methane (consisting of one atom of carbon and four atoms of hydrogen) combines with two molecules of oxygen (each molecule consists of two atoms) to yield one molecule of carbon dioxide (each molecule contains one carbon atom and two oxygen atoms) plus two molecules of water.

Table 4–1 lists the formulas for a number of compounds and the percentages of each of the elements they contain. Each of these compounds has its own unique formula; each has a definite composition.

TABLE 4–1 CHEMICAL COMPOSITION OF SOME COMMON COMPOUNDS

Compound	Formula	Mass percent of each element		
Table salt	$NaCl$	Na 39.3	Cl 60.7	
Sugar (sucrose)	$C_{12}H_{22}O_{11}$	C 42.2	H 6.4	O 51.4
Carbon dioxide	CO_2	C 27.3	O 72.7	
Alcohol (ethyl alcohol)	CH_3CH_2OH	C 52.2	H 13.0	O 34.8
Lye (sodium hydroxide)	$NaOH$	Na 57.5	O 40.0	H 2.5
Dimethyl ether	CH_3OCH_3	C 52.2	H 13.0	O 34.8
Washing soda (sodium carbonate)	Na_2CO_3	Na 43.4	C 11.3	O 45.3

Example 4–3: Law of definite composition *On analysis, it is found that a sample of sand contains 5.62 g of silicon and 6.40 g of oxygen. Another sample of sand contains 9.36 g of silicon and 10.64 g of oxygen. Show how these facts illustrate the law of definite composition.*

The ratio of silicon to oxygen in the first sample is

$$\frac{\text{Si}}{\text{O}} = \frac{5.62\text{ g}}{6.40\text{ g}} = \frac{5.62/6.40}{6.40/6.40} = \frac{0.879\text{ g Si}}{1.00\text{ g O}}$$

That is, there is 0.879 g of silicon for every 1.00 g of oxygen in the first sample. For the law to be illustrated, the same ratio must be obtained for the second sample. For this second sample,

$$\frac{\text{Si}}{\text{O}} = \frac{9.36\ \text{g}}{10.64\ \text{g}} = \frac{9.36/10.64}{10.64/10.64} = \frac{0.879\ \text{g Si}}{1.00\ \text{g O}}$$

Hence the law is illustrated.

Another method is to show that the percent composition is the same in both samples. The percent of oxygen in the first sample is

$$\text{percent of oxygen} = \frac{\text{mass of oxygen}}{\text{mass of compound}} \times 100$$

$$\frac{6.40\ \text{g O}}{5.62\ \text{g Si} + 6.40\ \text{g O}} \times 100 = 53.2\ \text{percent}$$

To illustrate the law, the same percentage must be obtained for the second sample:

$$\frac{10.64\ \text{g O}}{9.36\ \text{g Si} + 10.64\ \text{g O}} \times 100 = 53.2\ \text{percent}$$

Again, the law is found to be correct.

LAW OF MULTIPLE PROPORTIONS

The **law of multiple proportions** states that *when two elements combine to form more than one compound, the masses of one element that combine with a fixed mass of the other element in the various compounds are in the ratio of small whole numbers to one another.* Table 4–2 shows some examples of compounds that illustrate the law of multiple proportions.

TABLE 4–2 SOME COMPOUNDS THAT ILLUSTRATE THE LAW OF MULTIPLE PROPORTIONS

Compound	Formula	Mass ratios for atoms in compounds	
Nitrogen and oxygen:			
Nitrous oxide	N_2O	N:O	7:4
Nitric oxide	NO		7:8
Nitrogen trioxide	N_2O_3		7:12
Nitrogen dioxide	NO_2		7:16
Nitrogen pentoxide	N_2O_5		7:20
Carbon and oxygen:			
Carbon dioxide	CO_2	C:O	3:8
Carbon monoxide	CO		3:4
Carbon suboxide	C_3O_2		9:8
Carbon and hydrogen:			
Methane	CH_4	C:H	12:4
Ethane	C_2H_6		12:3
Ethene	C_2H_4		12:2
Ethyne	C_2H_2		12:1
Propane	C_3H_8		9:2

Water has the formula H_2O. Can there be compounds with formulas such as H_3O, HO_2, H_2O_2, and H_3O_2? The law of multiple proportions allows us to predict that there are. The compound H_2O_2, called hydrogen peroxide, contains two atoms of hydrogen and two atoms of oxygen in every molecule. The ratio of hydrogen to oxygen atoms is 1 to 1. You may ask why the formula is written as H_2O_2 rather than as HO. Experiments have shown that the **empirical formula** HO—*the simplest formula that shows the correct ratio of atoms*—does not represent the true structure of hydrogen peroxide and that the actual composition of the molecule is two hydrogen atoms united with two oxygen atoms. H_2O_2, then, is called the molecular formula. The **molecular formula** is the true formula for a compound; *it gives the actual number of each kind of atom present in the molecule.*

Example 4–4: Law of multiple proportions *An analysis of a compound composed of carbon and oxygen shows that it contains 27.3 percent carbon. Another compound of carbon and oxygen contains 42.8 percent carbon. Do these compounds illustrate the law of multiple proportions?*

To illustrate the law, we can compare the amount of carbon, C, in the first compound to a fixed mass of oxygen, O. In 100 g of the first compound, there are 27.3 g of carbon for every 72.7 g of oxygen. Hence,

$$\frac{27.3 \text{ g C}}{72.7 \text{ g O}} = \frac{x \text{ g C}}{1.00 \text{ g O}}$$

$x = 0.375$ g = the number of g of C for every g of O in the first compound.

Doing the same thing for the second compound gives

$$\frac{42.8 \text{ g C}}{57.2 \text{ g O}} = \frac{x \text{ g C}}{1.00 \text{ g O}}$$

$x = 0.748$ = the number of g of C for every g of O in the second compound.

Because the masses, 0.375 g and 0.748 g, bear a simple whole-number ratio to one another—1 to 2—the law is illustrated. The first compound has twice as much oxygen per carbon atom as does the second. The compounds are carbon dioxide, CO_2, and carbon monoxide, CO.

An important point is that atoms can have different arrangements in molecules. Table 4–1 above shows that the compounds sugar, ethyl alcohol, and dimethyl ether are all composed of the same elements; they differ only in the ratio in which these elements are combined. In fact, two of these compounds, ethyl alcohol and dimethyl ether, actually have the same percentage composition and the same empirical formula, C_2H_6O. This empirical formula is also the molecular formula, and it does indicate the correct composition; however, it does not take into account any possible variations in how the atoms are arranged in the molecule. The formulas given in Table 4–1—CH_3CH_2OH for ethyl alcohol and CH_3OCH_3 for dimethyl ether—are written to reflect different arrangements of the same atoms to produce different molecules, and thus are known as **structural formulas.** Such a phenomenon, in which the same percentages of the same elements are arranged in two or more different ways to produce two or more different compounds, is called **structural isomerism.**

Thus, when you see the formula for a compound written in what appears to be an awkward manner (for example, CH_3CH_2OH for ethyl alcohol rather than C_2H_6O), just remember there is a reason for it: It is an attempt to structurally distinguish that compound from others with the same molecular formula.

LAW OF CONSERVATION OF MASS-ENERGY

The **law of conservation of mass-energy** states that *matter and energy can be neither created nor destroyed, although it is possible to convert one to the other.* If you determine the combined masses of all the elements in a chemical reaction such as burning and then determine the mass of all the products produced by the reaction, you will find that there is no apparent loss of matter in the process; that is, the masses before and after are identical. For example, when a candle is burned, the masses of the gaseous products produced equal the masses of the original candle and the oxygen consumed by the burning.

In 1905, Albert Einstein (1879–1955) formulated his famous equation, $E = mc^2$. This means that the energy, E, is equivalent to the mass, m, times the square of the velocity of light, c^2. In a vacuum, light moves at 3.00×10^8 meters per second, or approximately 186,000 miles per second.

When heat, which is one form of energy, is liberated during a chemical reaction such as the burning of wood, the source of the energy is the reactants themselves. Such a process is called an **exothermic reaction,** since heat is given off in the course of the chemical change. During the reaction there is actually a loss of mass and energy, but in such ordinary (non-nuclear) chemical reactions, the loss is much too small to be measured, even by the most sensitive laboratory apparatus.

In a strict sense, the Einstein equation does not say that matter can be converted to energy and vice versa. It means that a loss in mass corresponds to a loss in energy. The energy and mass lost, however, are gained by some other matter in the universe. Thus there is an overall conservation of mass and energy.

THE MOLE CONCEPT APPLIED TO MOLECULES AND COMPOUNDS

The mass of a molecule is simply the sum of the masses of all the individual atoms in that molecule.

Example 4–5: Molecular masses *What is the molecular mass of the compound ethyl alcohol, which has the formula CH_3CH_2OH?*

The molecular mass is the sum of the atomic masses:

For 2 carbon atoms:	2×12.0 amu	= 24.0 amu
For 6 hydrogen atoms:	6×1.00 amu	= 6.0 amu
For 1 oxygen atom:	1×16.0 amu	= 16.0 amu
Molecular mass of C_2H_5OH		= 46.0 amu

Example 4–6: Moles *What is the mass of one mole of table sugar, $C_{12}H_{22}O_{11}$?*

Add the masses of each atom present in the molecule.

For 12 carbon atoms:	12×12.0 amu =	144 amu
For 22 hydrogen atoms:	22×1.0 amu =	22 amu
For 11 hydrogen atoms:	11×16.0 amu =	176 amu
One molecular mass	=	342 amu

Because the molecular mass is 342 amu, the mass of one mole is 342 g.

Example 4–7: Moles *How many grams of the compound $CaCl_2$ are needed to make one mole?*

To solve, simply add the atomic masses of each atom present in the formula $CaCl_2$ and express this in gram units.

For 1 calcium atom:	1×40.1 amu =	40.1 amu
For 2 chlorine atoms:	2×35.5 amu =	71.0 amu
Formula mass of $CaCl_2$:	=	111.1 amu

Hence, 111 g of $CaCl_2$ equals one mole. Later we will find that the compound $CaCl_2$ does not exist as $CaCl_2$ molecules but as calcium and chloride ions (charged particles), hence we use the term **formula mass** instead of molecular mass.

A **mole** can best be defined as the amount of material that contains Avogadro's number of particles—that is, 6.02×10^{23} particles per mole. One mole equals the atomic mass expressed in grams; for molecules, one mole equals the molecular mass express in grams. For example, one mole of ethyl alcohol, which has a molecular mass equal to 46.0 amu, is 46.0 g. Throughout the rest of this book, we will abbreviate Avogadro's number as *N*. For instance, consider the sugar molecule, $C_{12}H_{22}O_{11}$, which we studied above in Example 4–6. We may say that 144 g of carbon, which is 12 times the carbon's atomic mass of 12.0, contains $12N$ atoms of carbon; that is, it is 12 times Avogadro's number *N*. Similarly, 22 g of hydrogen contains $22N$ atoms of hydrogen, and 176 g of oxygen contains $11N$ atoms of oxygen. Hence the 342 g of sugar contains a total of $12N + 22N + 11N = 45N$ atoms. However, because each molecule of sugar contains 45 atoms, there must be $45N$ divided by 45 or *N* molecules of sugar in 342 g. The molecular mass expressed in grams, one mole, for any compound is the amount that contains Avogadro's number of molecules. (The definition of a mole applies not only to atoms and molecules, but also to any object such as subatomic particles and electrons.) Some mole relationships are illustrated in Table 4–3. The mole is undoubtedly the most important term in quantitative chemistry. It is used to express the number of particles or the amount of a compound in a chemical system.

TABLE 4–3 MOLE RELATIONSHIPS

Substance	Formula	Mass	Mass of one mole	Number of particles ($N = 6.02 \times 10^{23}$)
Helium	He	4.00 amu	4.00 g	N atoms
Oxygen	O_2	32.00 amu	32.00 g	N molecules (or $2N$ atoms)
Water	H_2O	18.0 amu	18.0 g	N molecules (or $3N$ atoms)
Mercury	Hg	200.6 amu	200.6 g	N atoms

Example 4–8: Moles *How many moles are contained in 100 g of the compound urea, $CO(NH_2)_2$?*

To find the number of atoms for the elements inside the parentheses of $CO(NH_2)_2$, we multiply everything inside the parentheses by the subscript outside and to the right of the parentheses, in this case, 2. In other words, the empirical formula for $CO(NH_2)_2$ is CON_2H_4.

To find the number of moles in 100 g of CON_2H_4, we first calculate the molecular mass of the compound.

For 1 carbon atom: 1×12.0 amu = 12.0 amu
For 1 oxygen atom: 1×16.0 amu = 16.0 amu
For 2 nitrogen atoms: 2×14.0 amu = 28.0 amu
For 4 hydrogen atoms: 4×1.0 amu = $\underline{4.0 \text{ amu}}$
Molecular mass of $CO(NH_2)_2$ = 60.0 amu

Hence, one mole equals 60.0 g. Because

$$\text{number of moles} = \frac{\text{mass in g}}{\text{molecular mass (g/mole)}}$$

we get for 100 g

$$\text{number of moles} = \frac{100 \not{g}}{60.0 \not{g}/\text{mole}} = 1.67 \text{ moles}$$

Example 4–9: Number of atoms per mole *How many atoms are in one mole of oxygen gas?*

The answer to this question is not 6.02×10^{23} atoms! Remember that oxygen gas is O_2—two atoms per molecule. One mole of oxygen gas contains 6.02×10^{23} molecules, and because there are two oxygen atoms in one O_2 molecule, the answer to the question is $2 \times 6.02 \times 10^{23}$ atoms or 1.20×10^{24} atoms.

Example 4–10: Converting moles to grams *How many grams of sodium chloride are needed to prepare a solution that contains 2.40 moles of sodium chloride, NaCl?*

The atomic mass for sodium is 23.0 amu and that for chlorine is 35.5 amu. The sum of both is 58.5 amu. Hence one mole of sodium chloride is 58.5 g. To find the number of grams needed for 2.40 moles, we multiply by the factor 58.5 g/mole:

$$2.40\ \cancel{\text{moles}} \times 58.5\ \text{g}/\cancel{\text{mole}} = 140\ \text{g}.$$

Example 4–11: Review problem *How many moles of water, how many molecules, and how many atoms of hydrogen and oxygen are contained in one cup, 8.00 fluid ounces, of water?*

This problem will review some of the topics we have been discussing. First, we must convert the given volume to metric units, remembering that there are 32 fluid ounces in a quart.

$$8.00\ \cancel{\text{oz}} \times \frac{1.00\ \cancel{\text{qt}}}{32.0\ \cancel{\text{oz}}} \times \frac{1.00\ \cancel{\text{l}}}{1.06\ \cancel{\text{qt}}} \times \frac{1000\ \text{ml}}{1.00\ \cancel{\text{l}}} = 236\ \text{ml}$$

Since the density of water is 1.00 g/ml, the mass of water is

$$236\ \cancel{\text{ml}} \times \frac{1.00\ \text{g}}{1.00\ \cancel{\text{ml}}} = 236\ \text{g}$$

Now we need to know the molecular mass in order to find the number of grams in 1.00 mole. We get this by adding the sum of the atomic masses.

For oxygen:	1×16.0	$= 16.0$ amu
For hydrogen:	$2 \times\ \ 1.00$	$= \underline{\ 2.0\ \text{amu}}$
Molecular mass		$= 18.0$ amu, hence there are 18.0 g/mole

The cup of water thus contains

$$236\ \cancel{\text{g}} \times \frac{1.00\ \text{mole}}{18.0\ \cancel{\text{g}}} = 13.1\ \text{moles}$$

By definition, one mole of a compound contains 6.02×10^{23} molecules, hence

$$13.1\ \cancel{\text{moles}} \times \frac{6.02 \times 10^{23}\ \text{molecules}}{1.00\ \cancel{\text{mole}}} = 78.9 \times 10^{23} = 7.89 \times 10^{24}\ \text{molecules}$$

Finally, since each molecule of water contains two atoms of hydrogen and one atom of oxygen, there are $2 \times 7.89 \times 10^{24} = 1.58 \times 10^{25}$ atoms of hydrogen and 7.89×10^{24} atoms of oxygen.

PROGRAMMED SEQUENCE:

Turn now to p. 314 and work through Set 4 of the Programmed Sequences. These problems review the concept of the mole, solidifying your grasp of the idea and testing your understanding. Since the mole is so crucial to quantitative chemistry, make sure that you take this opportunity to learn it well.

THE SIZE OF MOLECULES

In the preceding chapter, we learned that atoms have radii that vary from approximately 3.7×10^{-11} m to 23.5×10^{-11} m. Because molecules are composed of atoms, we would expect their sizes to depend entirely on the number of atoms they contain and the manner in which the atoms are joined together.

Theoretically, there is no upper limit to the size of molecules. Some molecules are so large that they can actually be seen and weighed on a laboratory balance. Carbon atoms, for example, are able to unite with other carbon atoms to form long chains containing thousands of atoms, and these chains can be connected together to form extremely large molecules. These large molecules are called **polymers,** and the science that studies them is called **macromolecular chemistry.** Natural rubber is an example of a polymer; an average rubber molecule might contain 26,000 atoms.

NEW TERMS

c: velocity of light, 3.00×10^8 m/s.

Chemical equation: a statement of a chemical reaction using formula notation.

Empirical formula: a formula that shows the correct ratio of atoms combined to form a molecule, but does not necessarily represent the true structure of the compound. HO is the empirical formula for hydrogen peroxide, H_2O_2.

Exothermic: a change in which heat is given off.

Formula mass: the sum of the atomic masses expressed in grams for each atom in the formula of a compound. The term is used in place of molecular mass for compounds like $CaCl_2$ that exists in ionic form as a combination of calcium and chloride ions.

Gram atomic mass: the atomic mass expressed in grams.

Gram molecular mass: the molecular mass expressed in grams.

Ion: a charged atom or molecule.

Law of conservation of mass-energy: matter and energy can be neither created nor destroyed, although it is possible to convert one to the other.

Law of definite composition: a given compound always contains the same elements in the same ratio by mass.

Law of multiple proportions: when two elements combine to form more than one compound, the masses of one element that combine with a fixed mass of the other element in the various compounds are in the ratio of small whole numbers to one another.

Mole: the amount of material that contains Avogadro's number of particles (6.02×10^{23} particles/mole).

Molecular formula: the true formula for a compound; it gives the actual number of each kind of atom present in the molecule.

Molecular mass: the sum of the masses of all the atoms that compose a molecule—expressed in amu.

Polymer: a high-molecular-mass compound.

Structural isomerism: the phenomenon of compounds with the same molecular formula but different structural formulas.

PROBLEM SET A

1. A sample of a chemical compound on analysis was found to contain 107.9 g of silver and 35.45 g of chlorine. What is the percent by mass of silver in the compound?
2. On analysis, it was found that a sample of salt contained 0.224 g of sodium and 0.346 g of chlorine. Another sample of salt contained 39.3 percent sodium. Show that the law of definite composition is illustrated.
3. A certain oxide of iron (a compound of iron and oxygen) contains 77.7 percent by mass of iron. A different oxide contains 69.9 percent by mass of iron. Show that the law of multiple proportions is illustrated by performing the following calculations:
   a. What is the mass of iron per 1.00 g of oxygen in the first compound?
   b. What is the mass of iron per 1.00 g of oxygen in the second compound?
   c. What is the ratio of the two numbers found in a and b? Express the ratio using two whole (integer) numbers.
4. A compound of carbon, hydrogen, and oxygen was analyzed for carbon and hydrogen only, and was found to contain 68.4 percent carbon and 13.6 percent hydrogen by mass. Calculate the mass of oxygen in a 1.50 g sample of this compound.
5. When a 0.503 g sample of magnesium reacted with oxygen, the product had a mass of 0.834 g. What is the percent by mass of oxygen in the product (magnesium oxide)?
6. What is the velocity of light expressed in ft/min?
7. What is the molecular or formula mass for the following compounds?
   a. O_2
   b. N_2O_5
   c. $CaSO_4$
   d. CH_3CO_2H
   e. $KClO_3$
8. What is the mass of one mole of each of the compounds of Problem 7?
9. How many molecules are in 3.00 moles of nitrogen gas; how many atoms?
10. How many grams are in one mole of the following compounds?
    a. $C_{12}H_{22}O_{11}$ (sucrose)
    b. H_2SO_4 (sulfuric acid)
    c. $C_2H_5OC_2H_5$ (ethyl ether)
    d. C_8H_{18} (octane)
    e. KNO_3 (potassium nitrate)
11. How many moles are in each of the following?
    a. 50.0g of H_2O
    b. 100 g of $ZnSO_4$
    c. 20.0 g of CH_3OH
    d. 50.0 g of CH_3CO_2H
    e. 90.0 g of UCl_4
12. How many grams are in each of the following?
    a. 15.0 moles of CO_2
    b. 2.50 moles of NH_3
    c. 0.670 mole of HCHO
13. Rubber consists of a large number of $CH_2C(CH_3)CHCH_2$ units. How many of these units are contained in a typical rubber molecule with a molecular mass of 400,000?
14. Complete the following table:

Formula of compound	Molecular or formula mass, amu	Mass of 1.00 mole of the compound, g	Number of moles in 100 g of compound	Number of moles of O atoms in 100 g
H_2O	___	___	___	___
H_2O_2	___	___	___	___
CH_3OH	___	___	___	___
CH_3CO_2H	___	___	___	___

15. Use some of the information from Problem 14 to complete the following table:

Formula of compound	Moles in 0.500 g	Molecules in 0.500 g	Moles of H atoms in 0.500 g	Number of H atoms in 0.500 g
H_2O	___	___	___	___
H_2O_2	___	___	___	___
CH_3OH	___	___	___	___
CH_3CO_2H	___	___	___	___

16. Elemental sulfur is sometimes in the form of S_8 molecules. Calculate the number of grams per mole of S_8 molecules.
17. Elemental chlorine is diatomic. How many atoms of chlorine are contained in a 500 g sample?

PROBLEM SET B

1. Analysis showed a 0.1037 g sample of a compound of gold and chlorine to contain 0.0673 g of gold. What is the percent by mass of chlorine in this compound?
2. Two different oxides of nitrogen, A and B, were analyzed. Oxide A contained 63.6 percent by mass of nitrogen and oxide B contained 30.4 percent of nitrogen by mass.
   a. What is the mass of nitrogen per 1.00 g of oxygen in compound A?
   b. What is the mass of nitrogen per 1.00 g of oxygen in compound B?
   c. What is the ratio of these two numbers from a and b? Give your answer in whole (integer) numbers.
3. Burning a 0.503 g sample of sulfur converted it into 1.005 g of an oxide of sulfur. What is the percent by mass of oxygen in this oxide?
4. What is the molecular or formula mass for the following compounds?
   a. ZnS
   b. C_4H_8
   c. MnO_2
   d. CaC_2
   e. $CaCO_3$
5. What is the mass of 3.00 moles of each of the compounds in Problem 4?
6. How many molecules are in 2.50 moles of carbon dioxide gas (CO_2)? How many atoms?
7. How many grams are in 2.00 moles of the following compounds?
   a. $NaNO_3$
   b. $BaSO_4$
   c. U_3O_8
   d. $C_{10}H_{22}$
8. How many moles are in the following?
   a. 250 g of $NaCl$
   b. 0.450 g of SO_2
   c. 100 g of $CH_3CH_2CH_2OH$
   d. 60.0 g of HNO_3

9. How many grams are in each of the following?
   a. 12.0 moles of H_2O
   b. 500 moles of CO
   c. 3.00 moles of KNO_3

10. Complete the following table:

Formula of compound	Molecular or formula mass, amu	Mass of 1.00 mole of the compound, g	Number of moles in 100 g of compound	Number of moles of N atoms in 100 g
NH_3	____	____	____	____
N_2H_4	____	____	____	____
NH_2OH	____	____	____	____

11. Use some of the information from Problem 10 to complete the following table:

Formula of compound	Moles in 0.500 g	Molecules in 0.500 g	Moles of H atoms in 0.500 g	Number of H atoms in 0.500 g
NH_3	____	____	____	____
N_2H_4	____	____	____	____
NH_2OH	____	____	____	____

12. Given the fact that elemental phosphorus is tetratomic, P_4, how many atoms of phosphorus are found in a 150 g sample of elemental phosphorus?

13. Oxygen is sometimes present in the triatomic form called ozone, O_3. Calculate the number of grams per mole for the compound ozone.

5

chemical formulas and equations

THIS WILL SAVE YOU TIME:

Refer to this list as you read through the chapter. In answering the questions, you will understand the important concepts you are most likely to be asked about later on.

After reading this chapter, you should be able to:

1. *Determine the empirical formula for a compound from percent composition data.*
2. *Determine the molecular formula for a compound from percent composition data and the molecular mass.*
3. *Calculate percent compositions from formulas.*
4. *Write and balance chemical equations. (Excluding the more complicated oxidation-reduction reactions, which are discussed in Chapter 16.)*
5. *Give examples of three types of chemical reactions.*

Every one of the more than a million known molecules has its own chemical formula. These formulas are a precise description of the number, and in some cases the arrangement, of atoms in the molecule. In this chapter, we will show how formulas are determined from experimental data. A formula, however, is much more than simply a label with numbers attached. We will see how formulas serve as keys to such calculations as percent composition by mass and what role they play in the writing and balancing of chemical equations.

DETERMINING EMPIRICAL FORMULAS

According to the law of definite composition, the percentage composition of any pure compound is constant. Knowing this and the relative masses of the atoms in the compound, we can calculate the simplest, or empirical, formula of a compound. The following examples show how.

Example 5–1: Determining an empirical formula *A 2.514 g sample of a pure compound consists of 2.020 g of zinc and 0.494 g of oxygen. What is the empirical formula for this compound?*

First, we find how many moles of each element are present. For zinc, Zn, this is

$$2.020 \not{g}\, \not{Zn} \times \frac{1 \text{ mole Zn}}{65.4 \not{g}\, \not{Zn}} = \frac{2.020 \text{ moles Zn}}{65.4} = 0.0309 \text{ mole Zn}$$

For oxygen, O, it is

$$0.494 \not{g}\, \not{O} \times \frac{1 \text{ mole O}}{16.0 \not{g}\, \not{O}} = \frac{0.494 \text{ mole O}}{16.0} = 0.0309 \text{ mole O}$$

Because the number of moles of zinc and oxygen are the same, the number of zinc and oxygen atoms must also be the same. Hence the simplest, or empirical, formula is Zn_1O_1, which is ZnO.

Example 5–2: Determining an empirical formula *What is the empirical formula for a compound that contains 56.4 percent phosphorus and 43.6 percent oxygen?*

By definition, percent means parts per hundred; therefore, 100 g of the compound contains 56.4 g of phosphorus and 43.6 g of oxygen. To find the number of moles,

$$\frac{56.4 \not{g}\, \not{P}}{100 \text{ g compound}} \times \frac{1.00 \text{ mole P}}{31.0 \not{g}\, \text{P}} = \frac{1.82 \text{ moles P}}{100 \text{ g compound}}$$

Similarly for oxygen,

$$\frac{43.6 \not{g}\, \not{O}}{100 \text{ g compound}} \times \frac{1.00 \text{ mole O}}{16.0 \not{g}\, \not{O}} = \frac{2.72 \text{ moles O}}{100 \text{ g compound}}$$

Because a mole of any element contains the same number of atoms as a mole of any other element, the two numbers 1.82 and 2.72 tell us the actual atomic ratio of the two elements in the compound. The formula $P_{1.82}O_{2.72}$ is incorrect, however, for only whole atoms can exist. To obtain small whole numbers, divide each by the smaller:

$$P_{1.82/1.82}O_{2.72/1.82} = P_1O_{1.5}$$

Finally, multiplying by 2 removes the fractions and gives the correct empirical formula P_2O_3. The compound is called phosphorus trioxide.

DETERMINING MOLECULAR FORMULAS

True, or molecular, formulas cannot be obtained from percent composition data alone. In the preceding problem we found that the empirical formula for a compound containing 56.4 percent phosphorus and 43.6 percent oxygen is P_2O_3; this does not happen to be the molecular formula. By separate experimental methods it has been found that one mole of this compound has a mass of 220 g. If the compound had the formula P_2O_3, the mass of one mole would be 110 g, half the experimentally determined value of 220. Obviously the molecular formula must be $(P_2O_3)_2$ or P_4O_6. The molecular formula is always equal to (empirical formula)$_n$, where n can be any integer from one to thousands. To find n,

$$n = \frac{\text{mass of one mole of compound determined by experiment}}{\text{mass of one mole calculated from empirical formula}}$$

Many experimental methods can be used to determine the molecular masses of compounds; two will be discussed in later chapters.

Example 5–3: Determining a molecular formula *An analysis of a pure compound of carbon and hydrogen showed that it contained 92.3 percent carbon by mass. A separate experiment showed that the mass of one mole was 78 g. What is the molecular formula for the compound?*

First find the empirical formula.
For carbon,

$$92.3\ \cancel{g\,C} \times \frac{1.00\text{ mole C}}{12.0\ \cancel{g\,C}} = 7.69\text{ moles C/100 g compound}$$

For hydrogen,

$$7.7\ \cancel{g\,H} \times \frac{1.00\text{ mole H}}{1.00\ \cancel{g\,H}} = 7.7\text{ moles H/100 g compound}$$

The empirical formula is thus

$$C_{7.7/7.7}H_{7.7/7.7} = C_1H_1 = CH$$

To find the molecular formula $(CH)_n$

$$n = \frac{78 \text{ g/mole (experimental value)}}{13 \text{ g/mole (calculated from empirical formula)}} = 6$$

The molecular formula is thus $(CH)_6$, or C_6H_6.

CALCULATING PERCENT COMPOSITION FROM FORMULAS

The molecular formula of a compound tells us the exact ratio in which atoms have combined to form molecules. The CCl_4 (carbon tetrachloride) molecule, for example, consists of four chlorine atoms combined with one carbon atom to form each molecule. The four chlorine atoms have a combined mass of 4×35.5 or 142 amu, and the carbon atom has an atomic mass of 12.0 amu; hence the molecular mass of the compound is 154 amu. The percentage of chlorine in the compound is obtained by dividing the total mass of the chlorine atoms by the mass of the molecule and multiplying by 100 to get a percent value:

$$\text{Percent chlorine} = \frac{4(Cl)}{CCl_4} \times 100 = \frac{142 \text{ amu}}{154 \text{ amu}} \times 100 = 92.2 \text{ percent}$$

Similarly for carbon,

$$\text{Percent carbon} = \frac{C}{CCl_4} \times 100 = \frac{12.0 \text{ amu}}{154 \text{ amu}} \times 100 = 7.8 \text{ percent}$$

Example 5–4: Percent composition *What is the percentage of sulfur, S, in the compound H_2SO_4?*

The single sulfur atom has a mass of 32.1 amu and the molecular mass of the compound is 2(1.00) for hydrogen, 1(32.1) for sulfur, and 4(16.0) for oxygen, to give a total of 98.1 amu. The percentage of sulfur is thus

$$\frac{S}{H_2SO_4} \times 100 = \frac{32.1 \text{ amu}}{98.1 \text{ amu}} \times 100 = 32.7 \text{ percent S}$$

Example 5–5: Percent composition *What is the percent composition of the compound SO_3, sulfur trioxide?*

The procedure is:

1. Find the molecular mass of the compound.
2. Count the number of atoms of a given element and multiply by its gram atomic mass.
3. Divide the result of step 2 by the result of step 1.
4. Multiply by 100 to put the answer in percent.
5. Repeat for each element in the compound.

For SO_3, the mass of one mole is 1(32.1) + 3(16.0) = 80.1 g. For sulfur, the sum of the gram atomic masses is 1(32.1) = 32.1 g. Then following steps 3 and 4 gives

$$\text{Percentage of S} = \frac{\text{S}}{\text{SO}_3} \times 100 = \frac{32.1 \text{ g}}{80.1 \text{ g}} \times 100 = 40.1$$

Repeating the same procedure for oxygen,

$$\text{Percentage of O} = \frac{3\text{ O}}{\text{SO}_3} \times 100 = \frac{3(16.0 \text{ g})}{80.1 \text{ g}} \times 100 = 59.9$$

The percentage of oxygen could have been obtained by simply subtracting the value for sulfur from 100. By making the independent calculation, however, you get an accurate check of your results, because the two values must add to 100.

In the preceding two problems, note that the symbols and formulas are used both to describe the composition of a substance and to specify an exact amount—one mole.

PROGRAMMED SEQUENCE:

Turn now to p. 317 and work Set 5 of the Programmed Sequences. These problems deal with percent compositions and empirical formulas. Such operations are important to many of the quantitative topics that follow, and you should be sure that you are at home with these calculations before going on.

WRITING AND BALANCING CHEMICAL EQUATIONS

If you know the reactants and products of a chemical change, you can write a balanced **chemical equation**—a symbolic statement used by chemists to show both qualitatively and quantitatively what has taken place. ("Qualitative" denotes a description of something; "quantitative" refers to a measurement of how much of something is present.) In the preceding chapter, the reaction of hydrogen with oxygen appeared as a preliminary example of an equation. Before considering additional examples, study the following rules.

To write a chemical equation

1. First determine the correct identity of all reactants and products. This step involves laboratory experimentation.
2. Find the correct molecular formulas for all the substances involved.
3. Write the molecular formulas for the reactants and separate them with a + sign.
4. Show the physical state of the reactants by using one of the following notations:
   - (*g*) gas
   - (*l*) liquid
   - (*s*) solid
   - (*aq*) aqueous, dissolved in water

   Note: These symbols are often omitted from equations when not needed for clarity. The notation (*aq*) is usually omitted for ions in aqueous solution; ($Na^+ = Na^+(aq)$)
5. Following the formulas for the reactants place one of the three symbols:

$\rightarrow$	means yields, produces, or gives
$=$ or $\rightleftharpoons$	used to indicate that the reaction can go in either direction, that is, the reactants produce products at the same time that the products are reacting to yield reactants.

6. Finally, write the molecular formulas for the products, along with the designation for their physical state, separated by + signs.
7. You can place some of the conditions needed to carry out the reaction above or below the arrow. You may need to specify temperature, pressure, heat, or other reaction conditions.
8. Other notations are sometimes required in place of those listed in step 4. These are: $\uparrow$ means a gas is given off; $\downarrow$ means a solid precipitates (literally, "falls out").

The chemical equation resulting from these steps must now be balanced. To observe the law of conservation of mass-energy, there must be as many atoms of each element on the reactant side of the equation as on the product side. Balancing chemical equations is mainly a matter of trial and error. It is good procedure, however, to balance the elements other than hydrogen and oxygen first, keeping the oxygen atoms for last. Some equations are difficult to balance, and special techniques are necessary.

Writing and balancing a chemical equation—an example

Consider the reaction of nitrogen with hydrogen to form ammonia. The first step is to identify the correct molecular formulas for all the substances involved and designate their physical state:

Nitrogen = $N_2(g)$
Hydrogen = $H_2(g)$
Ammonia = $NH_3(g)$

The second step is to place the formulas for the reactants on the left and the products on the right and separate them with a double arrow. A double arrow is used because it has been determined experimentally that the ammonia will partially decompose to nitrogen and hydrogen under the conditions used for the reaction.

$$H_2(g) + N_2(g) \rightleftharpoons NH_3(g)$$

Now the equation must be balanced. We first note that there is one more nitrogen atom on the left than on the right; the NH_3 must be multiplied by 2:

$$H_2(g) + N_2(g) \rightleftharpoons 2NH_3(g)$$

(Note that it is incorrect to multiply the number of nitrogen atoms in NH_3 by two, that is, N_2H_3; this multiplication changes the formula, making the compound something other than ammonia.) We now have the nitrogen atoms balanced, but there are only two hydrogens on the left compared to a total of six on the right. Multiplying the number of hydrogen molecules by 3 remedies this imbalance:

$$3H_2(g) + N_2(g) \rightleftharpoons 2NH_3(g)$$

We now have a balanced chemical equation—the number of atoms of each element in the reactants is the same as in the products.

Example 5–6: Writing balanced chemical equations *Oxygen can be prepared in the laboratory by heating potassium chlorate, $KClO_3(s)$.* The products are oxygen, $O_2(g)$, *and potassium chloride, $KCl(s)$. Write a balanced chemical equation for the reaction.*

The unbalanced, or skeletal, equation is

$KClO_3(s) \rightarrow KCl(s) + O_2(g)$

A single arrow is used because we know experimentally that the reverse reaction, the combination of $KCl(s)$ and $O_2(g)$, does not take place.

To balance the number of oxygen atoms, it is necessary to find the least common multiple for 3 and 2, which is 6. Then by dividing 3 into 6, we find that $KClO_3(s)$ must be multiplied by 2; similarly, dividing 2 into 6, we find that $O_2(g)$ must be multiplied by 3.

$2KClO_3(s) \rightarrow KCl(s) + 3O_2(g)$

The K and Cl atoms can be balanced by multiplying $KCl(s)$ by 2:

$2KClO_3(s) \text{ heat} \rightarrow 2KCl(s) + 3O_2(g)$

Many chemical equations are difficult to balance. An example is

$Cu(s) + HNO_3(aq) \rightarrow Cu(NO_3)_2(aq) + H_2O + NO(g)$

Try to balance it, but do not be discouraged if you fail. This equation is an example of a class of reactions called oxidation-reduction reactions. These equations, and the special techniques needed to balance them, will be considered in Chapter 16. (The numbers needed to balance the above equation are 3, 8, 3, 4, 2.)

THE TYPES OF CHEMICAL EQUATIONS

Most chemical reactions can be classified as one of four types. An example or two of each type is given below. Try balancing the skeletal equations. Check your results with the balanced equations shown.

1. *Combination reactions: $A + B \rightarrow AB$*
   The combination of hydrogen and oxygen to give water and of hydrogen and nitrogen to give ammonia are examples. Two more are:
   a. When calcium oxide is added to water, there is a vigorous reaction, with the liberation of heat to form calcium hydroxide.

      Word equation: calcium oxide plus water gives calcium hydroxide.
      Skeletal equation: $CaO(s) + H_2O(l) \rightarrow Ca(OH)_2(s)$
      Balanced equation: $CaO(s) + H_2O(l) \rightarrow Ca(OH)_2(s)$

   b. Magnesium burns in air to produce magnesium oxide.

      Word equation: magnesium plus oxygen gives magnesium oxide.
      Skeletal equation: $Mg(s) + O_2(g) \rightarrow MgO(s)$
      Balanced equation: $2Mg(s) + O_2(g) \rightarrow 2MgO(s)$

2. *Decomposition reactions:* $AB \rightarrow A + B$
   Two examples are:
   a. Calcium carbonate decomposes to give calcium oxide and carbon dioxide.

      Word equation: calcium carbonate gives calcium oxide and carbon dioxide
      Skeletal equation: $CaCO_3(s) \rightarrow CaO(s) + CO_2(g)$
      Balanced equation: $CaCO_3(s) \rightarrow CaO(s) + CO_2(g)$

   b. Mercury oxide decomposes on heating to give mercury metal and oxygen.

      Word equation: mercury oxide yields mercury metal plus oxygen
      Skeletal equation: $HgO(s) \rightarrow Hg(l) + O_2(g)$
      Balanced equation: $2HgO(s) \rightarrow 2Hg(l) + O_2(g)$

3. *Replacement, or substitution, reactions:* $A + BC \rightarrow AC + B$
   In this type of reaction, the element A replaces one of the elements in the other compound. An example is the reaction of iron metal with hot steam to give iron oxide and hydrogen.

   Word equation: iron plus water yields iron oxide plus hydrogen
   Skeletal equation: $Fe(s) + H_2O(g) \rightarrow Fe_3O_4(s) + H_2(g)$
   Balanced equation: $3Fe(s) + 4H_2O(g) \rightarrow Fe_3O_4(s) + 4H_2(g)$

4. *Exchange, or double replacement, reactions:* $AB + CD \rightarrow AD + CB$
   It is difficult to find a common inorganic (compounds containing no carbon) reaction that can be correctly classified as a double replacement reaction. Reactions such as

   $HCl(aq) + NaOH(aq) \rightarrow NaCl(aq) + H_2O(l)$

   and

   $NaCl(aq) + AgNO_3(aq) \rightarrow NaNO_3(aq) + AgCl(s)$

   are actually combination reactions involving ions, charged particles. The latter, for example, is actually*

   $Ag^+ (aq) + Cl^- (aq) \rightarrow AgCl(s)$

   Hence, no examples will be cited for exchange reactions.

In studying chemical equations, you should direct your attention primarily to the scheme for writing and balancing. Do not be discouraged if you do not understand the system for naming the various compounds or why calcium nitrate's formula is $Ca(NO_3)_2$ rather than CaN_2O_6 or $CaNO_3$—these topics are treated in the next chapters.

NEW TERMS

Combination reaction: $A + B \rightarrow AB$
Decomposition reaction: $AB \rightarrow A + B$
Double displacement reaction: $AB + CD \rightarrow AD + CB$

* The topic of net ionic equations is discussed in Chapter 15.

Ionic reaction: a reaction involving charged species.

Precipitate: a solid product that separates from a reaction mixture.

Replacement reaction: $A + BC \rightarrow AC + B$

Substitution reaction: $A + BC \rightarrow AC + B$

PROBLEM SET A

1. A pure compound contains 79.9 percent copper and 20.1 percent oxygen. What is the empirical formula for the compound?
2. Several pure compounds contain iron and oxygen, Fe_xO_y. Three of these have iron contents of 69.9, 72.4, and 77.8 percent, respectively. What are their empirical formulas?
3. A 5.000 g sample of uranium was heated in air and converted to 5.897 g of a pure compound containing uranium and oxygen, U_xO_y. What is the empirical formula for the compound?
4. An analysis of a compound showed it to consist of 19.3 percent sodium, 26.9 percent sulfur, and 53.8 percent oxygen. A molecular mass determination showed one mole to equal 238 g. What is the molecular formula?
5. What is the percentage composition of zinc sulfide, ZnS?
6. How much sulfur is contained in a 100 g sample of sodium sulfate, Na_2SO_4?
7. How much arsenic is in 100 g of an insecticide that contains 32.0 percent of the compound As_2O_5?
8. Calculate the percent of hydrogen in acetic acid, CH_3CO_2H.
9. Given the formula for tetraethyl lead, $Pb(C_2H_5)_4$. Calculate the percent composition for the compound.
10. A 50.0 g sample of the pure compound MnO_2 is converted into elemental manganese and elemental oxygen. How many grams of Mn can be produced if the conversion process is 100 percent?
11. Balance the following chemical equations:
    a. $NaOH(aq) + H_2S(g) \rightarrow Na_2S(aq) + H_2O(l)$
    b. $Fe(s) + O_2(g) \rightarrow Fe_2O_3(s)$
    c. $FeS_2(s) + O_2(g) \rightarrow Fe_2O_3(s) + SO_2(g)$
    d. $NH_3(g) + O_2(g) \rightarrow NO_2(g) + H_2O(g)$
    e. $C_4H_{10}(g) + O_2(g) \rightarrow CO_2(g) + H_2O(g)$
12. Balance the following reactions and label them by type:
    a. $Al(s) + Cr_2O_3(s) \rightarrow Al_2O_3(s) + Cr(s)$
    b. $NO(g) + O_2(g) \rightarrow NO_2(g)$
    c. $K_2CO_3(aq) + Sr(NO_3)_2(aq) \rightarrow KNO_3(aq) + SrCO_3 \downarrow$
    d. $C_7H_{14}(g) \rightarrow C_7H_8(g) + H_2(g)$
    e. $AgNO_3(aq) + NaI(aq) \rightarrow AgI(s) + NaNO_3(aq)$
13. Given the following balanced chemical equation:
    $2H_2 + O_2 \rightarrow 2H_2O$
    If 50.5 g of reactants are completely consumed, how many grams of products are formed? Why?

PROBLEM SET B

1. A pure compound, cobalt sulfide, is composed of 55.0 percent cobalt and 45.0 percent sulfur. Calculate the empirical formula for the compound.

2. Several different pure compounds contain only manganese and oxygen, Mn_xO_y. Three of these have manganese contents of 53.4, 63.2, and 72.1 percent, respectively. Calculate the empirical formula for each of the three compounds.
3. A 15.00 g sample of elemental mercury was converted to 17.40 g of a compound Hg_xS_y. What is the empirical formula for this compound?
4. A pure compound was found to have the following percent composition: 40.44 percent carbon, 7.92 percent hydrogen, 35.92 percent oxygen, and 15.72 percent nitrogen. An additional independent experiment found the mass of one mole to be equal to 89.1 g. What is the molecular formula for the compound?
5. Calculate the percentage composition of the compound C_2F_4. This compound, tetrafluoroethylene, is the basic unit in the commercial plasic Teflon.
6. How many grams of sulfur are contained in a one pound box of saccharin, empirical formula $C_7H_5O_3NS$?
7. Calculate how many grams of nitrogen are contained in a 1.00 kg sample of morphine, $C_{17}H_{19}NO_3$, a drug used as a pain killer.
8. Calculate the percentage of oxygen in lysergic acid, $C_{16}H_{16}N_2O_2$.
9. Ethylene glycol, $C_2H_6O_2$, is the primary ingredient in permanent-type antifreeze products. Calculate the percentage composition for this compound.
10. Balance the following chemical equations:
    a. $H_2(g) + N_2(g) \leftrightharpoons NH_3(g)$
    b. $CaCl_2(aq) + Na_2CO_3(aq) \rightarrow NaCl(aq) + CaCO_3(s)$
    c. $Al(OH)_3(s) \rightarrow Al_2O_3(s) + H_2O(g)$
    d. $B_2O_3(s) + H_2O \rightarrow H_3BO_3(aq)$
    e. $N_2O_5(g) \rightarrow NO_2(g) + O_2(g)$
11. Balance the following equations and label them by type:
    a. $Li(s) + N_2(g) \rightarrow Li_3N(s)$
    b. $BeF_2(aq) + H_2O \rightarrow Be(OH)_2(s) + HF(l)$
    c. $AgCN(s) \rightarrow Ag(s) + C_2N_2(g)$
    d. $P_4O_{10}(s) + H_2O \rightarrow H_3PO_4(aq)$
    e. $Al(s) + AgNO_3(aq) \rightarrow Al(NO_3)_3(aq) + Ag(s)$
12. Pyrite, FeS_2, is a common mineral. How many grams of pure iron metal could be obtained from an ore sample of 25.00 g that contained 37.0 percent by mass of FeS_2? Assume that the FeS_2 is the only source of iron in the ore and that the process is 100 percent efficient.

6

how to write formulas and name compounds

THIS WILL SAVE YOU TIME:

Refer to this list as you read through the chapter. In answering the questions, you will understand the important concepts you are most likely to be asked about later on.

After reading this chapter, you should be able to:

1. *Identify binary and ternary compounds from their formulas.*
2. *State the five basic rules for defining oxidation numbers.*
3. *Explain what is meant by the periodicity of the oxidation numbers and illustrate the concept.*
4. *Given the position of an element in the periodic table, predict its positive and negative oxidation numbers.*
5. *Name the common elements with two important oxidation numbers and tell what these numbers are.*
6. *List the Greek prefixes for the numbers 1 through 10.*
7. *State the basic rules for naming binary compounds.*
8. *Name binary compounds using both the recommended Stock system and the old -ous and -ic system.*
9. *Name binary acids.*
10. *Explain the use of the suffixes: -ide, -ate, -ous, and -ic.*
11. *List the names, formulas, and charges for the common polyatomic ions.*
12. *Name the common acids containing hydrogen combined with a polyatomic group.*
13. *Give the names and formulas for the important compounds listed in Table 6–8 that do not follow the general nomenclature rules.*
14. *Tell what is meant by oxyacids and salts.*
15. *Name hydrates.*

Ideally, the name given to a chemical substance should distinguish it clearly and unambiguously from all other known substances. Naming the elements presents no particular problem, simply because there are but 106 of them, but with more than one million compounds, naming becomes a gargantuan and difficult task. Over the centuries and years, chemists have developed a predictable system of chemical nomenclature that also serves as a form of classification.

The first basic distinction in nomenclature is between organic and inorganic compounds, the former term referring to most of those compounds containing the element carbon. In this chapter, we shall discuss some of the basic nomenclature for inorganic compounds. These compounds can be divided into three convenient categories: **binary compounds,** those containing two different elements; **ternary compounds,** those containing three different elements; and the remaining more complex compounds containing four or more elements. Our discussion of nomenclature begins with the study of oxidation numbers and the problem of predicting formulas for compounds.

PREDICTING FORMULAS FOR COMPOUNDS

What is the formula for the binary compound formed from aluminum and oxygen? By convention, the symbol for the more metallic element is written first in the formula. In comparing elements, the one that is farthest to the left or to the bottom of the periodic table is the more metallic. Thus the basic formula for the aluminum and oxygen compound is Al_xO_y. Now the problem is to find *x* and *y*. From our previous discussions we know that the actual formula can be derived only from percent composition data obtained in laboratory analyses plus an experimental determination of the molecular mass of the compound. However, some useful rules based on the concept of oxidation numbers enable us to predict formulas for a great number of compounds. In this example, knowing that the oxidation number for aluminum is $+3$ and -2 for oxygen allows us to predict the correct formula, Al_2O_3, for this compound.

OXIDATION NUMBERS

One of the most important periodic properties of elements is their oxidation state, or oxidation number. We shall define **oxidation number** by the following rules:

1. The oxidation number for any uncombined atom or free element—for example, N_2—is zero.
2. The oxidation number for oxygen is -2.
3. The oxidation number for hydrogen is $+1$.
4. For single ions, the oxidation number is equal to the charge on the ion.
5. The sum of the oxidation numbers for all atoms of all elements in a given formula equals zero; the sum equals the overall charge in the case of a polyatomic ion. (A polyatomic ion is a charged particle containing more than one atom.)

Example 6–1: Oxidation numbers *What is the oxidation number of arsenic in the compound As_2O_5?*

By definition, the oxidation number of each oxygen atom is −2 and the sum of the oxidation numbers of all the atoms must add to zero. Hence,

2(oxidation number of arsenic) + 5(oxidation number of oxygen) = 0
2(oxidation number of arsenic) + 5(−2) = 0
2(oxidation number of arsenic) − 10 = 0
2(oxidation number of arsenic) = +10
oxidation number of arsenic = +5

The formula for the compound carbon dioxide, CO_2, shows that the oxidation number for carbon is +4; for boron in the compound B_2O_3, it is +3, and so on. Examination of a series of known formulas for some binary compounds containing oxygen (Table 6–1) shows a definite correlation between the oxidation numbers and the groups in which the elements are located in the periodic table. All the elements of group 1A have a +1 oxidation number, the elements of group IIA have a +2 oxidation number, and so forth. In general, a positive oxidation number for any element is equal to the element's group number in the periodic table. For example, nitrogen is +5, carbon +4.

TABLE 6–1 FORMULAS FOR BINARY OXYGEN-CONTAINING COMPOUNDS OF SOME ELEMENTS

Group in periodic table						
I	II	III	IV	V	VI	VII
K_2O	CaO	B_2O_3	CO_2	N_2O_5	SO_3	Cl_2O_7
Cs_2O	BaO	Al_2O_3	SiO_2	As_2O_5	CrO_3	

The negative oxidation number for any element can be obtained by subtracting the group number from 8 and giving the difference a negative sign (Table 6–2): fluorine is −1, sulfur −2, and nitrogen −3.

TABLE 6–2 EXPECTED OXIDATION NUMBERS FOR ELEMENTS

Group in periodic table						
I	II	III	IV	V	VI	VII
+1	+2	+3	+4	+5	+6	+7
			−4	−3	−2	−1

Predicting a chemical formula simply involves joining elements with positive oxidation numbers to those with negative oxidation numbers, keeping it in mind that the sum of all numbers in the final formula must total zero.

PROGRAMMED SEQUENCE:

Turn to p. 322 and work Set 6 of the Programmed Sequences. These exercises review the handling of negative numbers, which is important to correct calculation of oxidation states. Make sure that you feel comfortable with negative numbers before going on.

Example 6–2: Predicting formulas *What is the predicted formula for a compound of calcium and fluorine?*

Calcium is in group IIA of the periodic table, so its oxidation number is +2. Fluorine is in group VIIA, and its oxidation number is +7 or −1. It is a general rule that the more metallic elements have the positive oxidation states and are written first in the formula. Hence the formula is CaF_2, because it takes two of the −1 fluorines to balance the +2 calcium.

Example 6–3: Predicting formulas *Predict the formula for the compound of the elements antimony, Sb, and sulfur.*

Antimony has an oxidation number of +5 or −3. Because it is to the left of sulfur in the periodic table, it is the more metallic of the two and will have the positive oxidation number. Sulfur has an oxidation number of +6 or −2, but in this compound, the −2 is the correct choice. The formula is thus Sb_2S_5; two +5 antimonys are needed to balance the five −2 sulfurs.

Unfortunately, there are many exceptions to this system of assigning oxidation numbers from the position of the elements in the periodic table. The major one is that many elements have more than one oxidation number, and therefore the oxidation numbers of these elements have to be memorized. Before memorizing a long list, however, check the group under which the element is located in the periodic table. If the oxidation number coincides with the group number, there is no reason to memorize it, simply because periodic tables are always available to chemistry students. Table 6–3 lists a number of common elements whose oxidation numbers should be memorized.

The series of compounds N_2O, NO, NO_2, N_2O_3, N_2O_4, and N_2O_5 illustrate some of the oxidation numbers exhibited by the element nitrogen. In these compounds nitrogen is +1, +2, +4, +3, +4, and +5, respectively. Only the formula N_2O_5 can be predicted from the periodic relationships. A compound can always be found that obeys the rule, however: CO_2, SO_3, and P_4O_{10} do, CO, SO_2, and P_2O_6 do not.

TABLE 6–3 OXIDATION NUMBERS OF SOME COMMON ELEMENTS THAT CANNOT BE PREDICTED FROM THE PERIODIC TABLE

Element	Oxidation number	Name
Cobalt	+2	Cobalt(II), cobaltous
	+3	Cobalt(III), cobaltic
Chromium	+3	Chromium(III), chromic
Copper	+1	Copper(I), cuprous
	+2	Copper(II), cupric
Iron	+2	Iron(II), ferrous
	+3	Iron(III), ferric
Lead	+2	Lead(II), plumbous
	+4	Lead(IV), plumbic
Manganese	+2	Manganese(II), manganous
Mercury	+1	Mercury(I), mercurous
	+2	Mercury(II), mercuric
Nickel	+2	Nickel(II), nickelous
Tin	+2	Tin(II), stannous
	+4	Tin(IV), stannic

NAMING COMPOUNDS

Binary compounds whose elements have only one oxidation number

Common table salt is a binary compound composed of the elements sodium and chlorine. Because sodium has a +1 oxidation number and chlorine a −1, the formula is NaCl. The familiar chemical name, sodium chloride, illustrates the scheme used for naming binary compounds. The more metallic element is named first and this is followed by a shortened form of the name of the second element to which an -ide ending has been added:

Name for NaCl =	name of metallic element	+ stem name of second element	+ -ide ending =	
	sodium	+ chlor-	+ -ide	= sodium chloride

Other examples are shown in Table 6–4. Note that binary compounds can have more than two atoms per molecule—for example, N_2O_5—but no more than two different kinds of atoms.

TABLE 6–4 NAMING BINARY COMPOUNDS

Formula for compound	Name of first element	Name of second element	Stem name of second element	Binary ending	Name of compound
LiF	lithium	fluorine	fluor-	-ide	lithium fluoride
HCl	hydrogen	chlorine	chlor-	-ide	hydrogen chloride
KBr	potassium	bromine	brom-	-ide	potassium bromide
CaI_2	calcium	iodine	iod-	-ide	calcium iodide
ZnO	zinc	oxygen	ox-	-ide	zinc oxide
BaS	barium	sulfur	sulf-	-ide	barium sulfide
NaH	sodium	hydrogen	hydr-	-ide	sodium hydride

Binary compounds containing an element with multiple oxidation numbers

Consider the well-known compounds carbon monoxide, CO, and carbon dioxide, CO_2. These are examples of binary compounds containing two nonmetallic elements, one of which exhibits multiple oxidation states: carbon in CO is $+2$; in CO_2 it is $+4$. If we followed the rules just presented for naming binary compounds, we would get the same name for both of them — carbon oxide. Obviously the same label cannot be used for two compounds, so the difference is indicated by greek prefixes attached to the basic name. CO is called carbon monoxide; the prefix mono-, meaning one, distinguishes it from CO_2, carbon dioxide, where the prefix di- designates two atoms of oxygen. The Greek prefixes from one to ten are:

1 mono-
2 di-
3 tri-
4 tetra-
5 penta-
6 hexa-
7 hepta-
8 octa-
9 nona-
10 deca-

Examples of binary compounds whose names include Greek prefixes are given in Table 6–5. The compound Al_2O_3 is correctly called dialuminum trioxide, but the name aluminum oxide is usually used. This shorter form is acceptable, because aluminum exhibits only the $+3$ oxidation state and there are no other binary compounds of aluminum and oxygen to create confusion. Calcium dichloride is another example, it is called simply calcium chloride. To make it easier to pronounce, the a of the prefixes tetra- and penta- is omitted when the first letter following the prefix is a vowel. For example, the correct form is pentoxide, not pentaoxide. Except when clarity demands it, the prefix mono- is usually omitted.

There are two methods besides the greek prefixes for designating the number of atoms in a given formula. If a metallic element has more than one oxidation state and forms more than one compound with the same nonmetallic element, the compounds can be distinguished by placing parentheses after the name of the metallic species and designating the oxidation number with a Roman numeral. For example, consider the two compounds formed from iron and chlorine, $FeCl_2$ and $FeCl_3$. The first of these is called iron(II) chloride and the second iron(III) chloride. This nomenclature scheme is called the Stock system, after Albert Stock (1876–1946), and is the recommended method for naming these compounds.

If a metallic element has two oxidation numbers and forms two different compounds with the same nonmetallic element, the compound with the lower oxidation number can be distinguished by adding the ending -ous to the stem name of the metal, and the higher oxidation number compound by an -ic ending. Thus $FeCl_2$ can be called ferrous chloride, and $FeCl_3$, ferric chloride (the ferr-stem derives from Latin *ferrum,* iron). The Stock system is preferred to this naming system. Unfortunately, however, it is still commonly used, and therefore you need to be familiar with it. Table 6–3 summarizes some of these names.

TABLE 6–5 NAMES FOR SOME BINARY COMPOUNDS

Formula	Name (parentheses indicate unused names)
Al_2O_3	Aluminum oxide (dialuminum trioxide)
CCl_4	Carbon tetrachloride
$CaCl_2$	Calcium chloride (calcium dichloride)
H_2O	Water (dihydrogen oxide)
NH_3	Ammonia (nitrogen trihydride)
N_2O	Dinitrogen oxide
NO	Nitrogen oxide
NO_2	Nitrogen dioxide
N_2O_3	Dinitrogen trioxide
N_2O_4	Dinitrogen tetroxide
N_2O_5	Dinitrogen pentoxide
PCl_3	Phosphorus trichloride
PCl_5	Phosphorus pentachloride
SO	Sulfur monoxide
SO_2	Sulfur dioxide
SO_3	Sulfur trioxide

Example 6–4: Naming compounds *Name the compound $CoCl_3$.*

From Table 6–3 we see that cobalt has two common oxidation numbers, +2 and +3. In this compound it is +3; hence, the name is cobalt(III) chloride. The older name cobaltic chloride is still in common use, but the Stock name is preferred.

Binary compounds containing hydrogen

Binary compounds of hydrogen and a nonmetal are named the same way as other binary compounds, HCl is hydrogen chloride, H_2S is hydrogen sulfide, and so on. When certain of these binary compounds of hydrogen and a nonmetal are mixed with water, the properties of the resulting solutions are quite different from those of the pure compounds. These aqueous solutions are called **acids.** To see how acids are named, consider HCl as an example.

$$HCl(g) + H_2O(l) \rightarrow HCl(aq)$$

hydrogen chloride + water → hydrochloric acid (a water solution of hydrogen chloride)

The name hydrochloric acid is formed by using the prefix for hydrogen, hydro-, followed by the stem name of the second element, chlor-, to which an -ic ending is added; this name is followed by the word acid.

Name for HCl(*aq*) =	prefix for hydrogen	+ stem name of second element	+ -ic ending	+ the word acid
	hydro-	+ chlor-	+ -ic	+ acid = hydrochloric acid

Additional examples of this naming system are:

$HF(g)$	hydrogen fluoride	$HF(aq)$	hydrofluoric acid
$HCl(g)$	hydrogen chloride	$HCl(aq)$	hydrochloric acid
$HBr(g)$	hydrogen bromide	$HBr(aq)$	hydrobromic acid
$H_2S(g)$	hydrogen sulfide	$H_2S(aq)$	hydrosulfuric acid
$H_2Se(g)$	hydrogen selenide	$H_2Se(aq)$	hydroselenic acid

Ternary compounds

Ternary compounds contain three different kinds of atoms. The most important ternary inorganic compounds are combinations of a metallic element and a polyatomic unit—that is, a unit of three or more atoms—consisting of another element and oxygen. The compound $ZnSO_4$ is an example. This ternary compound comprises the metallic element zinc combined with a polyatomic unit of one sulfur and four oxygen atoms. The SO_4 is a distinct structural entity. It is an ion with a −2 oxidation number or charge and it retains its identity as a unit during many chemical reactions. For example, note that the SO_4^{2-} ion remains unchanged in the reaction

$Zn(s)$ +	$H_2SO_4(aq)$ →	$H_2(g)$	+ $ZnSO_4(aq)$
zinc	sulfuric acid	hydrogen	zinc sulfate

Ternary compounds that contain the SO_4^{2-} ion are called sulfates. They are named by following the name of the metallic element with the name of the sulfate unit. The terms sulfate derives from adding the -ate ending to the stem name of the element other than oxygen in the unit. Some other examples of ternary -ate compounds are:

$Al_2(SO_4)_3$	aluminum sulfate	$KClO_3$	potassium chlorate
$CaCO_3$	calcium carbonate	Na_3PO_4	sodium phosphate
HNO_3	hydrogen nitrate	$PbCrO_4$	lead chromate

TABLE 6–6 NAMES, FORMULAS, AND CHARGES FOR SOME NEGATIVE POLYATOMIC IONS

Name	Formula
Arsenate	AsO_4^{3-}
Carbonate	CO_3^{2-}
Hydrogen carbonate (bicarbonate)	HCO_3^-
Chlorate	ClO_3^-
Chromate	CrO_4^{2-}
Nitrate	NO_3^-
Phosphate	PO_4^{3-}
Monohydrogen phosphate	HPO_4^{2-}
Dihydrogen phosphate	$H_2PO_4^-$
Sulfate	SO_4^{2-}
Hydrogen sulfate (bisulfate)	HSO_4^-

The formulas and oxidation numbers (equal to the charges on the ions) for the -ate polyatomic ions must be memorized. A list of the most important ones is given in Table 6–6.

Example 6–5: Naming compounds *What is the name for the compound $Al_2(SO_4)_3$?*

This compound is aluminum sulfate. Note the meaning of the parentheses and subscripts in the molecular formula.

sulfate unit

$Al_2(SO_4)_3$ ← 3 sulfate units. The 3 multiplies everything within the brackets

2 atoms of aluminum

(The empirical formula, $Al_2S_3O_{12}$, does not indicate the presence of the sulfate groups and is of no value for purposes of identification or naming.)

Example 6–6: Naming compounds *What is the name of the compound with the formula $Pb(NO_3)_2$?*

From formulas given previously, we know that the oxidation number for the NO_3 ion is -1. If we did not know it, we could predict it from the periodic table relationships; nitrogen is $+5$ and oxygen is -2, hence NO_3 is -1. Usually the periodic grouping indicates the oxidation number of the element other than oxygen in the ternary -ate compounds. Because there are two -1 nitrates combined with the lead, its oxidation number is $+2$. The name of the compound is either plumbous nitrate or lead(II) nitrate, with the latter name the preferred form.

Ternary compounds containing hydrogen and a polyatomic group

The compound HNO_3 is correctly called hydrogen nitrate, and H_2SO_4 is hydrogen sulfate. These and numerous other ternary compounds of hydrogen and a polyatomic ion have different properties when combined with water. A special nomenclature scheme is used for these aqueous acid solutions. At this point, we will consider only the aqueous solutions of hydrogen -ate compounds. For example,

$HNO_3(g)$	$+\ H_2O(l) \rightarrow$	$HNO_3(aq)$
hydrogen nitrate	water	nitric acid

The term nitric acid derives from adding an -ic ending to the stem name of the element other than hydrogen or oxygen. In this case, the element is nitrogen.

Name for $HNO_3(aq)$ = stem name of element other than H or O + -ic + the word acid

nitrø- + -ic + acid = nitric acid

Other examples are shown in Table 6–7.

TABLE 6–7 NAMES FOR SOME OXYACIDS: AQUEOUS SOLUTIONS OF HYDROGEN TERNARY -ATE COMPOUNDS

Formula	Acid name
$H_3BO_3(aq)$	Boric acid
$H_2CO_3(aq)$	Carbonic acid
$HClO_3(aq)$	Chloric acid
$HBrO_3(aq)$	Bromic acid
$HIO_3(aq)$	Iodic acid
$HNO_3(aq)$	Nitric acid
$H_2SO_4(aq)$	Sulfuric acid
$H_3PO_4(aq)$	Phosphoric acid

Additions and exceptions

A number of important exceptions and additions to our basic rules of nomenclature are listed in Table 6–8.

TABLE 6–8 NAMES AND FORMULAS FOR SOME IMPORTANT COMPOUNDS NOT PREVIOUSLY DISCUSSED

Formula	Name
$NaOH$	Sodium hydroxide
$Ca(OH)_2$	Calcium hydroxide
HCN	Hydrogen cyanide
KCN	Potassium cyanide
NH_4OH	Ammonium hydroxide
$(NH_4)_2SO_4$	Ammonium sulfate
$H(CH_3CO_2)$	Acetic acid
$Na(CH_3CO_2)$	Sodium acetate
$H_2C_2O_4$	Oxalic acid
$Na_2C_2O_4$	Sodium oxalate
$NaHCO_3$	Sodium bicarbonate (baking soda)
Na_2HPO_4	Sodium monohydrogen phosphate
$K_2Cr_2O_7$	Potassium dichromate
$KMnO_4$	Potassium permanganate

Compounds containing the OH^- polyatomic ion are called hydroxides, and those containing CN^- are cyanides. The -ide ending is used even though these compounds are ternary. Inorganic hydroxides are called **bases;** sodium hydroxide, NaOH, is also known by the common name lye. Sodium cyanide, NaCN, and other metal cyanides contain carbon but are considered inorganic compounds.

The polyatomic ion NH_4^+, ammonium, has a +1 oxidation number or charge. It is the only positive polyatomic ion we will have occasion to use in our discussions; it forms many common and important compounds. Ammonium behaves as a metallic element; hence it is named first in the name of the compounds. Ammonium chloride is NH_4Cl. Household ammonia is an aqueous solution of ammonia; it is called ammonium hydroxide, but its correct formula is $NH_3(aq)$.

A ternary compound important in discussions of both inorganic and organic chemistry is the organic acid $H(CH_3CO_2)$, acetic acid. Vinegar is an aqueous solution of this acid. Inorganic salts of this acid are called acetates. **Salts** are *compounds formed when the hydrogen ions of an acid are replaced by metal or ammonium ions.* Sodium acetate is $Na(CH_3CO_2)$. The polyatomic acetate ion is sometimes designated by the empirical formula $C_2H_3O_2^-$.

The polyatomic ion $C_2O_4^{2-}$, oxalate, forms organic compounds important to inorganic chemists. Oxalic acid is $H_2C_2O_4$, a poison found in many plants such as rhubarb. Fortunately, foods containing oxalic acid lose their toxicity when cooked. The salts of oxalic acid are oxalates; sodium oxalate is $Na_2C_2O_4$.

Table 6–8 lists a number of polyatomic -ate ions that contain hydrogen: PO_4^{3-} is phosphate, HPO_4^{2-} is monohydrogen phosphate, and $H_2PO_4^-$ is dihydrogen phosphate. The term bicarbonate is also used for the hydrogen carbonate ion, HCO_3^-.

Oxyacids and their salts

Many compounds contain the same elements as the ternary -ate compounds but differ in the number of oxygen atoms; for completeness, the scheme for naming them is included here. Table 6–9 explains the nomenclature for the series of polyatomic ions formed from chlorine and oxygen. These ions combine with hydrogen ions to form compounds known as **oxyacids** (in aqueous solution). The compounds formed by the combination of these ions with metal or ammonium ions are called **salts of oxyacids** (the -ate compounds are included in these groups). The names of a number of oxyacids and oxyacid salts are given in Table 6–10.

TABLE 6–9 NOMENCLATURE FOR POLYATOMIC IONS CONTAINING DIFFERENT NUMBERS OF OXYGEN ATOMS

Number of oxygen atoms compared to the **-ate** compound	Method for naming	Examples	
+1	per- + stem name of element + -ate	Perchlorate	ClO_4^-
0	stem name of element + -ate	Chlorate	ClO_3^-
−1	stem name of element + -ite	Chlorite	ClO_2^-
−2	hypo- + stem name of element + -ite	Hypochlorite	ClO^-

TABLE 6–10 NAMES AND FORMULAS FOR SOME OXYACIDS AND OXYACID SALTS

Oxyacid formula	Name	Oxyacid salt formula	Name
$HClO(aq)$	Hypochlorous acid	$NaClO$	Sodium hypochlorite
$HClO_2(aq)$	Chlorous acid	$NaClO_2$	Sodium chlorite
$HClO_3(aq)$	Chloric acid	$NaClO_3$	Sodium chlorate
$HClO_4(aq)$	Perchloric acid	$NaClO_4$	Sodium perchlorate
$HNO_2(aq)$	Nitrous acid	KNO_2	Potassium nitrite
$HNO_3(aq)$	Nitric acid	KNO_3	Potassium nitrate
$H_2SO_3(aq)$	Sulfurous acid	Na_2SO_3	Sodium sulfite
$H_2SO_4(aq)$	Sulfuric acid	Na_2SO_4	Sodium sulfate

Hydrates

The compound copper sulfate, $CuSO_4$, is a white powder that forms a blue solution when placed in water. When this solution evaporates, a deep-blue, crystalline compound is left behind. A percent composition analysis shows that this compound has the empirical formula $CuSO_9H_{10}$. The compound has all the properties of a sulfate: It shows tests for SO_4^{2-} ions, water is driven off when it is heated, and the nonvolatile residue is $CuSO_4$. This experimental evidence tells us that the formula for this compound is $CuSO_4 \cdot 5H_2O$. Five molecules of water are bound to a formula unit of copper(II) sulfate to form a single formula unit of copper(II) sulfate pentahydrate, or copper(II) sulfate 5-water. This compound is an example of a **hydrate,** *a compound that contains molecules of water (usually called* **water of crystallization**) *bound into its structure.* The species of copper sulfate that contains no bound water molecules, $CuSO_4$, is called anhydrous copper sulfate. Some important hydrate compounds are gypsum, $CaSO_4 \cdot 2H_2O$; epsom salts, $MgSO_4 \cdot 7H_2O$; and washing soda, $Na_2CO_3 \cdot 10H_2O$. Note that the formulas for hydrates reflect the fact that the water molecules act as units in the compound; a raised dot (·) separates the H_2O units from the rest of the atomic symbols.

CONCLUSIONS

In this chapter, we have discussed a number of rules and schemes for writing formulas and naming compounds. Those of you who are encountering chemical nomenclature for the first time are probably somewhat confused, and rightly so; familiarity with the names and formulas is best gained by working problems and by coming in contact with the actual compounds in the laboratory. With practice, nomenclature becomes predictable and straight forward.

Example 6–7: Naming compounds *What is the name of the compound with the formula* Fe_2O_3*?*

This compound occurs in nature as hematite, and when carefully refined for pigment, it is called Venetian red, a common name whose use in the chemical literature should be discouraged. Because Fe_2O_3 is a binary compound, the preferred name is iron oxide. To distinguish it from other binary compounds of iron and oxygen (FeO, Fe_3O_4), we could call it diiron trioxide. The name based on the oxidation state of iron is more acceptable, however. Because iron in this compound has a $+3$ oxidation number, it is called ferric oxide or iron(III) oxide, with the latter form preferred. To summarize,

hematite: common name

Venetian red: common name (also called rouge in cosmetics)

iron oxide: a correct name that does not distinguish it from other iron oxides

diiron trioxide: a cumbersome name that is not used

ferric oxide: an older name, correct but not recommended

iron(III) oxide: preferred name

Iron is a peculiarly Anglo-Saxon word not recognizable to people who do not speak English. There is much virtue in using the internationally acceptable Latin names rather than English. In the future, the compound Fe_2O_3 will probably be called ferrum(III) oxide.

NEW TERMS

Acid: aqueous solutions of some hydrogen-containing compounds such as HCl(*aq*) and H_2SO_4(*aq*). In aqueous solutions, acids give the hydronium, H_3O^+, ion.

Base: some inorganic compounds containing the OH^- ion. In aqueous solution, bases contain the hydroxide, OH^-, ion.

Binary compound: a compound containing two different elements.

Hydrate: a compound that contains water molecules bound into the structure in definite proportions.

Oxidation number: an arbitrary number assigned to each atom in a molecule. The sum of the oxidation numbers of all the atoms in a molecule is zero; the oxidation number of an ion is equal to its charge.

Oxidation state: a term used to indicate the oxidation number of an atom in a molecule.

Oxyacid: a ternary acid that contains hydrogen, oxygen, and a third element.

Polyatomic ion: a charged particle that contains more than one element. The sulfate ion, SO_4^{2-}, is an example.

Salt: a compound in which the H ions of an acid have been replaced by metal ions or ammonium ions.

Ternary compound: a compound containing three different elements.

PROBLEM SET A

1. What is the oxidation number of each element in the following compounds: FeO, H_2S, Sb_2S_3, $KMnO_4$, $Ca(NO_3)_2$?
2. A compound has the formula $M_x^{2+}A_y^{3-}$. What is the least common multiple (disregarding the negative sign) of the two charges, +2 and −3? By what number must 3 be multiplied to give the least common multiple? By what number must 2 be multiplied? These numbers are the x and y in the formula. What is the correct formula?
3. Many—but not all—compounds are made up of positive and negative ions. In these compounds, the sum of the positive charges must equal the sum of the negative charges (disregarding the negative signs). If we invent three positive ions, A^+, B^{2+}, and C^{3+}, they could form a total of nine different compounds with the three negative ions, X^-, Y^{2-}, and Z^{3-}. Derive the formulas for the nine possibilities and list them in tabular form by filling out the following chart.

	X^-	Y^{2-}	Z^{3-}
A^+	____	____	____
B^{2+}	____	____	____
C^{3+}	____	____	____

4. On the basis of their position in the periodic table, predict a chemical formula for the oxide of each of the following elements: Se, Cu, Cr, Li, Ce, Zr.
5. Give the names and formulas for the compounds formed from iodine and the following elements: rubidium, calcium, tin, antimony.
6. Complete the following table. Give both the name and the formula for each compound formed from the indicated species.

	S^{2-}	Cl^-	SO_4^{2-}	NO_3^-	PO_4^{3-}
Al^{3+}	________	________	________	________	________
Fe^{2+}	________	________	________	________	________
Hg^{2+}	________	________	________	________	________
Na^+	________	________	________	________	________

7. Give an acceptable name for each of the following binary compounds:
   a. MgO
   b. HgO
   c. HBr
   d. CF_4
   e. KH
   f. ZnS
   g. CS_2
   h. OF_2
8. Give an acceptable name for each of the following ternary compounds:
   a. $BeCrO_4$
   b. $Zn_3(PO_4)_2$
   c. $Mg(ClO_3)_2$
   d. K_2SO_4
   e. Na_2CO_3
   f. $Ca(NO_3)_2$
9. Give an acceptable name for the following compounds:
   a. NH_4Br
   b. $H(CH_3CO_2)$
   c. $Sn(CN)_2$
   d. KOH
   e. $Fe(OH)_2$
   f. $Al(OH)_3$
10. The following are oxyacid salts other than those of the -ate acids. Name them.
   a. $ZnSO_3$
   b. $KClO$
   c. $NaNO_2$
   d. $Al(ClO_4)_3$
   e. $Ca(ClO_2)_2$
11. Give the correct names for both the compound and the aqueous acid solution of the following:
   a. HCl
   b. H_2S
   c. HF
   d. H_2SO_4
12. Name the following oxyacids:
   a. $HClO(aq)$
   b. $H_2SO_3(aq)$
   c. $HNO_2(aq)$
   d. $HClO_4(aq)$
13. Write formulas for the following compounds:
   a. cobalt(II) chloride
   b. iron(III) sulfate
   c. mercury(II) bromide
   d. zinc arsenate
   e. lead(II) chromate
   f. cupric phosphate
   g. copper(I) oxide
   h. lead(IV) oxide
14. Give the correct formula for:
   a. sodium hydroxide
   b. ammonium carbonate
   c. stannous oxalate
   d. sodium permanganate
   e. lithium cyanide
   f. zinc acetate
15. Give the correct formula for:
   a. calcium hypochlorite
   b. manganese sulfite
   c. potassium chlorite
   d. barium nitrite
   e. ferric perchlorate
   f. sulfuric acid 2-water

PROBLEM SET B

1. What is the oxidation number of each element in the following compounds: FeS, $NaOH$, K_2CrO_3, $Na_2Cr_2O_7$?
2. On the basis of their position in the periodic table, predict a chemical formula for the sulfide of each of the following elements: Hg, K, La, Al.
3. Give the names and formulas for the compounds formed from bromine and the following elements: cesium, potassium, strontium, and lead.
4. Complete the following table. Give both the name and formula for each of the compounds formed from the indicated species.

	F^-	CO_3^{2-}	ClO_3^-	PO_4^{3-}
K^+	____	____	____	____
Sn^{4+}	____	____	____	____
Pb^{2+}	____	____	____	____
Al^{3+}	____	____	____	____

5. Give an acceptable name for each of the following binary compounds:
   a. NiS
   b. NO_2
   c. PbO_2
   d. $TiCl_3$
   e. $CoCl_2$
   f. SnF_4
   g. CBr_4
   h. $CuCl_2$
6. Give an acceptable name for each of the following ternary compounds:
   a. $Sn(CO_3)_2$
   b. $CuSO_4$
   c. $Co(NO_3)_3$
   d. $Pb(CrO_4)_2$
   e. $Ca_3(AsO_4)_2$
   f. $KMnO_4$
7. Give an acceptable name for the following compounds:
   a. K_2HPO_4
   b. $KHCO_3$
   c. $K_2C_2O_4$
   d. $NaHSO_4$
   e. $Ca(CH_3CO_2)_2$
   f. $(NH_4)_2SO_4$
8. Name the following:
   a. KNO_2
   b. $Al(ClO_3)_3$
   c. $FeSO_3$
   d. $NaClO_4$
9. Give the correct names for both the compound and the aqueous acid solution for the following:
   a. HNO_3
   b. $HClO_3$
   c. H_2SO_5
   d. $HClO_2$
10. Write formulas for the following compounds:
   a. tin(II) fluoride
   b. calcium carbonate
   c. ferrous nitrate
   d. chromium sulfate
   e. calcium bicarbonate
   f. potassium dichromate
   g. sodium hypochlorite
   h. ammonium oxalate
   i. potassium acetate
   j. copper(II) permanganate
   k. sodium carbonate decahydrate
   l. magnesium sulfate 7-water

7

gases and the kinetic-molecular theory

THIS WILL SAVE YOU TIME:

Refer to this list as you read through the chapter. In answering the questions, you will understand the important concepts you are most likely to be asked about later on.

After reading this chapter, you should be able to:

1. *State the postulates of the kinetic-molecular theory.*
2. *Distinguish between heat and temperature.*
3. *Define temperature in terms of the kinetic-molecular theory.*
4. *Explain why the molecules of two different gases move with different velocities even though they are both at the same temperature and pressure.*
5. *Explain the meaning of the "per unit area" part of the definition of pressure and tell how it is possible for a small girl walking across a floor to exert a greater pressure than a large man.*
6. *Tell what is meant by atmospheric pressure, how it is measured, and what units are used to designate it.*
7. *Explain what a manometer is and how it works.*
8. *Explain the meaning of pressure in terms of the kinetic-molecular theory.*
9. *Using the kinetic-molecular theory, explain the laws of nature discovered by Boyle, Charles, Avogadro, and Dalton.*
10. *State the mathematical expressions for Boyle's, Charles's, Gay-Lussac's, and Dalton's laws and be able to solve gas-law problems using these relationships.*
11. *Explain the significance of the absolute zero of temperature.*
12. *Convert °C to °K and vice versa.*
13. *Tell what is meant by an ideal gas and explain why gases deviate from ideal behavior.*
14. *Work gas-law problems using the equation of state, $PV = nRT$.*
15. *Explain what is meant by the molar gas volume.*
16. *Calculate the volume occupied by a given mass (or number of moles) of a gas.*
17. *Calculate molecular masses of a gas given the density or the mass of a given volume.*

Gases easily qualify as the most fascinating of the three states of matter because of their phenomenal properties. We know that gases have mass even though the most powerful microscopes reveal no visible particles of gas. Gases can be compressed easily—forced into a smaller volume at a greater pressure—as in an automobile tire. But when the valve is removed from the tire, the air escapes, showing that gases expand. Gases can be liquefied, even solidified; dry ice is the solid state of carbon dioxide, which is a gas at room temperature. Gases occupy the whole volume of their containers; they cannot congregate in some smaller area. Gases diffuse—mix with one another—without any energy added to them; the odor of a rotten egg broken in one corner of the kitchen will soon fill the whole room as the ill-smelling hydrogen sulfide gas diffuses through the air.

All these phenomena and more can be explained by the **kinetic-molecular** theory of gases. The basic idea of this theory is that gases consist of molecules moving randomly and unceasingly. In this chapter, we will examine the kinetic-molecular theory and the related concepts of pressure and temperature.

THE KINETIC-MOLECULAR THEORY

The basic statement of the kinetic molecular theory is that *gas molecules move constantly and randomly.* Seven basic postulates can be drawn from this statement:

1. All gases are made up of minute particles called molecules or atoms.
2. Gas molecules are widely separated from one another (unless the gas is under high pressure). The space between molecules is a vacuum.
3. Gas molecules move constantly. Their speed is fast, their line of travel straight, but their direction is random.
4. The velocity of gas molecules increases as temperature rises, decreases as temperature falls.
5. Gas molecules move at different speeds, but at a given temperature the average kinetic energy of all gas molecules is the same.
6. Gas molecules collide frequently with one another and with the walls of their container.
7. When gas molecules collide, there is no loss of energy.

These statements introduce three fundamental concepts: temperature, molecular motion, and kinetic energy. We will discuss each idea in some detail and then consider how the kinetic-molecular theory applies to the fundamental laws of gas phenomena.

TEMPERATURE

Temperature is *a measure of the intensity of heat,* and heat, in turn, is a form of energy we can measure in calorie units. When a hot body is placed next to a cold one, heat flows from the hot body to the cold one. This heat transfer is shown by a decrease in the temperature of the hot body and an increase in the temperature of the cooler one. If there is no heat transfer, the two bodies are at the same temperature.

Kinetic energy (KE) is defined as *the energy a body has because it is moving.* Mathematically, kinetic energy can be defined as

$$KE = \frac{mv^2}{2}$$

where m is the mass and v the velocity of the object. Because gases are composed of very large numbers of molecules moving at various speeds, the kinetic-energy equation for gases is usually written as

$$KE = \frac{m\bar{v}^2}{2}$$

where the overbar ($\bar{v}$) designates the average velocity of the collection of molecules.

It is a law of nature that the average kinetic energy of gas molecules is constant at a given temperature. At any one temperature, all gases have the same average kinetic energy. This law allows us to define temperature theoretically: *The temperature of a gas is proportional to the average kinetic energy of the gas molecules.* In other words, the greater the average kinetic energy of the gas molecules, the higher the temperature, and vice versa.

MOLECULAR MOTION

At a given temperature, the average kinetic energy of a collection of gas molecules is constant. This does not mean, however, that all the molecules are moving at the same velocity, or speed. The mathematical definition of kinetic energy, $\frac{1}{2}m\bar{v}^2$, contains two factors, the velocity and the mass. Thus molecules with a large mass and a low velocity can have the same kinetic energy as smaller molecules traveling at high speed. In a container of a gas at a constant temperature, some molecules move relatively slowly, while others move with velocities two or three times the average. The average kinetic energy of oxygen molecules at 0°C is the same as the average kinetic energy of hydrogen molecules at the same temperature; the average hydrogen molecule, however, moves four times as fast as the average oxygen molecule. At 0°C, the molecular velocity of oxygen molecules averages 4.61×10^2 m/s.

Gas molecules collide constantly with each other and with the walls of the containing vessel. One of the basic postulates of the kinetic-molecular theory says that these collisions entail no net loss of energy. In the jargon of science, these collisions are said to be perfectly elastic. A given molecule may lose energy on collision and the molecule it strikes may gain energy, but the average kinetic energy of both molecules is the same both before and after. Gas molecules collide very frequently. Oxygen molecules at 0°C and normal atmospheric pressure average 1.6×10^{28} collisions/cc/s.

PRESSURE

Pressure is defined as *force per unit area.* The "per unit area" portion of the definition is very important to understanding the concept. This point is illustrated in Figure 7–1. This figure shows a block 3 cm × 3 cm × 10 cm resting on a table in two different positions. In both position **A** and position **B,** the block exerts the same

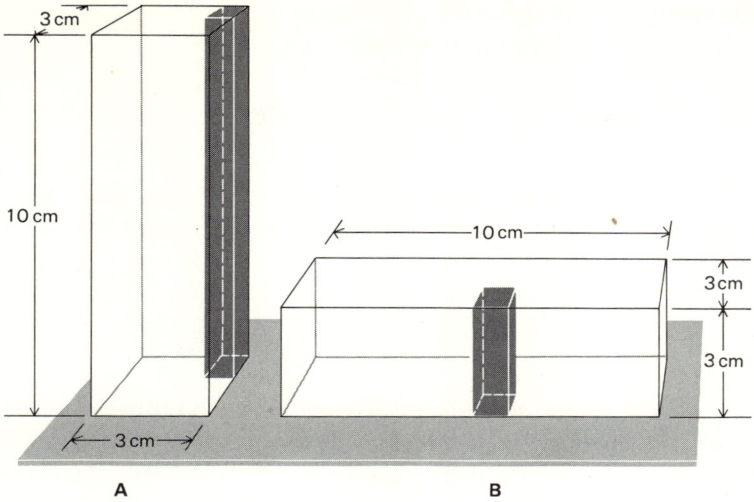

Figure 7–1 A comparison of the pressures exerted by a block resting in two different positions. In position **A,** the block exerts 10/3 times the pressure it does in position **B.** The total force exerted by the block is the same in both positions.

total force; the pressures, however, are different in each position. The pressure exerted by the block in position **A** is equal to the weight of the shaded section, which has a height of 10 cm and a base area of 1 cm^2. The block in position **B** exerts a pressure equal to the weight of the shaded section, which is 3 cm high and has a base area of 1 cm^2. Thus the pressure in position **A** is 10/3 times as great as the pressure in position **B.**

An everyday example of the "per unit area" aspect of pressure is the damage done to floors by women's spike-heeled shoes. All the woman's weight (force) is exerted over a very small area to create a tremendous pressure. For example, a 120-lb woman in spike heels with a $\frac{1}{4}$ in. diameter exerts a force of approximately 2450 lb/in^2 or 352,000 lb/ft^2. When the same weight is distributed over the larger area of a flat heel, the pressure is much less, only 20 lb/in^2 for an average-sized heel, and the floor is not damaged.

Evangelista Torricelli (1608–1647) was the first person to measure the pressure of gases in the atmosphere, a feat he accomplished by inventing the **barometer.** His experiment is illustrated in Figure 7–2. A glass tube approximately 1 m long with one open end is filled with mercury. A finger is placed over the open end to exclude air, and the tube is then inverted into a reservoir of mercury. The mercury drops to a level that is independent of the size or shape of the tube. The height of the mercury is a measure of the atmospheric pressure.

Why doesn't all the mercury flow out of the barometer tube? The kinetic-molecular theory tells us that gas molecules are in motion and that they are constantly bombarding anything they come into contact with. These molecular bombardments are the source of gas pressure. In the case of the barometer, the gas molecules of the atmosphere exert a downward pressure indicated as P_a in Figure 7–2. This pressure pushes the mercury up the tube. An opposite pressure exerted by the weight of the mercury, P_m, forces the mercury out of the tube. When these

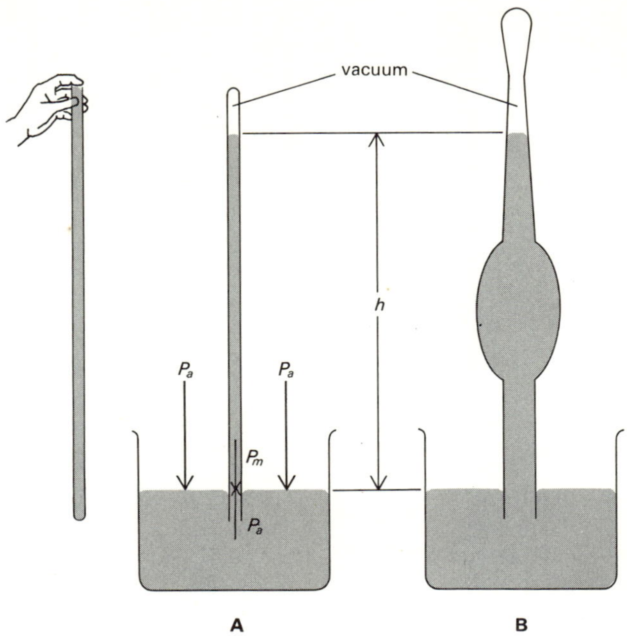

Figure 7–2 A mercury barometer. Drawing **B** illustrates the principle that the height of the mercury column is independent of the size or shape of the barometer tube.

two forces balance one another, $P_a = P_m$, the height of the mercury in the tube remains constant, and h is a measure of the atmospheric pressure.

Atmospheric pressure varies with the weather. The pressure drops when a storm approaches and stays down throughout the storm period. This phenomenon makes the barometer a very reliable tool for predicting the weather. Atmospheric pressure also varies with altitude; the higher one ascends into the atmosphere, the lower the pressure. At an altitude of 3.5 miles the pressure is about half that at sea level, at 11 miles it is only one-tenth, and at 19 miles the pressure drops to only 1 percent of the sea-level value.

Two units are used in measurements of gas pressure. The first is the **atmosphere** (abbreviation, atm), approximately equal to the atmospheric pressure at sea level on a clear day. More precisely, *one atmosphere is the pressure needed to support a column of mercury 760 mm high.* Another pressure unit is called the **torr,** named in honor of Torricelli. *One torr equals the pressure needed to support a column of mercury 1 mm high.* Thus one atmosphere equals 760 torr. Pressures are still sometimes stated in mm or cm of Hg, but the torr is the preferred unit.

In the laboratory, gas pressures are commonly measured with a device known as a **manometer,** illustrated in Figure 7–3. The U tube is filled with mercury. The gas whose pressure is to be measured exerts a pressure against one end of the U tube, and the atmosphere exerts a pressure on the other end. Measuring the difference in height (h) of the columns gives the pressure difference between the two. If, for example, h is 200 mm and the atmosphere is measured at 760 torr, the pressure of the gas in the vessel is 960 torr.

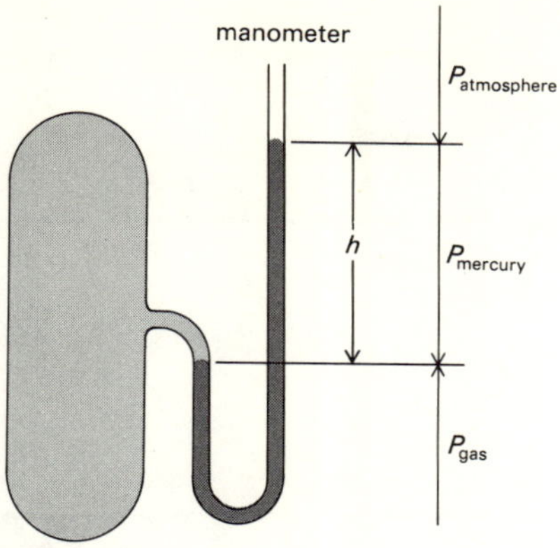

Figure 7–3 A mercury manometer.

STANDARD CONDITIONS OF TEMPERATURE AND PRESSURE

Since the volume of a gas varies with the temperature and pressure, some sort of standard is needed for purposes of comparison. These conditions, called **standard conditions of temperature and pressure** and abbreviated to **STP,** are set at 0°C and 1.00 atmosphere.

BOYLE'S LAW

Robert Boyle (1627–1691) formulated the famous gas law that carries his name. The law states: *At a constant temperature, the volume of a fixed mass of gas is inversely proportional to the pressure.* Stated mathematically, Boyle's law is the equation

$$P = \frac{k}{V} \quad \text{or} \quad PV = k$$

where k is a constant.

Figure 7–4 illustrates Boyle's law. A known mass of gas is put into a cylinder fitted with a moveable piston and connected to a manometer. Moving the piston alters the volume of the gas, and any changes in the pressure of the gas can be seen on the manometer. It is found that doubling the volume reduces the pressure to one-half its initial value, tripling the volume lowers the pressure to one-third, and so on. The mathematical formula for Boyle's Law can be restated as

$$P_1V_1 = P_2V_2 \text{ (temperature constant)}$$

where the subscript 1 refers to the gas before expansion or compression and the subscript 2 indicates the gas after the change.

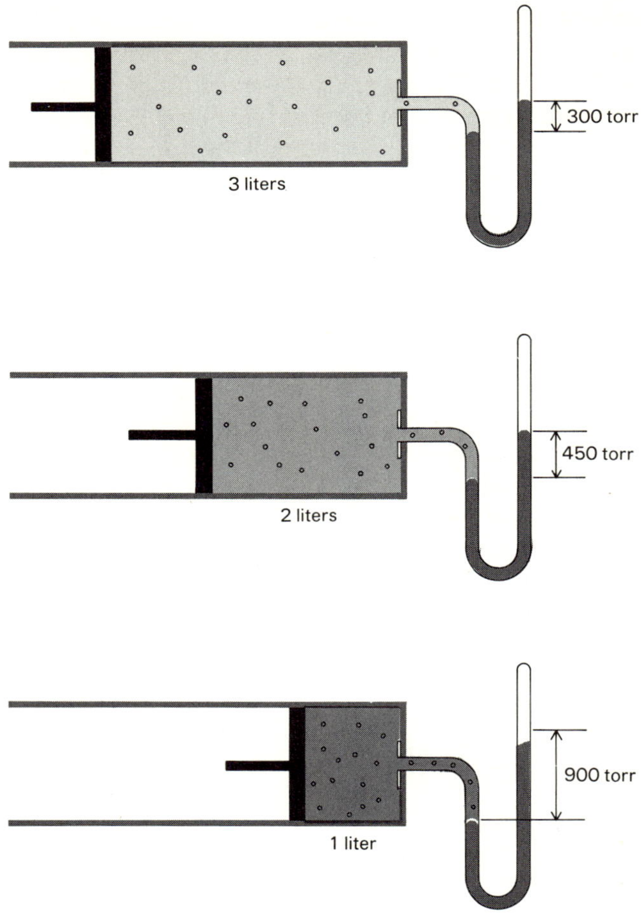

Figure 7–4 Apparatus to illustrate Boyle's law. At constant temperature, the volume of a fixed mass of gas is inversely proportional to the pressure.

Example 7–1: Boyle's law *A mass of nitrogen gas occupies 4.00 liters at a pressure of 720 torr. What volume will this same mass of gas occupy at a pressure of 1 atm (760 torr), assuming that the temperature remains constant?*

The pressure increases, so the volume of the gas must decrease. One way to solve the problem is to apply a correction factor to the initial volume:

$$4.00 \text{ liters} \times \frac{720 \text{ torr}}{760 \text{ torr}} = 3.79 \text{ liters}$$

Another way is to substitute the values into the $P_1V_1 = P_2V_2$ expression:

$$V_2 = \frac{P_1V_1}{P_2} = \frac{(720 \text{ torr})(4.00 \text{ liters})}{760 \text{ torr}} = 3.79 \text{ liters}$$

As we have seen, molecular bombardments cause pressure. The number of molecules striking a given area of surface depends on the number of molecules present in a given volume and on the velocities of those molecules. In the Boyle's law experiment, the molecular velocities do not change, for the temperature is held constant. Therefore, the observed pressure changes must result from a change in the concentration of bombarding molecules. Figure 7–4 shows that this explanation is correct. When the piston of the apparatus is in the 1-liter position, the molecules have only half as much space to move in as they did in the 2-liter position. As a result, they strike the walls twice as often, thus increasing the pressure.

DALTON'S LAW

In 1801, John Dalton performed an experiment in which he added water vapor to dry air and noticed an increase in the pressure of the resulting mixture. To account for this change, Dalton formulated the law that bears his name, also known as the law of partial pressures. The **partial pressure** of a gas in a mixture is *the pressure the gas would have if it alone occupied the total volume of the container.* Dalton's **law of partial pressures** states: *The total pressure of a mixture of gases is equal to the sum of the partial pressures of the individual gases in the mixture.* Thus in any nonreacting gas mixture, the pressure exerted by any one gas is unaffected by the kind or quantity of the other gases mixed with it.

Example 7–2: Partial pressures *Two 1-liter containers are connected by a closed valve. One container has a pressure of 400 torr of nitrogen and the second a pressure of 800 torr of oxygen. Both gases are at the same temperature. What happens when the valve is opened?*

If we assume that there is no volume or heat effect when the gases diffuse and mix with one another and that the gases do not react, then the final total pressure will equal the sum of the partial pressures of the two gases. If the nitrogen occupied the total volume, it would have a pressure of 200 torr, because the volume is doubled. By the same reasoning, oxygen would have a partial pressure of 400 torr. Dalton's law then tells us that the total pressure is

$$P_{\text{total}} = P_{N_2} + P_{O_2} = 200 \text{ torr} + 400 \text{ torr} = 600 \text{ torr}$$

CHARLES'S LAW

The volume of a fixed mass of gas depends on the temperature as well as the pressure. According to the kinetic-molecular theory, a rise in temperature increases the average kinetic energy of gas molecules. Molecules move faster at higher temperatures; they are able to hit harder and bombard the container's walls more frequently in a given period of time, thus increasing the pressure. If the volume of the gas is held constant, pressure increases with increasing temperature; alternatively, if the pressure is held constant, the volume must increase with a rise in temperature. **Charles's law,** named in honor of Jacques Charles (1746–1823), describes the relationship between gas volume and temperature.

The apparatus pictured in Figure 7–5 can be used to illustrate the principle of Charles's law. The calibrated gas chamber is connected to a mercury leveling bulb. When the level of mercury inside the container is the same as the level in the movable bulb, the confined gas is at atmospheric pressure. The gas can be kept at a

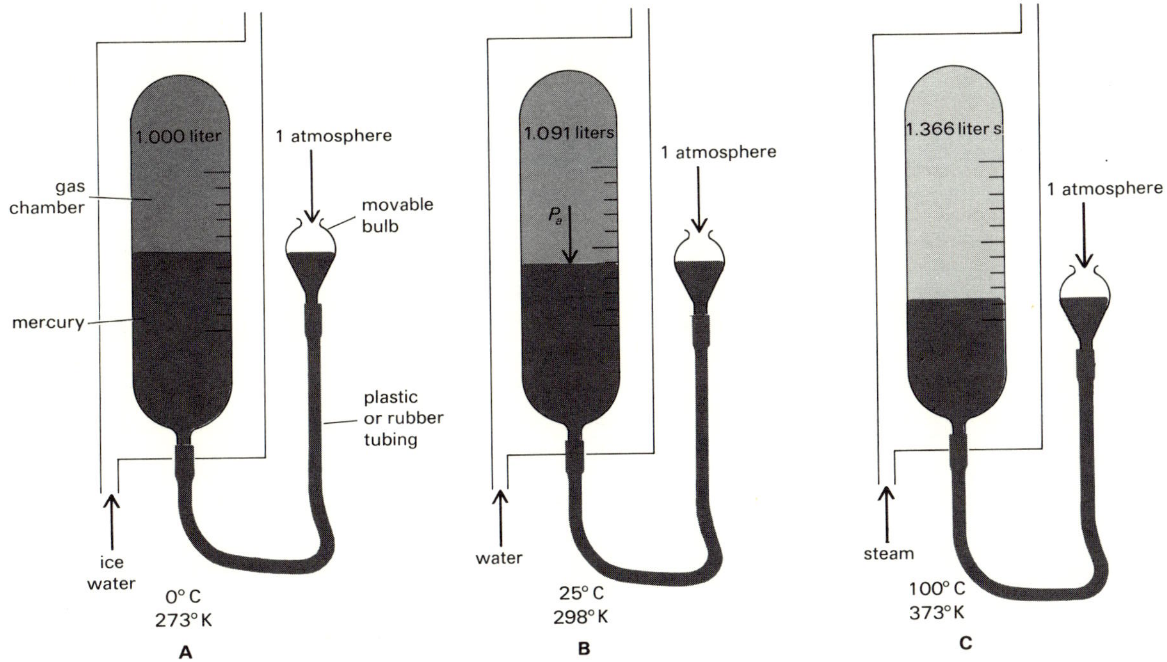

Figure 7–5 Apparatus to illustrate Charles's law. At constant pressure, the volume of a fixed mass of gas increases with rising temperature.

constant pressure by simply raising or lowering the leveling bulb, and it can be heated or cooled by pumping liquids of various temperatures into the jacket surrounding the chamber. In Figure 7–5, the gas is shown with an initial volume of 1 liter, a temperature of 0°C, and a pressure of 1 atm, or 760 torr. If you change the temperature to 25°C by passing water at that temperature through the surrounding jacket, the volume increases to 1.091 liters after the mercury level is adjusted to keep the gas at atmospheric pressure. When steam is passed through the jacket, Figure 7–5C, the volume at 100°C, and 1 atm is found to be 1.366 liters. The results of this type of experiment are plotted in Figure 7–6; you can see that there is a straight-line relationship between volume and temperature.

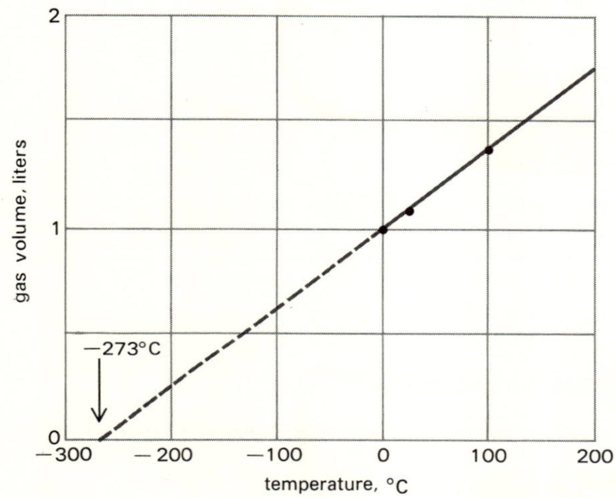

Figure 7–6 Graph illustrating Charles's law. The volume of a fixed mass of gas is proportional to the temperature when the pressure is kept constant.

THE ABSOLUTE TEMPERATURE SCALE

The straight-line of Figure 7–6 has been extrapolated to zero volume. If the relationship of Charles's law were valid at low temperatures, a gas at −273°C would have no volume! This is not the case, though, because a gas changes into a solid or liquid before it reaches this low temperature. The −273°C temperature is of great theoretical and practical importance, however. The kinetic energy of a gas, $m\bar{v}^2/2$, is proportional to the temperature. If the temperature of a gas were lowered to a point where the velocities of its molecules were zero, the temperature would also be zero. This zero temperature is calculated to be −273.15°C and is called the **absolute zero** of temperature; it is the temperature at which all molecular motion ceases. Absolute zero has not been attained in the laboratory, but scientists have succeeded in getting to within a fraction of a degree of it.

Much laboratory work employs a convenient standard of temperature measurement known as the **absolute, or Kelvin, temperature scale.** The Kelvin scale assigns 0°K to absolute zero, and has the same interval between degrees as the Celsius scale. Since 0°K is absolute zero, the Kelvin scale has no negative temperatures. Figure 7–7 shows a few relationships among the Kelvin, Celsius, and Fahrenheit scales. °C and °K are related by the following expression:

°K = °C + 273.15 (273 for our purposes)

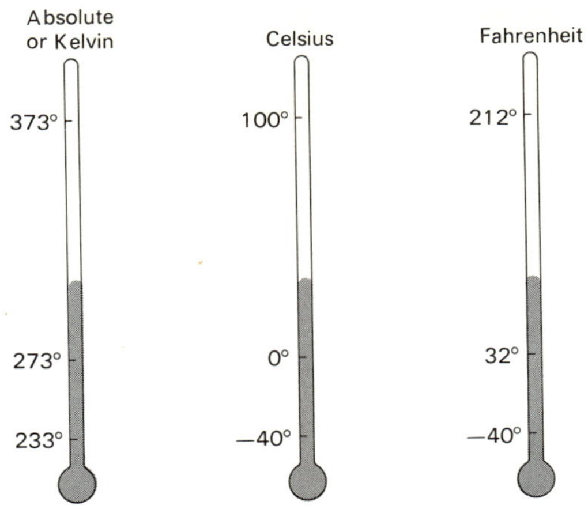

Figure 7–7 Comparison of the three temperature scales.

Example 7–3: Converting °C to °K *Convert 240°C to °K.*

Simply add 273° to 240° to get 513°K.

The Kelvin temperature scale is the only one suitable for working gas-law problems. If the temperature of a gas is changed from 100°C to 200°C (pressure constant), the volume of the gas does not double. On the other hand, if the temperature of a fixed mass of gas is changed from 100°K to 200°K (pressure constant), the volume does double.

We can now state Charles's law: *With pressure held constant, the volume of a given mass of gas is directly proportional to the Kelvin temperature.* In mathematical form, the law is

$$V_1 = kT_1 \quad \text{or} \quad \frac{V_1}{T_1} = \frac{V_2}{T_2}$$

where T is the Kelvin temperature.

Example 7–4: Charles's law *The initial volume of a balloon is 3.00 liters at 25°C and 760 torr. What is the volume if the temperature is increased to 100°C with no change in pressure?*

The temperature must be expressed in °K. Volume increases as temperature rises; therefore, the correction factor in this problem must be greater than 1.

$$3.00 \text{ liters} \times \frac{373\text{°K}}{298\text{°K}} = 3.76 \text{ liters}$$

Alternatively, using the equation $V_1/T_1 = V_2/T_2$

$$\frac{3.00 \text{ liters}}{298°\cancel{K}} = \frac{V_2}{373°\cancel{K}} \qquad V_2 = 3.76 \text{ liters}$$

BOYLE'S AND CHARLES'S LAWS COMBINED

The mathematical expressions for Boyle's and Charles's laws can be combined to form a general gas law relating *P, V,* and *T*. This is

$$\frac{PV}{T} = k \quad \text{or} \quad \frac{P_1V_1}{T_1} = \frac{P_2V_2}{T_2}$$

If T_1 equals T_2, the expression reduces to Boyle's law. If $P_1 = P_2$, the equation gives Charles's law. The third possibility, where V_1 equals V_2, gives a mathematical expression for **Gay-Lussac's law,** named in honor of Joseph Gay-Lussac (1778–1850). Gay-Lussac's law states that *the pressure of a fixed mass of gas at a constant volume is directly proportional to the temperature.*

Example 7–5: Combined gas laws *The measured volume of a gas is 1000 ml at a temperature of 60°C and a pressure of 720 torr. What volume will the gas occupy at STP?* (Remember: STP means 0°C and 760 torr.)

First tabulate the data:

Initial conditions		Final conditions	
V_1	1000 ml	V_2	?
P_1	720 torr	P_2	760 torr
T_1	333°K	T_2	273°K

Substituting the data into the equation gives

$$\frac{(720 \text{ torr})(1000 \text{ ml})}{333°\cancel{K}} = \frac{(760 \text{ torr})(V_2)}{273°\cancel{K}}$$

$$V_2 = \frac{(720 \text{ torr})(1000 \text{ ml})(273°\text{K})}{(333°\text{K})(760 \text{ torr})}$$

$$V_2 = 777 \text{ ml}$$

Note that the general equation can be rearranged to show two corrections applied to the volume: one for pressure and one for temperature change.

$$V_2 = \frac{P_1}{P_2} \times \frac{T_2}{T_1} \times V_1$$

AVOGADRO'S LAW

In 1811, Avogadro formulated a law that can be explained in terms of the kinetic-molecular theory. **Avogadro's law** states: *Equal volumes of gases at the same temperature and pressure contain equal numbers of molecules.* The drawing in Figure 7–8 illustrates the principle. Three gases, helium, nitrogen, and carbon dioxide, are compared. If temperature and volume are the same and if an equal number of molecules of each of the gases is put into similar containers, all the gases will exert equal pressures. At first, it may seem unreasonable to say that a big molecule of carbon dioxide with a mass of 44 exerts the same pressure as a little mass-4 helium atom. However, when we calculate pressure, we must take into account both the velocity

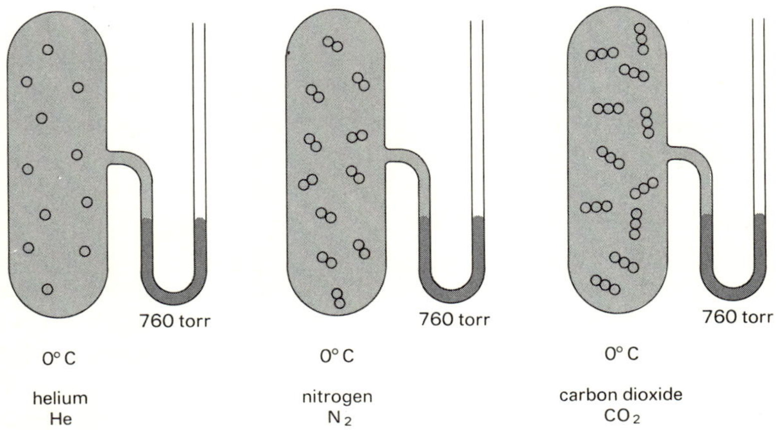

Figure 7–8 Schematic representation of Avogadro's law. Equal volumes of gases at the same temperature and pressure contain equal numbers of molecules.

of the atoms (or molecules) and their mass. In a given period of time, the faster moving helium atoms collide almost five times as often with the walls of the vessel as do the slower moving carbon dioxide molecules. As a result, both gases exert equal pressures.

Example 7–6: Comparing the number of gas molecules *Compare the number of molecules in a 2-liter cylinder of oxygen gas at 10 atm pressure and 25°C with the number of hydrogen molecules in a 4-liter cylinder at 5 atm and 25°C.*

Compare the volumes of the two gases at the same temperature and pressure. The volumes at STP, for example, will be directly proportional to the number of molecules present. For oxygen:

$$\frac{(10\ \cancel{\text{atm}})(2\ \text{liters})}{298\cancel{°\text{K}}} = \frac{(1\ \cancel{\text{atm}})(V_2)}{273\cancel{°\text{K}}}$$

$$V_2\ (\text{oxygen}) = 18.3\ \text{liters at STP}$$

For hydrogen:

$$\frac{(5\ \cancel{atm})(4\ \text{liters})}{298\cancel{°K}} = \frac{(1\ \cancel{atm})(V_2)}{273\cancel{°K}}$$

V_2 (hydrogen) = 18.3 liters at STP

The cylinders contain the same number of molecules.

IDEAL GASES

At the beginning of this text, we stated that a law is a statement of the behavior of nature and that a law falling short of a complete description of a phenomenon must be altered or rejected. The gas laws we have described require some revision. The reason is that as pressures increase and temperatures fall, gases deviate considerably from the behavior described in the gas laws. At lower temperatures and higher pressures, new forces come into play. The crowded molecules may attract each other and the space occupied by the molecules can significantly decrease the free volume in which they can move. To overcome these difficulties, the word *gas* in the statements of the gas laws is replaced with the term **ideal gas.** An ideal gas, by definition, is *a gas that behaves exactly as described by the gas laws;* that is, one in which there are no forces of attraction or repulsion between molecules and in which the molecular volumes are negligible. In most laboratory experiments, deviations from the gas laws are small and corrections can easily be made. At lower pressures and higher temperatures, however, deviations from ideal-gas behavior are quite negligible, and no corrections need be made.

EQUATION OF STATE FOR AN IDEAL GAS

The combined gas-law expression

$$\frac{P_1V_1}{T_1} = \frac{P_2V_2}{T_2}$$

can be written in another general form as

$$\frac{PV}{T} = \text{a constant}$$

The constant in this equation depends on the amount of gas present. Thus we can write it as

$$\frac{PV}{T} = nR$$

where n is the number of moles of gas and R is a constant called the **ideal gas constant.**

This equation, $PV = nRT$, is called **the equation of state for an ideal gas,** and is one of the most useful relations in chemistry. The value of the ideal gas constant, R, depends on the units chosen for P, V, and T. When pressure is expressed in atmospheres, volume in liters, and temperature in degrees Kelvin, the ideal gas constant has the following value:

$$R = 0.0820575 \quad \text{or} \quad \frac{0.0821 \text{ liter atm}}{\text{mole °K}}$$

Example 7–7: Using the equation of state for an ideal gas *What is the volume occupied by 3.00 moles of nitrogen at 1520 torr pressure and 0°C?*

This problem can be solved easily by using the gas-law equation, $PV = nRT$. First, the data must be expressed in the proper units.

$$n = 3.00 \text{ moles}$$
$$P = \frac{1520}{760} \text{ atm} = 2.00 \text{ atm}$$
$$T = 273\text{°K}$$
$$R = \frac{0.0821 \text{ liter atm}}{\text{mole °K}}$$

Substituting the data into the equation gives

$$(2.00 \cancel{\text{atm}})(V) = (3.00 \cancel{\text{moles}}) \left(\frac{0.0821 \text{ liter} \cancel{\text{atm}}}{\cancel{\text{mole}} \cancel{\text{°K}}}\right) (273\cancel{\text{°K}})$$

$$V = \frac{(3.00)(0.0821 \text{ liter})(273)}{2.00} = 33.6 \text{ liters}$$

Example 7–8: Using the equation of state for an ideal gas *How many moles of oxygen gas are needed to fill a 10.0 liter cylinder to a pressure of 8.00 atmospheres at 25°C?*

First, express the given data in units corresponding to the value of R.

$$R = \frac{0.0821 \text{ liter atm}}{\text{mole °K}}$$
$$V = 10.0 \text{ liters}$$
$$P = 8.00 \text{ atm}$$
$$T = 298\text{°K}$$
$$n = ?$$

Substituting the data into the $PV = nRT$ equation gives

$$(8.00 \text{ atm})(10.0 \text{ liters}) = n \left(\frac{0.0821 \text{ liter atm}}{\text{mole °K}}\right) (298\text{°K})$$

$$n = \frac{(8.00 \text{ atm})(10.0 \text{ liters})}{(0.0821 \text{ liter atm/mole °K})(298 \text{°K})}$$

$n = 3.27$ moles

MOLAR GAS VOLUME

None of the factors in the ideal-gas equation, $PV = nRT$, depends on the nature of the gas in question. That is, the equation applies to any gas, whether it is light, heavy, pure, or a mixture. If we have two equal volumes of two different gases, A and B, at the same temperature and pressure, then we can write:

$$P_A V_A = n_A R T_A$$

and

$$P_B V_B = n_B R T_B$$

However, since $P_A = P_B$, $V_A = V_B$, and $T_A = T_B$, it must follow that $n_A = n_B$. This, of course, is Avogadro's law: equal volumes of gases at the same temperature and pressure contain the same number of moles (or molecules).

Using the ideal-gas equation, we can calculate the volume occupied by 1.00 mole of any gas. This volume, measured at 273°K and 1 atmosphere pressure, STP, is a constant called the **molar gas volume.** At STP, it equals 22.4 liters per mole. With the molar gas volume, the molecular mass of a gas can be calculated when its mass, volume, temperature, and pressure are known.

Example 7–9: Calculating molar gas volume *Calculate the volume occupied by 1.00 mole of any gas at 0°C and 1.00 atmosphere pressure. This volume by definition is the molar gas volume.*

To solve, substitute the given values into the ideal gas equation. Thus

$$PV = nRT$$

$$(1.00 \text{ atm})(V) = 1.00 \text{ mole} \left(\frac{0.0821 \text{ liter atm}}{\text{mole °K}}\right)(273 \text{°K})$$

$V = 22.4$ liters

Example 7–10: Calculating the mass of one mole of a gas *10.0 g of a gas has a volume of 8.00 liters at STP. Calculate the mass of one mole of the gas.*

We know that 22.4 liters(STP) of any gas has a mass equal to the molecular mass in grams. Thus solving the problem means finding the mass of 22.4 liters(STP) of the gas.

$$M = \frac{22.4 \text{ liters(STP)}}{1.00 \text{ mole}} \times \frac{10.0 \text{ g}}{8.00 \text{ liters(STP)}}$$

$$M = \frac{28.0 \text{ g}}{\text{mole}}$$

Alternatively, using ratios,

$$\frac{8.00 \text{ liters(STP)}}{10.0 \text{ g}} = \frac{22.4 \text{ liters(STP)}}{M} \qquad M = \frac{28.0 \text{ g}}{\textit{mole}}$$

This problem can also be solved using the ideal-gas equation. It is necessary, however, to remember the definition of a mole:

$$\text{number of moles} = \frac{\text{mass of substance in grams}}{\text{molecular mass of the substance}} = \frac{m}{M} = n$$

Substituting this value into the gas-law equation yields

$$PV = \frac{m}{M} RT$$

Writing the given values into the equation gives

$$(1.00 \text{ atm})(8.00 \text{ liters}) = \frac{10.0 \text{ g}}{M} \left(\frac{0.0821 \text{ liter atm}}{\text{mole °K}}\right) (273\text{°K})$$

$$M = \frac{10.0 \text{ g}(0.0821 \text{ liter atm/mole °K})(273\text{°K})}{(1.00 \text{ atm})(8.00 \text{ liters})}$$

$$M = \frac{28.0 \text{ g}}{\text{mole}} = \text{the molecular mass of the gas}$$

Example 7–11: Calculating the molecular mass of a gas *The density of a gas at 730 torr and 40°C is 2.39 g/liter. What is the molecular mass of the gas?*

The problem states that 1.00 liter measured at 730 torr and 40°C has a mass of 2.39 g. To find the molecular mass, we need to know the mass of 22.4 liters measured at STP. The first step in solving the problem, therefore, is to express both volumes at the same conditions of temperature and pressure. In other words, we must convert the 1.00 liter at 730 torr and 40°C to STP. This can be done with the combined gas law. The known values are

$P_1 = 730$ torr	$P_2 = 760$ torr
$V_1 = 1.00$ liter	$V_2 = V_2$
$T_1 = 40 + 273 = 313$°K	$T_2 = 273$°K

Substituting into the equation gives

$$\frac{(730 \text{ torr})(1.00 \text{ liter})}{313\text{°K}} = \frac{(760 \text{ torr})(V_2)}{273\text{°K}}$$

$$V_2 = \frac{(730)(1.00 \text{ liter})(273)}{(313)(760)} = 0.838 \text{ liter}$$

We now know that if 0.838 liter(STP) of the gas has a mass of 2.39 g, 22.4 liters(STP) would have a mass of

$$M = \frac{22.4 \text{ liters(STP)}}{1.00 \text{ mole}} \times \frac{2.39 \text{ g}}{0.838 \text{ liter(STP)}}$$

$$M = \frac{63.9 \text{ g}}{\text{mole}} = \text{the molecular mass of the gas}$$

Perhaps the easiest solution to the problem is to use the $PV = nRT$ expression directly. The data are

$$P = 730 \text{ torr} = \frac{730}{760} \text{ atm} = 0.960 \text{ atm}$$

$$V = 1.00 \text{ liter}$$

$$R = \frac{0.0821 \text{ liter atm}}{\text{mole °K}}$$

$$T = 273 + 40 = 313\text{°K}$$

$$n = \frac{m}{M} = \frac{2.39 \text{ g}}{M}$$

Substituting these values and solving for M gives

$$(0.960 \text{ atm})(1.00 \text{ liter}) = \frac{2.39 \text{ g}}{M}\left(\frac{0.0821 \text{ liter atm}}{\text{mole °K}}\right)(313\text{°K})$$

$$M = \frac{2.39 \text{ g}(0.0821/\text{mole})(313)}{(0.960)(1.00)}$$

$$M = \frac{64.0 \text{ g}}{\text{mole}}$$

(Both answers are correct, 63.9 and 64.0. The difference is due to rounding the various numbers to three significant figures. Should this be unclear, refer back to Programmed Sequence 3 and to Chapter 2 for discussion of significant figures.)

NEW TERMS

Absolute zero of temperature: 0°K, or −273°C.

Atmosphere: a pressure unit equal to 760 torr.

Avogadro's law: equal volumes of gases at the same temperature and pressure contain equal numbers of molecules.

Barometer: a device for measuring atmospheric pressures.

Boyle's law: at constant temperature, the volume of a given mass of gas is inversely proportional to the pressure.

Charles's law: at constant pressure, the volume of a given mass of gas is directly proportional to the Kelvin temperature.

Dalton's law: the total pressure of a mixture of gases is equal to the sum of the partial pressures of the individual gases in the mixture.

Diffusion: the spontaneous process whereby particles forming a mixture tend to become completely mixed.

Equation of state for an ideal gas: $PV = nRT$.

Gas constant: R equals 0.0821 liter atmosphere/mole °K.

Gay-Lussac's law: at constant volume, the pressure of a fixed mass of gas is directly proportional to the Kelvin temperature.

Ideal gas: a gas that behaves exactly as described by the gas laws.

Kelvin temperature: absolute temperature, °K = °C + 273.

Kinetic energy: the energy of motion; $KE = \frac{1}{2}m\bar{v}^2$.

Manometer: a device for measuring pressures of gases.

Molar gas volume: the volume of 1.00 mole of any gas at STP: 22.4 liters (STP).

Partial pressure: the pressure exerted by a gas in a mixture. It is the pressure the gas would exert if it alone occupied the total volume.

Pressure: force per unit area.

STP: standard temperature and pressure, 0°C (or 273°K) and 1.00 atmosphere (or 760 torr).

Temperature: a measure of the intensity of heat. The temperature of a gas is proportional to its kinetic energy.

Thermometer: any device used for temperature measurements.

Torr: a pressure unit equal to the pressure needed to support a column of mercury 1 mm in height. 760 torr = 1 atmosphere.

Vapor: any substance in the gaseous state—usually used to refer to gases like water vapor that are likely to return to the liquid or solid state under normal conditions.

PROBLEM SET A

1. A 3.00-liter cylinder of gas at room temperature has a pressure of 20.0 atm. What would the volume of the gas be at 1.00 atm and the same temperature?
2. Gas in a 300-ml pressurized shaving-cream can has a pressure of 1200 torr at 27°C. What pressure would the gas exert if the can were heated to 300°C?
3. Convert 60°C to °K.
4. On the ground, a balloon has a volume of 100 kl at a temperature of 27°C and atmospheric pressure. What volume will the balloon have when it reaches an altitude where the temperature is −10°C and the pressure 300 torr?
5. Given 700 ml of gas at 700 torr and 100°C. What volume does the gas occupy at STP?
6. If 550 ml of oxygen gas at 25°C is heated to a temperature of 198°C, what will be the new volume of the gas? Assume constant pressure.
7. A given mass of nitrogen gas has a volume of 750 liters at a pressure of 750 torr. To what value in atmospheres must the pressure be changed if the volume is to be reduced to 3.50 liters? Temperature remains constant.
8. The measured volume of a gas was 16.0 liters at a temperature of 37°C and a pressure of 2.00 atmospheres. What volume would this gas occupy at STP?
9. Air is 79 percent nitrogen. What is the partial pressure of nitrogen at normal atmospheric pressure?
10. Container A has a volume of 9.00 liters and a pressure of 800 torr at 20°C. Container B

has a volume of 12.0 liters and a pressure of 600 torr at −10°C. Which container has the greater number of molecules?

11. A 10.0 liter cylinder contains oxygen at a temperature of 25°C and a pressure of 780 torr. At a constant temperature, nitrogen gas is pumped into this cylinder until the total pressure is 1400 torr. What is the partial pressure of the oxygen gas in the resulting gas mixture?

12. What is the volume occupied by 4.50 moles of oxygen gas at STP?

13. How many moles of carbon dioxide gas are needed to fill a 500 ml flask at 25°C and 750 torr pressure?

14. A 125 ml sample of a gas at 96°C and 1.20 atmospheres has a mass of 0.760 g. Calculate the molecular mass of the gas.

15. Calculate the density of chlorine gas (Cl_2) at STP. Report your answer in grams/liter.

16. At STP, what will be the volume in liters of 25.0 g of CO_2?

17. Flask A contains 15.0 liters of He gas. Flask B contains 25.0 liters of N_2 gas. Each volume is measured at the same temperature and pressure. If flask A contains 5.25 moles of He, how many moles of N_2 are in flask B?

PROBLEM SET B

1. A given sample of gas occupies a volume of 12.7 liters at a pressure of 655 torr. To what pressure will you have to go to change the volume to 20.0 liters? Assume constant temperature.

2. A 150 ml capsule of H_2S gas at 23°C is under a pressure of 2000 torr. What volume will this gas occupy at 15°C and a pressure of 0.93 atm?

3. Given a balloon with a volume of 5.00 liters at 750 torr and 25°C. What would the pressure be if the balloon were compressed to a volume of 3.80 liters and heated to a temperature of 40°C?

4. Given a gas cylinder filled with a gas at a pressure of 120 atmospheres. The cylinder is filled and stored at 25.0°C and has a safety factor of 50 atm—that is, beyond 170 atm of pressure the cylinder will explode. How high a temperature would the gas in the cylinder have to reach before the cylinder exploded? (Hint: the volume before explosion remains constant, so you can assume some convenient volume like 1.00 liter.)

5. Convert the following Celsius values to degrees Kelvin: 0°, 25°, and 100°.

6. 15.0 liters of hydrogen gas at 25°C is reduced to a new volume of 8.5 liters. If the pressure is held constant, what is the new temperature of the hydrogen?

7. In a laboratory experiment, 3.60 liters of dry oxygen are collected at 20°C and 720 torr pressure. What volume would this gas occupy at STP?

8. A volume of gas at STP is 580 ml. What volume would this gas occupy at 100°C and 530 torr?

9. A given mass of Ne gas has a volume of 25.0 liters at a pressure of 0.75 atm. If the pressure is changed to 300 torr and the temperature remains constant, what will be the volume of the Ne?

10. Given the partial pressures of the following three gases in a 15.0 liter cylinder: He = 88 atm; N_2 = 25 atm; Ne = 50 atm. What is the total pressure?

11. If the cylinder of Problem 10 is at 900°C, how many grams of neon does it contain?

12. A volume of 1.32 liters of a gas, measured at 19°C and 0.97 atm, has a mass of 2.55 g. Calculate the molecular mass of the gas.

13. Calculate the density of H_2S gas at the common laboratory conditions of 22°C and 755 torr. Express your answer in g/liter. (Hint: one mole of H_2S, 34.1 g, has a volume of 22.4 liters at STP.)

14. Given 150 ml of Ar gas at 20°C and 1.10 atm pressure. How many grams of Ar are present in this sample?

15. In a gaseous mixture of equal masses of CH_4 and C_2H_6, the partial pressure of CH_4 is 0.56 atm. What is the partial pressure of C_2H_6?

16. Calculate the mass of 20.0 liters of oxygen at 22°C and 750 torr.

17. Calculate the volume of 15.0 g of NH_3 gas at 780 torr and 50°C.

8

liquids, solids, and the kinetic-molecular theory

THIS WILL SAVE YOU TIME:

Refer to this list as you read through the chapter. In answering the questions, you will understand the important concepts you are most likely to be asked about later on.

After reading this chapter, you should be able to:

1. *Distinguish among solids, liquids, and gases on the basis of the kinetic-molecular theory.*
2. *Distinguish between amorphous and crystalline solids.*
3. *Describe the structure of a sodium chloride crystal.*
4. *Explain the phenomena of diffusion, surface tension, evaporation, and sublimation in terms of the kinetic-molecular theory.*
5. *Explain what is meant by vapor pressure and dynamic equilibrium.*
6. *Draw a graph showing how the vapor pressure of water varies with temperature.*
7. *Define boiling point and explain how it varies with pressure.*
8. *Explain the physical principle behind cooking with a pressure cooker.*
9. *Explain in detail what happens in terms of state changes and temperature when a substance gains or loses heat.*
10. *Draw a graph for water showing the temperature and state changes occurring when a given mass is heated or cooled (Figure 8–12).*
11. *Explain the cooling effect produced by evaporation using the principles of the kinetic-molecular theory.*
12. *Work problems involving energy and state changes.*

Gases behave as the equations of the kinetic-molecular theory predict only if we assume that the molecules do not attract one another. Actually this assumption is unwarranted. The molecules of real gases always have an attraction for one another, but at lower pressures or at higher temperatures the attractive forces are outweighed by the great distances between molecules or by the increased kinetic energy of the particles. When the temperature of a gas drops or the pressure increases, or both, molecules begin to attract one another and they start to form clusters. These attractive forces are weak, but they have enough strength to limit the molecules' complete freedom of movement and to form what we call the liquid state. In the liquid state, the molecules retain some ability to move, both as individuals and as clusters. When temperature drops further or pressure increases even more, or both, the motion of the molecules and clusters slows, and the forces of attraction between them become much stronger. At some eventual point of temperature and pressure, the molecules and clusters of molecules are held in definite positions. This is the solid state.

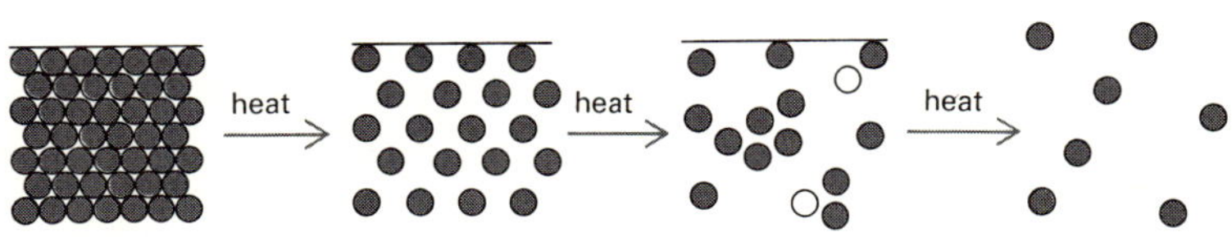

Figure 8–1 States of molecular aggregation.

The distinction among solids, liquids, and gases can be illustrated by observing what happens when heat is added to a solid (Figure 8–1). The molecules of the solid cannot move and are located in definite positions with respect to one another. As heat is added, the energy of the molecules increases, and they begin to vibrate in their fixed positions more and more vigorously. Eventually they acquire enough energy so that clusters of molecules break apart and begin to move about freely. This condition is the liquid state. Adding more heat breaks the clusters into smaller and smaller units, and eventually some molecules get enough energy to leave the group and escape to the gas state. When all the molecules are able to move independently and the attractive forces between molecules are negligible, the state of an ideal gas is reached.

In this chapter, we will see how energy changes are involved in state changes, and we will apply the principles of the kinetic-molecular theory to the liquid and solid states.

CRYSTALLINE SOLIDS

The forces of attraction between structural units (atoms, molecules, or ions) in a solid are much greater than the forces between the same species in the gas or liquid state. As a result, the structural units cannot change position with respect to other units, but they do vibrate about their fixed equilibrium positions. The arrangement of atoms, molecules, or ions in these equilibrium positions can be either a symmetrical formation called a **crystal,** or a nonsymmetrical distribution called an **amorphous solid.** A crystal is defined as *a solid in which the structural units are arranged in a repeating pattern in three dimensions.* Crystals can vary in size from microscopic samples to large specimens weighing several thousand pounds.

Crystalline solids melt at a precise temperature, while the melting point for amorphous solids can vary over a limited temperature range. True amorphous solids, however, do not exist; all solids have at least some structural organization.

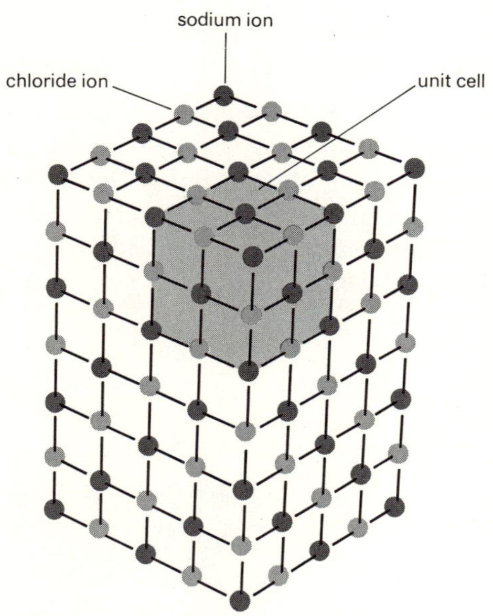

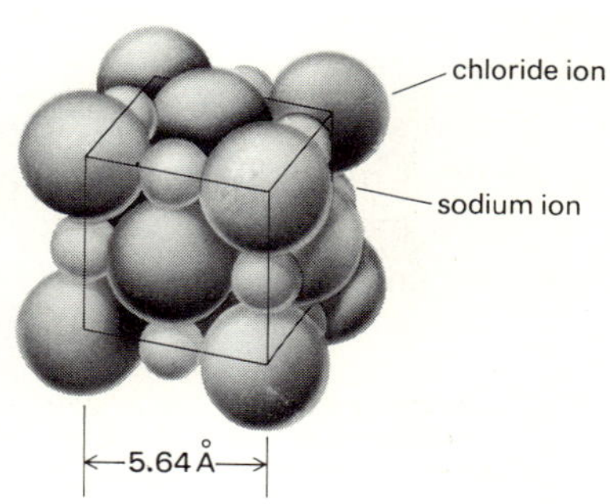

Figure 8–2 Sodium chloride crystal. Each sodium ion is surrounded by six chloride ions, and each chloride ion by six sodium ions.

Figure 8–2 illustrates the solid structure of the sodium chloride crystal. Every sodium ion is surrounded by six chloride ions: four in one plane, one on top, and one below. Each chloride ion is likewise surrounded by six sodium ions. The structure is cubic, with all the faces—flat surfaces of the crystal—meeting at 90° angles. The shaded volume on the left side of the figure represents this basic cubic structure. The right illustration takes the relative sizes of the ions into account and thus presents a more accurate picture.

DIFFUSION

Molecules in a liquid are randomly arranged and move constantly. There is sufficient cohesiveness among the molecules to keep the substance within well-defined boundaries and maintain its volume even though it does not assume the characteristic shape of a solid. Unlike gases, molecules in the liquid state can move only in a given volume. Molecules in a liquid are more tightly packed, and they cannot move very far before colliding with other molecules.

Molecular motion in liquids can be illustrated by a simple diffusion experiment. Place a drop of ink or a drop of aqueous potassium permanganate, $KMnO_4$, a solution with an intense purple color, on the surface of water in a test tube. Observe the distribution of color, but do not shake the tube. At first, the drop forms a distinct region, but then the color slowly diffuses throughout the water until the whole volume of liquid eventually has the same uniform shade. The kinetic-molecular theory explains the phenomenon; the experiment shows that liquids consist of molecules in a state of constant random motion.

SURFACE TENSION

The cohesion of molecules in the liquid state gives rise to the property of liquids called **surface tension.** Liquid surfaces behave as if a thin rubber membrane were stretched over them. A molecule in the bulk of the liquid is surrounded by similar molecules. Thus equal attractive forces pull on it from all directions. The forces acting on a surface molecule, though, are unbalanced. One side is in contact with the identical molecules of the liquid, but the other side is in contact with the dissimilar gas molecules of the atmosphere. This imbalance of forces on the surface molecules creates the membrane effect (Figure 8–3).

Figure 8–3 Surface tension. The drawing illustrates the unbalanced forces acting on molecules at the surface of a liquid.

Surface tension can be dramatically illustrated by placing a sewing needle on the surface of water. If you do it carefully and the resting needle is not disturbed, it will float, even though the needle has much greater density than the water. To sink, the needle would have to break through the layer of surface molecules, the imaginary membrane. The attractive forces holding the surface molecules are strong enough to counteract gravity's downward pull on the needle.

Another experiment illustrating surface tension is shown in Figure 8–4**A,** which shows water rising in a capillary tube. Water is pulled up the tube because the attractive forces between water and glass are greater than the attractive forces between water molecules. The opposite is true for mercury (Figure 8–4**B**). The cohesive forces between mercury atoms are greater than those between mercury and glass, so the mercury column in a capillary tube is depressed. We say that water wets the glass (attraction) and that mercury does not (repulsion). Note the shape differences between the **meniscus**—*the curved upper surface of the liquid column where it touches the walls of its container*—for water and that for mercury. The water meniscus is concave upward, while the mercury meniscus is concave downward.

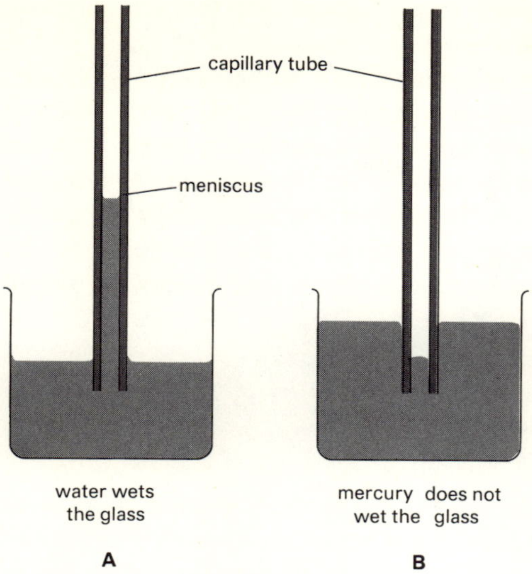

Figure 8–4 Capillary attraction. **A** Water rises in a capillary tube because of the attractive forces between water and glass molecules. **B** The same experiment using mercury, a liquid which does not wet the glass. The mercury level in the capillary tube is lower than in the reservoir because mercury atoms have a greater attraction for one another than for glass.

EVAPORATION AND SUBLIMATION

Molecules in the liquid state are in constant, chaotic motion. The average kinetic energy of the molecules depends on the temperature, but at any given temperature velocities do vary: some molecules move faster than others. A molecule with a greater than average amount of energy, possibly acquired by collision with another molecule, can, if it is near the surface of the liquid, break through the surface barrier and enter the gas state (Figure 8–5**A**). Once the fast-moving escaped molecule enters the atmosphere, it collides with air molecules, and its chance of returning to the liquid lessens as it moves away from the liquid surface (Figure 8–5**B**). This is **evaporation,** *the process by which a molecule leaves the liquid state and enters the gas state.* Any given volume of liquid will continue to evaporate until all the liquid disappears.

The rate of evaporation increases with temperature. As the temperature of a liquid rises, the percentage of molecules with a greater than average amount of energy grows. Hence there is a greater probability of a larger number of molecules breaking through the surface-tension barrier.

Liquids evaporate more rapidly in a vacuum than in air (Figure 8–5**C**). In the air, escaping molecules collide with molecules of oxygen and nitrogen, deflecting them. The air keeps the molecules close to the surface of the liquid, increasing the chance that some will return to the liquid. In a vacuum, there are no molecules of air for the escaping molecules to strike, and evaporation proceeds more quickly.

Molecules in the solid state are not free to move past one another, but as the temperature of a solid increases, molecular vibrations become more active. This vibrational energy can be transferred from one molecule to another. If a particular molecule gains enough energy, *it can break away from the rigid solid structure and enter the liquid or gas state*. This process, called **sublimation,** occurs with many solids. Wet clothes hung outdoors in freezing weather will dry; the water goes directly from ice to water vapor. A more familiar example is dry ice (solid carbon dioxide), which readily makes the transition from solid to gas at atmospheric

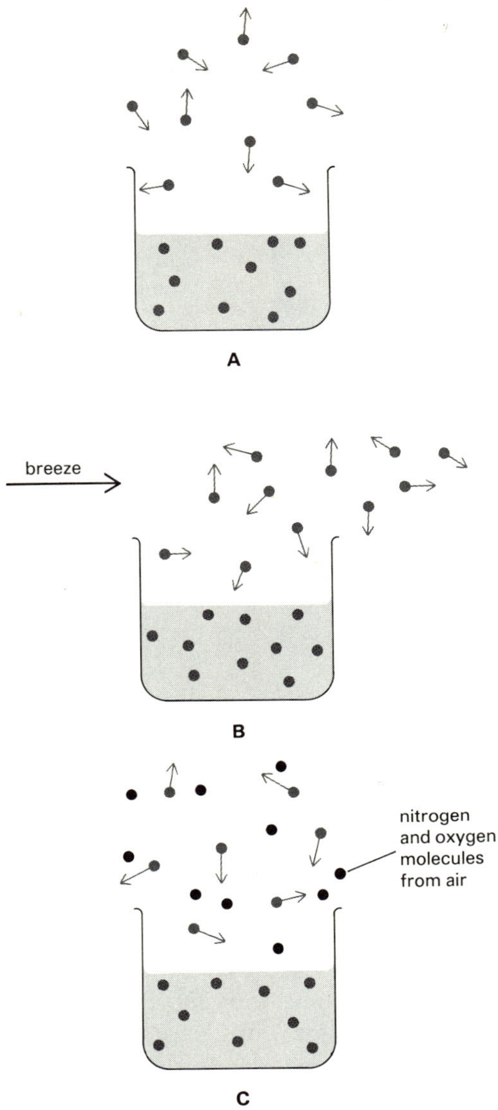

Figure 8–5 How evaporation proceeds. **A** Molecules escape from the surface; most cannot return to the liquid state because equilibrium is not established. **B** Air currents speed evaporation by pushing escaping molecules away from the surface of the liquid, thus lessening the chance of their return to the liquid. **C** Liquids evaporate more slowly into air than into a vacuum because collisions of escaping molecules with oxygen and nitrogen molecules in the air delays their movement away from the surface of the liquid.

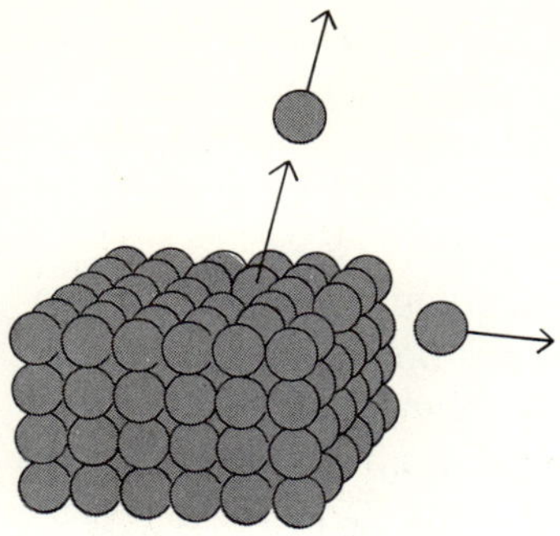

Figure 8–6 Sublimation. Molecules leave the surface of a solid and enter directly into the gas state.

pressure without the appearance of a liquid state. In effect, sublimation is the same thing as evaporation, except that it occurs in solids rather than liquids. Figure 8–6 illustrates the process.

VAPOR PRESSURE

If a liquid is placed in a closed container (Figure 8–7), the molecules leaving the liquid state and entering the gas state have no place to go. As a result, they remain near the liquid surface. Under these conditions, many molecules strike the surface and are recaptured by the liquid. An equilibrium is soon established; the rate at which molecules leave the surface equals the rate at which they reenter the liquid. This steady state is called a **dynamic equilibrium**—that is, *a balance of two opposing forces. The pressure exerted by the vapor molecules* is called the **equilibrium vapor pressure.**

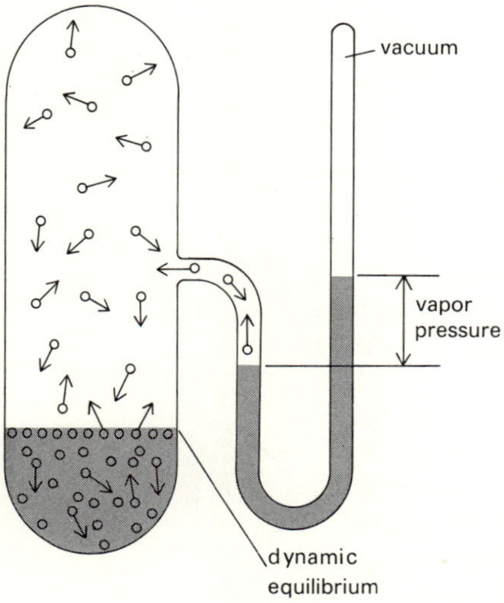

Figure 8–7 Equilibrium vapor pressure. In a closed container at a fixed temperature, the rate at which molecules leave the surface of a liquid equals the rate of their return.

The magnitude of the vapor pressure for a compound at a given temperature depends on the nature of the interactions between molecules of the substance. Different liquids have different vapor pressures. A compound such as water has a lower vapor pressure than does ether, $CH_3CH_2OCH_2CH_3$, because water molecules have greater forces of intermolecular attraction than do ether molecules.

Figure 8–8 shows how the vapor pressures of several liquids vary with temperature. Table 8–1 shows how temperature affects the vapor pressure of water. The vapor pressure for a given liquid is a constant at a given temperature. It does not depend on the size or shape of the container or the amount of liquid present. This fact can be demonstrated with the experiment shown in Figure 8–9. The key to the experiment is the different proportions of gas and liquid in the two containers. Container **A** has a large volume of liquid with a large surface area compared to container **B,** where the volume and the surface area of the liquid are small. When both containers are at the same temperature, the manometers record the same vapor pressure. Remember, pressure is a per-unit-area phenomenon. The rate at which molecules leave and return to a given unit area in container **A** is the same as in container **B.** Naturally, container **B** has more vapor molecules because of its greater volume and surface area, but as long as the two states, liquid and vapor, are both present, the vapor pressure will remain constant at a given temperature.

Figure 8–10 illustrates a method for measuring the vapor pressures of liquids. A drop of liquid inserted into a barometer tube will cause the mercury level to drop because of the pressure exerted by the vapor molecules in the confined space at the top of the tube. A small amount of liquid must be present so that equilibrium conditions are established. The difference in the height of the mercury before and after the liquid is added directly measures the equilibrium vapor pressure at the temperature of the experiment. The vapor pressure of liquid mercury is only 10^{-3} torr at room temperature; it does not add significantly to the observed measurements.

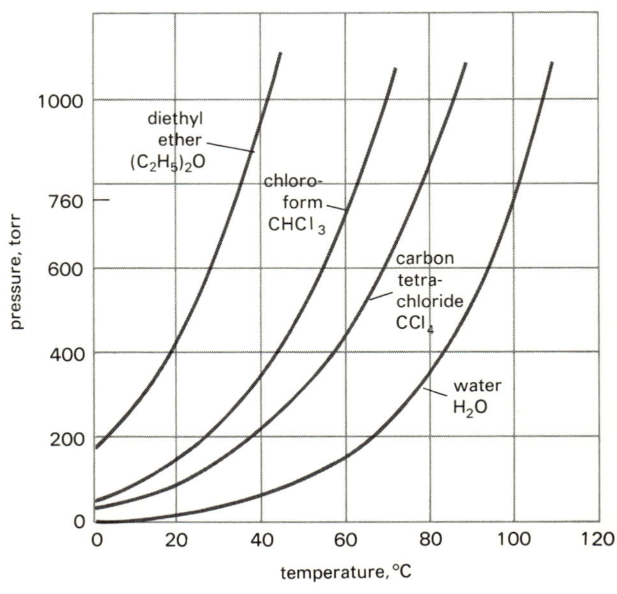

Figure 8–8 Equilibrium vapor pressures for several compounds.

TABLE 8-1 VAPOR PRESSURE OF WATER

Temperature °C	Vapor pressure, torr
−10	2.1
0	4.6
5	6.5
10	9.2
15	12.8
16	13.6
17	14.5
18	15.5
19	16.5
20	17.5
21	18.6
22	19.8
23	21.1
24	22.4
25	23.8
26	25.2
27	26.7
28	28.3
29	30.0
30	31.8
35	40.2
40	55.3
45	71.9
50	92.5
55	118.0
60	149.4
65	187.5
70	233.7
75	289.1
80	355.1
85	433.6
90	525.8
95	633.9
100	760.0
110	1074.6

BOILING POINT

The **boiling point** of a liquid is defined as *the temperature at which its vapor pressure equals the external pressure*. When heat is applied to a liquid, its vapor pressure increases until it equals atmospheric pressure. At this point where one pressure equals the other, the liquid can evaporate freely, and bubbles of vapor can be seen forming throughout the liquid.

The boiling point varies with the external pressure present above the surface

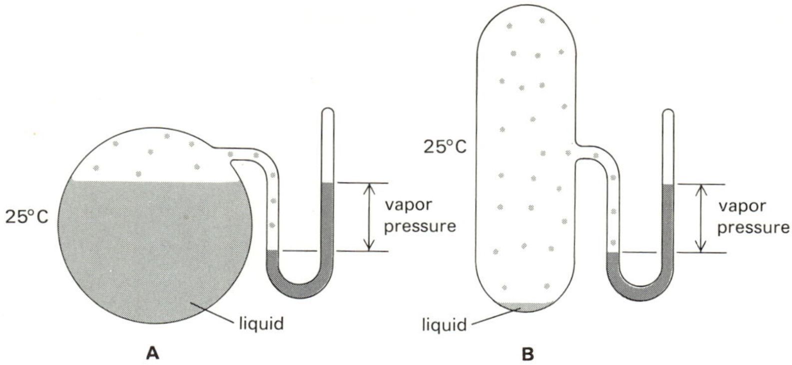

Figure 8–9 The equilibrium vapor pressure is the same in containers **A** and **B.** The vapor pressure of a given liquid at a fixed temperature is independent of the size or shape of the containing vessel and does not depend on the amount of liquid present, as long as at least some is present.

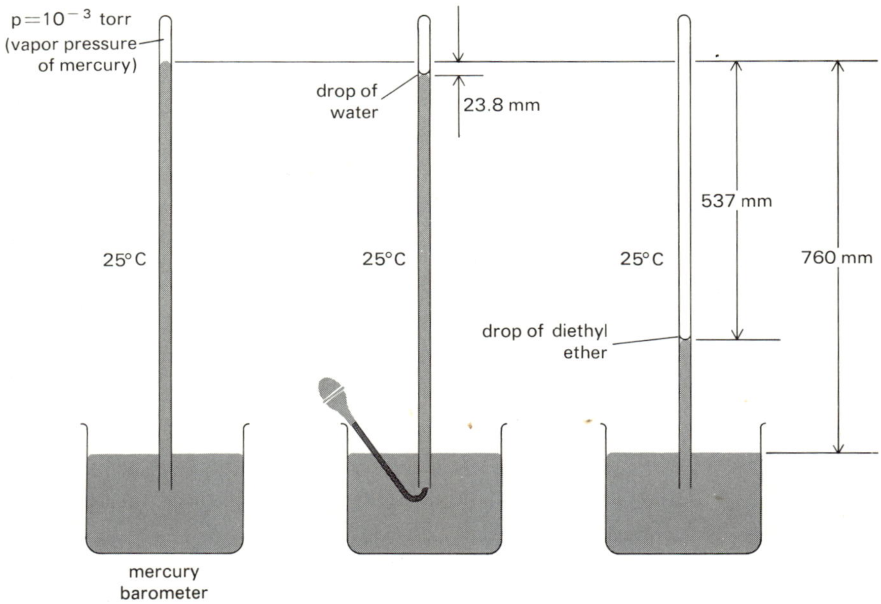

Figure 8–10 Determining equilibrium vapor pressures with a barometer tube.

of the liquid. The normal boiling point for water is 100°C at 760 torr pressure. At lower pressures, the boiling point decreases; it is 96°C at 650 torr and 83°C at 420 torr. Conversely, an increase in pressure raises the boiling point; for example, water boils at 101.4°C at 800 torr and 179.0°C at 10 atm.

The time required to cook foods depends on the temperature of the food. If water is present, the maximum cooking temperature that can be obtained is the boiling point of water. Since the boiling point of water varies with atmospheric pressure, it follows that cooking time is also a function of external pressure. At high altitudes, where the atmospheric pressure is lower, the boiling point of water is under 100°C, and cooking takes longer than at sea level. Cooking time, however, can be greatly shortened with a pressure cooker. In these pots, a tightly sealed lid prevents the steam from escaping, and the pressure within the vessel rapidly increases. This increased pressure raises the boiling point of water, and the food can be cooked much more rapidly at these higher temperatures. All pressure cookers have a pressure valve designed to let excessive pressure escape and keep the pot from exploding.

Example 8–1: Boiling point *What is the boiling point of chloroform at 600 torr?*

To answer this question, you must refer to the tabulated data in the literature. Many references are available for this type of information. Most frequently used is the *Handbook of Chemistry and Physics.**

Look at the handbook data plotted in Figure 8–8 to get the answer to the problem, 54°C.

ENERGY AND CHANGES OF STATE

Figure 8–11 shows the scientific terminology used to describe the energy-transfer processes that accompany changes from one state to another. The conversion of a liquid to a gas or of a solid to a liquid is an **endothermic** process: it requires the addition of heat. In the reverse processes, heat is given off, making them **exothermic.**

When heat is added to a solid near the absolute zero of temperature, the molecules begin to vibrate about their fixed positions in the solid structure. As the molecules get more energy, their vibration increases. The temperature of the solid rises until it reaches the melting point. The temperature remains constant at the melting point until all the solid is converted to a liquid. This effect is illustrated in

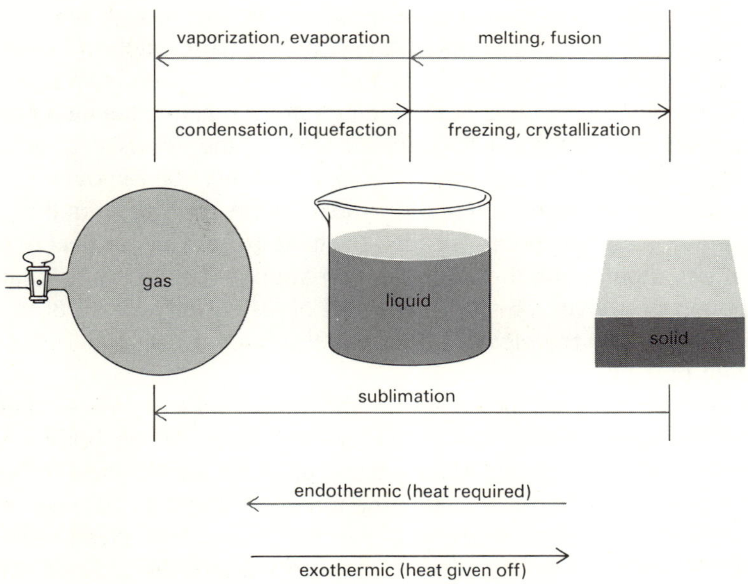

Figure 8–11 Nomenclature for energy-transfer processes in changes of state.

* Published by CRC Press Inc., Cleveland, OH.

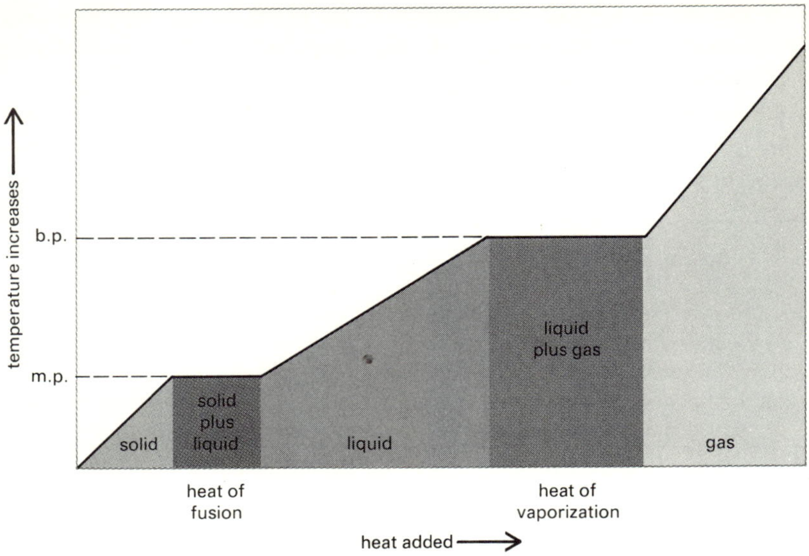

Figure 8–12 Temperature and state changes that occur when a substance is heated.

Figure 8–12, where the temperature of a given mass of a substance is plotted as a function of the amount of heat added at constant pressure.

The temperature at which solid and liquid states coexist in equilibrium at 760 torr pressure is called the **melting point** (abbreviated m.p.). At the melting point (Figure 8–12), the substance absorbs additional heat, but the temperature does not increase: all the added energy is used to overcome the forces of attraction holding the molecules in their fixed solid-state positions. When enough energy has been added to convert all the solid to liquid, the continual absorption of heat does raise the temperature. *The amount of heat needed to convert a given mass of a substance from the solid to the liquid state at the melting point* is called the **heat of fusion,** or **heat of melting.** Conversely, if the process goes in the opposite direction and a liquid changes into a solid, *the amount of heat that must be removed to convert a given mass of the substance from liquid to solid at the freezing point* (f.p., the same temperature as the melting point) is called the **heat of freezing,** or **heat of solidification.** The melting point and the heat of fusion are physical constants characteristic of a compound of a given pressure. The heat of fusion for water is 80 cal/g at the melting point, 0°C; for the element bromine, the heat of fusion is 16.1 cal/g at its melting point of −7°C.

Adding more heat to molecules in the liquid state increases their kinetic energy, and the temperature rises to the boiling point. At the boiling point, the absorption of more heat energy does not increase the temperature (Figure 8–12). All the energy is consumed in breaking the forces of attraction between molecules in the liquid state. The amount of energy needed to convert a given mass of liquid to vapor at the boiling point is called the **heat of vaporization,** and for the reverse process, changing from gas to liquid, the same constant is called the **heat of condensation,** or **heat of liquefaction.** When enough energy has been added to vaporize all the liquid molecules, additional heat simply increases the kinetic energy of the gas molecules and raises the temperature. The specific value for the heat of vaporization is a characteristic of a given compound. The value for water is 540 cal/g at the 100°C boiling point, and for bromine at its boiling point of 59°C, the heat of vaporization or condensation is only 45.5 cal/g.

A simple experiment illustrates some of the principles involved. If a paper cup full of water is placed above a hot gas flame, the paper will not burn until all the water has evaporated. As long as liquid water remains in the cup, all the heat absorbed by the system goes to increase the kinetic energy of the water molecules and to break the forces of attraction between those molecules so that they can evaporate. When the water is all gone, however, the temperature of the paper cup increases rapidly, and it soon bursts into flame. The principle behind this experiment has a practical application in the daily task of cooking dinner. If you are in a hurry to cook a pot of vegetables, do not turn up the gas or electric burner once water is boiling. Regardless of how much heat you apply, the temperature will remain the same. The water will boil faster, but its temperature will remain at 100°C. One way to speed up the process is to increase the boiling point of water by applying pressure; that is, use a pressure cooker.

Example 8–2: Energy and changes of state *How many calories of heat are needed to convert 10.0 g of ice at 0°C to 10.0 g of steam at 100°C?*

The problem involves three steps: the first is to calculate the amount of heat needed to melt the ice, a state change at 0°C; the second step is to calculate the heat needed to bring the liquid water to 100°C; the third step is to calculate the heat needed to convert the water to steam, a state change at 100°C. To solve the problem, you need to know the heat of fusion and the heat of vaporization for water.

Step 1. 10.0 g of ice to 10.0 g water at 0°C

$$10.0\,\cancel{g}\left(\frac{80\text{ cal}}{1.00\,\cancel{g}}\right) = 800\text{ cal} \qquad (80\text{ cal/g is the heat of fusion})$$

Step 2. 10.0 g of water at 0°C to 10.0 g water at 100°C

$$10.0\,\cancel{g}\left(\frac{1.00\text{ cal}}{1.00\,\cancel{g}\cancel{^\circ}}\right)(100\cancel{^\circ}) = 1000\text{ cal}$$

(1.00 cal/1.00 g H_2O/° is the definition of a calorie)

Step 3. 10.0 g of water at 100° to 100 g steam at 100°

$$10.0\,\cancel{g}\left(\frac{540\text{ cal}}{1.00\,\cancel{g}}\right) = 5400\text{ cal} \qquad (540\text{ cal/g is the heat of vaporization})$$

The total number of calories needed is therefore

$800 + 1000 + 5400 = 7200$ cal, or 7.2 kcal

COOLING BY EVAPORATION

The temperature of an evaporating liquid decreases. The evaporating molecules are those with greater than average kinetic energy, enough to overcome the forces of attraction between molecules in the liquid state. Molecules with less energy are not free to evaporate. Essentially, the hot molecules evaporate, and the cooler ones remain behind.

NEW TERMS

Boiling point: the temperature at which the vapor pressure of a liquid equals the external pressure.

Capillary attraction: the phenomenon in which a liquid rises in a tube of very small diameter because the liquid has wetted the walls of the tube.

Crystal: a solid in which the structural units are arranged in a repeating, symmetrical pattern in three dimensions.

Dynamic equilibrium: a state of balance in which opposing actions occur at the same rate

Endothermic: a process requiring the addition of heat energy.

Evaporation: vaporization—the change from the liquid to the gaseous state.

Heat of condensation: the quantity of heat released when a given mass of gas is condensed to the liquid state at the boiling point. For water, the heat of condensation is 540 cal/g.

Heat of fusion: the quantity of heat required to convert a given mass of solid to the liquid state at the melting point. For water, the heat of fusion is 80 cal/g.

Heat of solidification: the quantity of heat released when a given mass of liquid is changed to solid at the freezing point. It is numerically equal to the heat of fusion.

Heat of vaporization: the quantity of heat required to convert a given mass of liquid to the gaseous state at the boiling point. It is numerically equal to the heat of condensation.

Melting point: the temperature at which the solid and liquid phases coexist in equilibrium.

Meniscus: the curved upper surface of a liquid column where it touches the walls of the containing vessel.

Sublimation: the process of going directly from the solid to the gaseous state with no intervening change to liquid.

Surface tension: the intermolecular attraction that causes a liquid surface to become as small as possible, and causes it to behave as if a thin membrane were stretched over it.

Vapor pressure: the pressure exerted by a vapor in equilibrium with its liquid or solid phase.

PROBLEM SET A

1. Devise a method for determining the relative surface tensions of various liquids.
2. By reference to Figure 8–8, estimate the boiling point of water at 400 torr.
3. By reference to Figure 8–8, estimate the normal boiling point of diethyl ether.
4. The vapor pressures for a number of compounds at 20°C follow. Arrange the compounds in order of increasing rate of evaporation, and identify the substances with the highest and the lowest boiling points: mercury, 0.0012 torr; water, 17.5 torr; benzene, 74.7 torr, carbon tetrachloride, 91 torr; acetic acid, 11.7 torr.

5. At what temperature is the vapor pressure of neon 760 torr?
6. Explain why fish are able to survive in a lake when the outdoor temperature drops to 30° below zero. (Hint: water has its maximum density at 3.98°C—that is, ice floats.)
7. How many calories are required to heat 25.0 g of water from 0°C to 20°C?
8. How many calories of heat are required to convert 25.0 g of ice at 0°C to water at the same temperature?
9. How many calories of heat are required to convert 25.0 g of ice at 0°C to 25.0 g of steam at 100°C?
10. Given the heat of vaporization of bromine as 45.5 cal/g, calculate the heat of vaporization of a mole of bromine. (Remember that bromine is diatomic, Br_2.)

PROBLEM SET B

1. With reference to Figure 8–4, suppose that something dissolved in water keeps it from rising as high in the capillary as pure water does. Which has the higher surface tension, pure water or the solution?
2. By reference to Figure 8–8, estimate the temperature at which water has the same vapor pressure as chloroform at 40°C.
3. At what temperature is the vapor pressure of sodium chloride 760 torr?
4. Estimate the boiling point of water at 1000 torr pressure (refer to Figure 8–8).
5. At what pressure will water have a boiling point of only 55°C (refer to Figure 8–8)?
6. Explain how a barometer tube can be used to measure vapor pressure.
7. How many calories of heat must be removed from 1.00 mole of water to cool it from 100°C to 25°C?
8. How many calories of heat must be removed from 1.00 mole of steam at 100° to convert it to ice at 0°C?
9. The heat of fusion for bromine is 16.1 cal/g at its melting point of −7°C. Calculate the number of calories that would have to be removed from 30.0 g of liquid bromine at −7°C to convert it to solid bromine.
10. Liquid ether poured on the hand feels very cold. Explain why this is the case.

9

stoichiometry: calculations from chemical equations

THIS WILL SAVE YOU TIME:

Refer to this list as you read through the chapter. In answering the questions, you will understand the important concepts you are most likely to be asked about later on.

After reading this chapter, you should be able to:

1. *Explain the quantitative meaning of chemical equations.*
2. *Write balanced equations for all the reactions involved in preparing sulfuric acid from elemental sulfur by the contact process.*
3. *Work stoichiometry problems using the mole-factor method.*
4. *List some of the chemical and physical properties of sulfuric acid.*
5. *Tell how sulfuric acid reacts with water to produce hydrates.*
6. *Work stoichiometry problems involving limiting reagents.*
7. *Calculate percent yields.*
8. *Work stoichiometry problems involving molar volumes.*
9. *Explain what is meant by thermochemical reactions and the meaning of ΔH.*
10. *Work stoichiometry problems involving ΔH.*

In our original treatment of chemical equations, we saw that the numbers of atoms on the reactant side must equal the number on the product side. Balanced equations are of interest not only in terms of what reacts with what to produce what, but also in terms of how much reacts. The branch of chemistry that deals with calculations of mass implied in chemical equations is known as **stoichiometry.** To lend continuity to our discussion of stoichiometry, many of the formulas and equations in this chapter are drawn from the chemistry of sulfuric acid.

SULFURIC ACID

Sulfuric acid, $H_2SO_4(aq)$, is the most commonly used industrial chemical. United States production in 1971 amounted to 26.6 teragrams (26.6×10^{12} g, or 58.6 billion pounds). About one-third this total went into the production of fertilizers. Other major uses include the refining of petroleum, the manufacture of chemicals, processing in the iron and steel industry, and the production of paints, rayon, and a host of other products.

Sulfuric acid has been known since ancient times. Thirteenth-century Byzantine manuscripts describe its preparation from the roasting of vitriols—metal sulfates. Originally sulfuric acid was called oil of vitriol.

WHAT CHEMICAL EQUATIONS TELL US

Every balanced chemical equation contains a large amount of quantitative information. We can take a close look at the equation for the preparation of sulfur dioxide and see what the expression tells us. Sulfur dioxide, SO_2, occurs in nature in volcanic gases and as a dissolved compound in many water sources. It is also one of the undesirable components of smog. It is a colorless gas, 2.2 times heavier than air, with an irritating odor (the choking odor sometimes attributed to sulfur actually comes from SO_2; sulfur is odorless). One method for preparing sulfur dioxide is the burning of sulfur, S_8.

$$S_8(s) + 8O_2(g) \rightarrow 8SO_2(g)$$

With this statement as an example, we can detail the quantitative information a balanced equation provides:

1. *What the reactants are and what products result.*
   In our example, solid sulfur reacts with oxygen gas to give sulfur dioxide gas.
2. *The formulas for each reactant and each product.*
   The formula for solid sulfur is $S_8(s)$, for oxygen gas $O_2(g)$, and for sulfur dioxide gas $SO_2(g)$.
3. *The relative number of molecules of each reactant in the reaction and the number of molecules of products formed.*
   One molecule of sulfur reacts with 8 molecules of oxygen to give 8 molecules of sulfur dioxide.
4. *The relative number of atoms of each element in the reaction.*
   Eight atoms of sulfur react with 16 atoms of oxygen to give 8 molecules of sulfur dioxide comprising a total of 24 atoms.

5. *The relative number of molecular masses of each reactant and product.*
   One molecular mass of S_8 reacts with 8 molecular masses of O_2 to give 8 molecular masses of SO_2.
6. *The relative number of grams (or other mass units) of each substance reacted or produced.*
   256.8 g of S_8 react with 256.0 g of O_2 to give 512.8 g of SO_2.
7. *The relative number of moles of each substance reacted or produced.*
   One mole of S_8 reacts with 8 moles of O_2 to give 8 moles of SO_2.

To summarize, this equation contains the following quantitative information:

$S_8(s)$	$+ 8O_2(g)$	$\rightarrow 8SO_2(g)$
1 molecule	+ 8 molecules	→ 8 molecules
8 atoms	+ 16 atoms	→ 24 atoms
1 molecular mass	+ 8 molecular masses	→ 8 molecular masses
8(32.1) amu	+ 8(32.0) amu	→ 8(64.1) amu
1 mole	+ 8 moles	→ 8 moles
256.8 g	+ 256.0 g	→ 512.8 g
256.8 kg	+ 256.0 kg	→ 512.8 kg

From these data, we can calculate how much of a given reactant is needed to produce a given amount of product, or conversely, how much product can be obtained from a given amount of reactant.

THE MOLE-FACTOR CALCULATION METHOD

The recommended method for solving stoichiometry problems involves the use of mole factors. The method is best illustrated by carefully following the steps used to solve a typical problem. The following problem will serve as our example: *What mass of sulfur dioxide can be prepared from the complete combustion of 100 g of sulfur?*

The problem is solved with these steps:

1. *Determine the correct formulas for the reactants and products.*
   The formula for sulfur is S_8, for oxygen O_2, and for sulfur dioxide SO_2.
2. *Write and balance the chemical equation for the reaction.*
   We know this equation already. It is

   $$S_8(s) + 8O_2(g) \rightarrow 8SO_2(g)$$

3. *Identify the mole data needed to solve the problem.*
   The problem is to find out how many moles of SO_2 can be formed from how many moles of S_8. From the equation, we see that 1 mole of S_8 will produce 8 moles of SO_2.
4. *Determine the mole factor.*
   We will define **the mole factor** as

   $$\text{mole factor} = \frac{\text{number of moles of desired substance indicated by the balanced equation}}{\text{number of moles of starting substance indicated by the balanced equation}}$$

   For our problem, the mole factor is

   $$\text{mole factor} = \frac{8 \text{ moles of } SO_2}{1 \text{ mole of } S_8}$$

5. *Determine the number of moles of starting substance.*
   In this problem, we have 100 g of the starting substance, S_8. The molecular mass of S_8 is 8(32.1) g = 256.8 g = 257 g. The conversion factor is therefore*

$$\frac{257 \text{ g } S_8}{1.00 \text{ mole } S_8} \quad \text{or} \quad \frac{1.00 \text{ mole } S_8}{257 \text{ g } S_8}$$

Hence

$$100 \cancel{\text{g}}\ \cancel{S_8}\left(\frac{1.00 \text{ mole } S_8}{257 \cancel{\text{g}}\ \cancel{S_8}}\right) = 0.389 \text{ mole } S_8$$

6. *Multiply the number of moles of the starting substance (step 5) by the mole factor (step 4).*

$$0.389 \cancel{\text{mole}}\ \cancel{S_8}\left(\frac{8.00 \text{ moles } SO_2}{1.00 \cancel{\text{mole}}\ \cancel{S_8}}\right) = 3.11 \text{ moles } SO_2$$

We have now determined how many moles of SO_2 can be produced from 100 g of S_8. To determine the production of SO_2 in grams, we have to convert from moles to grams.

7. *Convert the mole data to the desired units with the proper conversion factor.*
   The problem asks for the answer in grams.

$$1.00 \text{ mole of } SO_2 = [32.1 + 2(16.0)] \text{ g} = 64.1 \text{ g}$$

The conversion factor is

$$\frac{64.1 \text{ g } SO_2}{1.00 \text{ mole } SO_2} = \frac{1.00 \text{ mole } SO_2}{64.1 \text{ g } SO_2}$$

After choosing the correct factor—the one allowing the units to be canceled—we can multiply to obtain the answer.

$$3.11 \cancel{\text{moles}}\ \cancel{SO_2}\left(\frac{64.1 \text{ g } SO_2}{1.00 \cancel{\text{mole}}\ \cancel{SO_2}}\right) = 199 \text{ g } SO_2$$

Thus 199 g SO_2 can be prepared by the complete combustion of 100 g of S_8.

Example 9–1: Mole-factor calculations *Another method of producing sulfur dioxide is the roasting of sulfide ores. How many moles of SO_2 can be produced by roasting (heating in the presence of air) 3.00 kg of zinc sulfide?*

Step 1. The formulas for the reactants are ZnS(s) and O_2(g). The products are zinc oxide, ZnO(s), and sulfur dioxide, SO_2(g).

Step 2. The skeletal equation is:

$$ZnS(s) + O_2(g) \rightarrow ZnO(s) + SO_2(g)$$

The balanced equation is:

$$2ZnS(s) + 3O_2(g) \rightarrow 2ZnO(s) + 2SO_2(g)$$

* Note that the numbers in the mole fraction, obtained from the balanced chemical equation, are pure numbers. That is, when we say that 8 moles of SO_2 can be produced from 1 mole of S_8, the numbers 8 and 1 can be expressed to as many significant figures as desired. Since we have been using three-significant-figure accuracy in all our examples, these numbers are written as 8.00 and 1.00.

Step 3. From the balanced equation, we see that 2 moles of SO_2 are produced from 2 moles of ZnS.

Step 4. The mole factor is:

$$\frac{2.00 \text{ moles } SO_2}{2.00 \text{ moles ZnS}}$$

Step 5. The number of moles of starting material is:

1 mole of ZnS $= 65.4 + 32.1 \text{ g} = 97.5 \text{ g}$

Hence the conversion factor from grams to moles is

$$\frac{1.00 \text{ mole ZnS}}{97.5 \text{ g ZnS}}$$

We also need to convert kg to g with the conversion factor

$$\frac{1000 \text{ g}}{1.00 \text{ kg}}$$

Therefore, 3.00 kg of ZnS equals

$$3.00\ \cancel{\text{kg ZnS}}\left(\frac{1000\ \cancel{\text{g ZnS}}}{1.00\ \cancel{\text{kg ZnS}}}\right)\left(\frac{1.00 \text{ mole ZnS}}{97.5\ \cancel{\text{g ZnS}}}\right) = 30.8 \text{ moles ZnS}$$

Step 6. Now the number of moles of SO_2 produced from 30.8 moles of ZnS is found by multiplying with the mole factor.

$$30.8\ \cancel{\text{moles ZnS}}\left(\frac{2.00 \text{ moles } SO_2}{2.00\ \cancel{\text{moles ZnS}}}\right) = 30.8 \text{ moles } SO_2 = \text{answer}$$

Step 7. No further conversions are necessary since the problem asked for the answer in moles.

As you gain skill in solving these stoichiometry problems, you will find it convenient to apply all the factors in one mathematical operation. In this case,

$$3.00\ \cancel{\text{kg ZnS}}\left(\frac{1000\ \cancel{\text{g ZnS}}}{1.00\ \cancel{\text{kg ZnS}}}\right)\left(\frac{1.00\ \cancel{\text{mole ZnS}}}{97.5\ \cancel{\text{g ZnS}}}\right)\left(\frac{2.00 \text{ moles } SO_2}{2.00\ \cancel{\text{moles ZnS}}}\right) = 30.8 \text{ moles } SO_2$$

THE CONTACT PROCESS FOR SULFURIC ACID

Two commercial processes are used to convert sulfur dioxide to sulfuric acid. The more important one is the contact process. Sulfur dioxide will react with oxygen from the air to form sulfur trioxide, SO_3, when the clean dry gases are passed over a suitable catalyst. A **catalyst** is a *substance that changes the rate of a chemical reaction but does not itself appear in the reaction products.* One important commercial catalyst is the solid vanadium pentoxide, V_2O_5.

$$2SO_2(g) + O_2(g) \xrightarrow[V_2O_5 \text{ catalyst}]{400°C} 2SO_3(g)$$

The SO_3 will react with water to form sulfuric acid.

$$SO_3(g) + H_2O(l) \rightarrow H_2SO_4(aq)$$

Because SO_3 tends to form another sulfur oxide, S_2O_6, which is less reactive with water, sulfuric acid is made commercially by first reacting the SO_3 with sulfuric acid to form pyrosulfuric acid, $H_2S_2O_7$.

$$SO_3(g) + H_2SO_4(l) \rightarrow H_2S_2O_7(l)$$

Adding pyrosulfuric acid to water produces sulfuric acid.

$$H_2S_2O_7(l) + H_2O(l) \rightarrow 2H_2SO_4(l)$$

Pure sulfuric acid can be prepared by this technique.

PROPERTIES OF SULFURIC ACID

Sulfuric acid is a colorless, syrupy liquid with a density of 1.84g/me at 20°C; it boils at 330°C. Adding water to sulfuric acid liberates enough heat to make the acid splatter and boil. Never, never should water be added to sulfuric acid. Sulfuric acid is very corrosive and will produce serious burns. (Acid should always be added to water, never vice versa.)

The following example details an important property of sulfuric acid—its reaction with certain metals.

Example 9–2: Properties of sulfuric acid *Dilute sulfuric acid reacts with certain metals to produce hydrogen. What mass of iron(II) sulfate is formed when 20.0 g of iron completely reacts with sulfuric acid?*

The balanced equation for this reaction is

$$Fe(s) + H_2SO_4(aq) \rightarrow FeSO_4(aq) + H_2(g)$$

One mole of Fe produces one mole of $FeSO_4$. Hence the mole factor is

$$\frac{1.00 \text{ mole } FeSO_4}{1.00 \text{ mole } Fe}$$

The gram-to-mole conversion factors are

$$\frac{55.8 \text{ g Fe}}{1.00 \text{ mole Fe}} \quad \text{and} \quad \frac{152 \text{ g } FeSO_4}{1.00 \text{ mole } FeSO_4}$$

Combining all the factors into one mathematical manipulation and multiplying by the 20.0 g Fe gives

$$20.0 \text{ g } \cancel{Fe}\left(\frac{1.00 \cancel{\text{ mole Fe}}}{55.8 \text{ g } \cancel{Fe}}\right)\left(\frac{1.00 \cancel{\text{ mole } FeSO_4}}{1.00 \cancel{\text{ mole Fe}}}\right)\left(\frac{152 \text{ g } FeSO_4}{1.00 \cancel{\text{ mole } FeSO_4}}\right) = 54.5 \text{ g } FeSO_4$$

Remember that these conversion factors all equal 1. Thus

$$\frac{55.8 \text{ g Fe}}{1.00 \text{ mole Fe}} = \frac{1.00 \text{ mole Fe}}{55.8 \text{ g Fe}}$$

Use the factor that permits you to cancel units and gives the correct answer.

Sulfuric acid is an excellent dehydrating agent—a substance that has strong affinity for water. The acid reacts with water to form hydrates. As you may recall from Chapter 6, hydrates are compounds with water molecules chemically bound into their structures in definite proportions.

$H_2SO_4 + H_2O \rightarrow H_2SO_4 \cdot H_2O$
sulfuric acid monohydrate or
sulfuric acid 1-water

Sulfuric acid also forms other hydrates. The general formula is $H_2SO_4 \cdot xH_2O$, where x is a value from 1 to 4.

Example 9–3: Formation of hydrates *What mass of sulfuric acid dihydrate is formed if 50.0 g of sulfuric acid react with water according to the equation*

$H_2SO_4 + 2H_2O \rightarrow H_2SO_4 \cdot 2H_2O$

Note that the term dihydrate indicates two molecules of water bound in the molecule. This compound is different from the monohydrate. Sulfuric acid dihydrate is also called sulfuric acid 2-water.

The mole factor is

$$\frac{1.00 \text{ mole } H_2SO_4 \cdot 2H_2O}{1.00 \text{ mole } H_2SO_4}$$

The gram-to-mole conversion factors are

$$\frac{98.1 \text{ g } H_2SO_4}{1.00 \text{ mole } H_2SO_4} \quad \text{and} \quad \frac{134 \text{ g } H_2SO_4 \cdot 2H_2O}{1.00 \text{ mole } H_2SO_4 \cdot 2H_2O}$$

To solve

$$50.0 \text{ g } \cancel{H_2SO_4}\left(\frac{1.00 \cancel{\text{mole } H_2SO_4}}{98.1 \text{ g } \cancel{H_2SO_4}}\right)\left(\frac{1.00 \cancel{\text{mole } H_2SO_4 \cdot 2H_2O}}{1.00 \cancel{\text{mole } H_2SO_4}}\right)\left(\frac{134 \text{ g } H_2SO_4 \cdot 2H_2O}{1.00 \cancel{\text{mole } H_2SO_4 \cdot 2H_2O}}\right)$$

$$= 68.3 \text{ g } H_2SO_4 \cdot 2H_2O$$

The dehydrating action of sulfuric acid is so intense that it will actually pull water out of substances such as table sugar, paper, cellulose (including pants and dresses), fingers, and so on. The reaction with sugar is

$C_{12}H_{22}O_{11} + 11 \; H_2SO_4 \rightarrow 12C + 11H_2SO_4 \cdot H_2O$

ADDITIONAL EXAMPLES OF STOICHIOMETRY PROBLEMS

Mole-mole calculations

The simplest kind of stoichiometry problems are those in which one calculates the number of moles of a substance that has reacted with, or is produced from, a given number of moles of another substance. The next two examples illustrate this type of problem.

Example 9–4: Mole-mole calculations *How many moles of oxygen can be obtained from the decomposition of 2 moles of hydrogen peroxide?*

First write and balance the equation for the reaction.

Skeletal equation: $H_2O_2 \rightarrow H_2O + O_2$
Balanced equation: $2H_2O_2 \rightarrow 2H_2O + O_2$
2 moles → 2 moles + 1 mole

From the balanced equation, we see that every 2 moles of decomposing hydrogen peroxide produces 1 mole of oxygen.

Example 9–5: Mole-mole calculations *How many moles of oxygen are needed to react completely with 6 moles of hexane, C_6H_{14}? The products are carbon dioxide and water.*

First write and balance the equation for the reaction.

Skeletal equation: $C_6H_{14} + O_2 \rightarrow CO_2 + H_2O$
Balanced equation:
step 1: $C_6H_{14} + 9\frac{1}{2}O_2 \rightarrow 6CO_2 + 7H_2O$
step 2: $2C_6H_{14} + 19O_2 \rightarrow 12CO_2 + 14H_2O$
2 moles + 19 moles → 12 moles + 14 moles

The balanced equation tells us that 2 moles of hexane react with 19 moles of oxygen; hence 6 moles of hexane will react with 3(19), or 57, moles of oxygen.

Mole-mass calculations

In this type of problem, we wish to calculate the number of moles of a substance produced from, or reacted with, a given mass of another substance, or vice versa—that is, given mass, calculate moles.

Example 9–6: Mole-mass calculations *Iron(III) oxide reacts with coke (carbon) in a blast furnace to produce carbon monoxide and molten iron. How many moles of iron can be produced from 50.0 kg of pure iron(III) oxide?*

The first step in any stoichiometry problem is to write and balance the equation for the reaction.

Skeletal equation: $Fe_2O_3 + C \rightarrow Fe + CO$
Balanced equation: $Fe_2O_3 + 3C \rightarrow 2Fe + 3CO$

The mole factor is

$$\frac{2.00 \text{ moles Fe}}{1.00 \text{ mole } Fe_2O_3}$$

The molecular mass of Fe_2O_3 is $2(55.8) + 3(16.0) = 160$ g. Therefore, the needed conversion factor is

$$\frac{1.00 \text{ mole } Fe_2O_3}{160 \text{ g } Fe_2O_3}$$

To convert from kg to g, we also need the factor

$$\frac{1000 \text{ g}}{1.00 \text{ kg}}$$

The solution is then

$$50.0 \cancel{\text{kg}} \cancel{Fe_2O_3}\left(\frac{1000 \cancel{\text{g}} \cancel{Fe_2O_3}}{1.00 \cancel{\text{kg}} \cancel{Fe_2O_3}}\right)\left(\frac{1.00 \cancel{\text{mole}} \cancel{Fe_2O_3}}{160 \cancel{\text{g}} \cancel{Fe_2O_3}}\right)\left(\frac{2.00 \text{ moles Fe}}{1.00 \cancel{\text{mole}} \cancel{Fe_2O_3}}\right) = 625 \text{ moles Fe}$$

Example 9–7: Mole-mass calculations *How many grams of zinc metal are needed to completely react with 20.6 moles of hydrochloric acid?*

Let us review the steps in the mole-factor calculation method.

Step 1. The formulas for the reactants are $Zn(s)$ for zinc and $HCl(aq)$ for hydrochloric acid. The products are zinc chloride, $ZnCl_2$, and hydrogen, H_2.

Step 2. *Skeletal equation:* $Zn(s) + HCl(aq) \rightarrow ZnCl_2(aq) + H_2(g)$
Balanced equation: $Zn(s) + 2HCl(aq) \rightarrow ZnCl_2(aq) + H_2(g)$

Step 3. One mole of $Zn(s)$ reacts with 2 moles of $HCl(aq)$.

Step 4. Note that in this problem we are concerned only with the amount of the reactant. In this case, the quantity we seek is the mass of Zn. HCl is the starting substance. Therefore, the mole factor is

$$\frac{1.00 \text{ mole Zn}}{2.00 \text{ moles HCl}}$$

Step 5. The number of moles of starting substance is 20.6 moles of HCl.

Step 6. Calculate the number of moles of Zn reacted.

$$20.6 \cancel{\text{moles}} \cancel{HCl}\left(\frac{1.00 \text{ mole Zn}}{2.00 \cancel{\text{moles}} \cancel{HCl}}\right) = 10.3 \text{ moles Zn}$$

Step 7. Finally, to convert to grams, we need the conversion factor

$$\frac{65.4 \text{ g}}{1.00 \text{ mole Zn}}$$

$$10.3 \cancel{\text{moles}} \cancel{Zn}\left(\frac{65.4 \text{ g Zn}}{1.00 \cancel{\text{mole}} \cancel{Zn}}\right) = 674 \text{ g Zn}$$

Mass-mass calculations

The next two examples illustrate mass-mass calculations.

Example 9–8: Mass-mass calculations *How many grams of NO(g) result from the oxidation of 36.0 g of ammonia gas?*

The formulas for the reactants are $NH_3(g)$ for ammonia and $O_2(g)$ for oxygen. The products are nitric oxide, NO(g), and water, H_2O.

Skeletal equation: $NH_3(g) + O_2(g) \rightarrow NO(g) + H_2O$

Balanced equation: $4NH_3(g) + 5O_2(g) \rightarrow 4NO(g) + 6H_2O$

The mole factor is

$$\frac{4.00 \text{ moles NO}}{4.00 \text{ moles } NH_3}$$

Since

$$1 \text{ mole } NH_3 = (14.0 + 3(1.00)) \text{ g} = 17.0 \text{ g}$$

and

$$1 \text{ mole NO} = 14.0 + 16.0 = 30.0 \text{ g}$$

the needed conversion factors are

$$\frac{1.00 \text{ mole } NH_3}{17.0 \text{ g}} \quad \text{and} \quad \frac{30.0 \text{ g}}{1.00 \text{ mole NO}}$$

Solving, we get

$$36.0 \text{ g } NH_3 \left[\frac{1.00 \text{ mole } NH_3}{17.0 \text{ g } NH_3}\right]\left[\frac{4.00 \text{ moles NO}}{4.00 \text{ moles } NH_3}\right]\left[\frac{30.0 \text{ g NO}}{1.00 \text{ mole NO}}\right] = 63.5 \text{ g NO}$$

Example 9–9: Mass-mass calculations *Sodium hydroxide can be prepared by reacting sodium carbonate with calcium hydroxide (slaked lime). If 42.0 kg of sodium hydroxide is prepared by this reaction, how many kilograms of calcium carbonate are produced as a by-product?*

The reactants are sodium carbonate, Na_2CO_3, and calcium hydroxide, $Ca(OH)_2$. The products are calcium carbonate, $CaCO_3$, and sodium hydroxide, NaOH.

Skeletal equation: $Na_2CO_3 + Ca(OH)_2 \rightarrow CaCO_3 + NaOH$

Balanced equation: $Na_2CO_3 + Ca(OH)_2 \rightarrow CaCO_3 + 2NaOH$

Two moles of NaOH are formed for 1 mole of $CaCO_3$. The mole factor is

$$\frac{1.00 \text{ mole } CaCO_3}{2.00 \text{ moles NaOH}}$$

1 mole of $CaCO_3 = 40.1 + 12.0 + 3(16.0)\ g = 100\ g$

1 mole of $NaOH = 23.0 + 16.0 + 1.00\ g = 40.0\ g$

The needed conversion factors are

$$\frac{1.00 \text{ mole } CaCO_3}{100 \text{ g } CaCO_3} \qquad \frac{1.00 \text{ mole NaOH}}{40.0 \text{ g NaOH}} \qquad \frac{1000 \text{ g}}{1.00 \text{ kg}}$$

Solving, we get

$$42.0 \cancel{\text{kg NaOH}} \left[\frac{1000 \cancel{\text{g NaOH}}}{1.00 \cancel{\text{kg NaOH}}}\right]\left[\frac{1.00 \cancel{\text{mole NaOH}}}{40.0 \cancel{\text{g NaOH}}}\right]\left[\frac{1.00 \cancel{\text{mole } CaCO_3}}{2.00 \cancel{\text{moles NaOH}}}\right]\left[\frac{100 \text{ g } \cancel{CaCO_3}}{1.00 \cancel{\text{mole } CaCO_3}}\right]\left[\frac{1.00 \text{ kg } CaCO_3}{1000 \cancel{\text{g } CaCO_3}}\right] = 52.5 \text{ kg } CaCO_3$$

LIMITING-REAGENT PROBLEMS

Many laboratory preparations commonly employ an excess of one of the reactants. Therefore, calculations to determine the amount of product expected should be based on the reagent not present in excess—that is, the one that will be completely used in the reaction. This reagent is called the **limiting reagent.**

Example 9–10: Limiting reagent *10.0 g of NaOH(aq) are reacted with 10.0 g of HCl(aq) to give water and NaCl(aq). How much NaCl is formed?*

Balanced equation:	NaOH(*aq*) +	HCl(*aq*) →	NaCl(*aq*) +	H_2O
Mole data:	1 mole	1 mole	1 mole	

Which mole factor should we use to solve the problem?

$$\frac{1.00 \text{ mole NaCl}}{1.00 \text{ mole NaOH}} \quad \text{or} \quad \frac{1.00 \text{ mole NaCl}}{1.00 \text{ mole HCl}}$$

We first have to decide which reactant is not present in excess.

1 mole of NaOH = 40.0 g
1 mole of HCl = 36.5 g.

These data show that it takes more grams of NaOH than HCl to produce one mole of NaCl. Thus the HCl is present in excess, and our calculations must be based on the 10.0 g of NaOH.

$$10.0 \cancel{\text{g NaOH}} \left[\frac{1.00 \cancel{\text{mole NaOH}}}{40 \cancel{\text{g NaOH}}}\right]\left[\frac{1.00 \cancel{\text{mole NaCl}}}{1.00 \cancel{\text{mole NaOH}}}\right]\left[\frac{58.5 \text{ g NaCl}}{1.00 \cancel{\text{mole NaCl}}}\right] = 14.6 \text{ g NaCl}$$

What answer would you get if you used the other mole factor?

PERCENT YIELD

The amount of product actually obtained from a chemical reaction is often less than the amount calculated from the stoichiometrical relationships. The lower yield may be due to any of a number of causes. For example, some of the reactants may not react completely, there may be side reactions giving different products, and some of the products may react to re-form the original reactants. In any case, you will probably get less product from the reaction than predicted by calculation. It is therefore important to know what percentage yield you have obtained. Percent yield is defined as follows:

$$\text{Percent yield} = \frac{\text{actual yield}}{\text{theoretical yield}} \times 100$$

Percent-yield data are very important to industrial chemists. For a reaction to be commercially important, it must have a good percent yield of the desired product.

Example 9–11: Percent yield *The gas ethene (common name, ethylene), C_2H_4, reacts with oxygen to form the commercially important compound ethylene oxide, C_2H_4O, according to the equation*

$$2C_2H_4(g) + O_2(g) \rightarrow 2C_2H_4O(g)$$

If 180 g of ethylene oxide is obtained from 120 g of ethene, what is the percent yield?

The theoretical yield is obtained by working a standard stoichiometry problem to find how much ethylene oxide can be formed from 120 g of ethene.

The mole factor is

$$\frac{2.00 \text{ moles of } C_2H_4O}{2.00 \text{ moles } C_2H_4}$$

$$1 \text{ mole } C_2H_4 = 2(12.0) + 4(1.00) = 28.0 \text{ g}$$

$$1 \text{ mole } C_2H_4O = 2(12.0) + 4(1.00) + 16.0 = 44.0 \text{ g}$$

The conversion factors are

$$\frac{1.00 \text{ mole } C_2H_4}{28.0 \text{ g } C_2H_4} \quad \text{and} \quad \frac{1.00 \text{ mole } C_2H_4O}{44.0 \text{ g } C_2H_4O}$$

The theoretical yield is

$$120 \text{ g } C_2H_2 \left[\frac{1.00 \text{ mole } C_2H_4}{28.0 \text{ g } C_2H_4}\right]\left[\frac{2.00 \text{ moles } C_2H_4O}{2.00 \text{ moles } C_2H_4}\right]\left[\frac{44.0 \text{ g } C_2H_4O}{1.00 \text{ mole } C_2H_4O}\right] = 189 \text{ g } C_2H_4O$$

The percent yield is

$$\frac{\text{actual yield}}{\text{theoretical yield}} \times (100) = \frac{180 \text{ g}}{189 \text{ g}} (100) = 95.2\%$$

PROBLEMS INVOLVING GAS VOLUMES

In many reactions, the reactants and products are gases. These gases are usually measured by volume and not by mass. However, we know from our discussion in Chapter 7 that, given the volume of a gas at STP, we can easily calculate the number of moles of gas by using the molar gas volume (22.4 liters (STP) of any gas = 1.00 mole), or the ideal gas law formula ($PV = nRT$). These relationships permit us to work stoichiometry problems involving gas volumes.

Example 9–12: Stoichiometry and gas volumes *Show that the volume of oxygen produced from the decomposition of water is half the volume of hydrogen produced.*

The volume changes accompanying a chemical reaction can be deduced from Avogadro's law, which says that equal volumes of gases contain equal numbers of molecules at a given temperature and pressure. A mole, by definition, contains 6.02×10^{23} molecules; hence the equation for the reaction tells us

$2H_2O$	$\rightarrow$	$2H_2$	$+ O_2$
2 moles	$\rightarrow$	2 moles	+ 1 mole
$2(6.02 \times 10^{23})$	$\rightarrow$	$2(6.02 \times 10^{23})$	$+ 1(6.02 \times 10^{23})$
molecules H_2O		molecules H_2	molecules O_2

The volume of a gas that contains 1 mole, or 6.02×10^{23} molecules, is a universal constant equal to 22.4 liters at STP. Thus the equation says

$$2(22.4 \text{ liters}_{STP}\ H_2O) \rightarrow 2(22.4 \text{ liters}_{STP}\ H_2) + (22.4 \text{ liters}_{STP}\ O_2)$$

or simply

$$2 \text{ volumes } H_2O(g) \rightarrow 2 \text{ volumes } H_2 + 1 \text{ volume } O_2$$

In any balanced chemical equation, the relative volumes of gaseous reactants and products are given by the numbers preceding the molecular formulas, assuming that reactants and products are at the same temperature and pressure.

Example 9–13: Stoichiometry and gas volumes *What volume of hydrogen (at STP) can be obtained from the decomposition of 10 g of water?*

Balanced equation:	$2H_2O(l) \rightarrow$	$2H_2(g)$	$+ O_2(g)$
Mole data:	2 moles	2 moles	

The coefficients of the balanced equation show that 2 moles of H_2O produce 2 moles of H_2; hence the mole factor is

$$\frac{2.00 \text{ moles } H_2}{2.00 \text{ moles } H_2O}$$

The needed conversion factors are

$$\frac{1.00 \text{ mole } H_2O}{18.0 \text{ g } H_2O} \qquad \frac{1.00 \text{ mole } H_2}{22.4 \text{ liters}_{STP} \ H_2} \quad \text{or} \quad \frac{22.4 \text{ liters}_{STP} \ H_2}{1.00 \text{ mole } H_2}$$

The solution is

$$10.0 \ \cancel{g} \ \cancel{H_2O} \left[\frac{1.00 \ \cancel{\text{mole}} \ \cancel{H_2O}}{18.0 \ \cancel{g} \ \cancel{H_2O}}\right]\left[\frac{2.00 \ \cancel{\text{moles}} \ \cancel{H_2}}{2.00 \ \cancel{\text{moles}} \ \cancel{H_2O}}\right]\left[\frac{22.4 \text{ liters}_{STP} \ H_2}{1.00 \ \cancel{\text{mole}} \ \cancel{H_2}}\right] = 12.4 \text{ liters}_{STP} \ H_2$$

What we have done is to solve for the number of moles of hydrogen in the usual fashion and then multiply this answer by the molar-volume conversion factor.

Example 9–14: Stoichiometry and gas volumes *What volume of oxygen (at STP) can be prepared by heating 100 g of potassium chlorate?*

Balanced equation: $2 \ KClO_3(s) \rightarrow 2KCl(s) + 3O_2(g)$
Mole data: 2 moles — 3 moles

Mole factor: $\dfrac{3.00 \text{ moles } O_2}{2.00 \text{ moles } KClO_3}$

Conversion factors: $\dfrac{1.00 \text{ mole } KClO_3}{123 \text{ g } KClO_3} \qquad \dfrac{22.4 \text{ liters}_{STP} \ O_2}{1.00 \text{ mole } O_2}$

Solution:

$$100 \ \cancel{g} \ \cancel{KClO_3} \left[\frac{1.00 \ \cancel{\text{mole}} \ \cancel{KClO_3}}{123 \ \cancel{g} \ \cancel{KClO_3}}\right]\left[\frac{3.00 \ \cancel{\text{moles}} \ \cancel{O_2}}{2.00 \ \cancel{\text{moles}} \ \cancel{KClO_3}}\right]\left[\frac{22.4 \text{ liters}_{STP} \ O_2}{1.00 \ \cancel{\text{mole}} \ \cancel{O_2}}\right] = 27.3 \text{ liters}_{STP} \ O_2$$

Example 9–15: Stoichiometry and gas volumes *What volume of hydrogen, collected in the laboratory at 25°C and 750 torr, is formed from the reaction of 10.0 g of zinc metal with hydrochloric acid?*

Balanced equation: $Zn(s) + 2HCl(aq) \rightarrow ZnCl_2(aq) + H_2(g)$
Mole data: 1 mole — 1 mole

Mole factor: $\dfrac{1.00 \text{ mole } H_2}{1.00 \text{ mole } Zn}$

Conversion factors: $\dfrac{1.00 \text{ mole } Zn}{65.4 \text{ g } Zn} \qquad \dfrac{22.4 \text{ liters}_{STP} \ H_2}{1.00 \text{ mole } H_2}$

We first solve for the volume of hydrogen gas at STP.

$$10.0 \ \cancel{g} \ \cancel{Zn} \left[\frac{1.00 \ \cancel{\text{mole}} \ \cancel{Zn}}{65.4 \ \cancel{g} \ \cancel{Zn}}\right]\left[\frac{1.00 \ \cancel{\text{mole}} \ \cancel{H_2}}{1.00 \ \cancel{\text{mole}} \ \cancel{Zn}}\right]\left[\frac{22.4 \text{ liters}_{STP} \ H_2}{1.00 \ \cancel{\text{mole}} \ \cancel{H_2}}\right] = 3.42 \text{ liters}_{STP} \ H_2$$

Finding the volume at 25°C and 750 torr means solving a standard gas-law problem.

The data are:

$P_1 = 760$ torr
$V_1 = 3.42$ liters
$T_1 = 273°K$

$P_2 = 750$ torr
$V_2 = ?$
$T_2 = 298°K$

$$\frac{(760\ \cancel{\text{torr}})(3.42\ \text{liters})}{273°K} = \frac{(750\ \cancel{\text{torr}})(V_2)}{298°K}$$

$$V_2 = \frac{(760)(3.42\ \text{liters})(298)}{(273)(750)} = 3.78\ \text{liters}\ H_2\ \text{(measured at 25°C and 750 torr)}$$

THERMOCHEMICAL EQUATIONS

Chemical reactions are accompanied by energy changes. As we have seen, a reaction that gives off heat is exothermic, one that absorbs it is endothermic.

Heat can be included as a reactant or as a product in balanced chemical equations called **thermochemical equations.** An example is the equation for the reaction of the gas methane, CH_4 (natural gas is approximately 80 percent CH_4), with oxygen. This reaction occurs when natural gas is burned in home furnaces or kitchen stoves. It can be written in two ways:

$$CH_4(g) + 2O_2(g) \rightarrow CO_2(g) + 2H_2O(l) + 213\ \text{kcal}$$

or

$$CH_4(g) + 2O_2(g) \rightarrow CO_2(g) + 2H_2O(l) \qquad \Delta H = -213\ \text{kcal}$$

The heat change that occurs when a reaction is carried out at constant pressure is called the **enthalpy change** and is designated by the notation ΔH. The Δ symbol in mathematics symbolizes the difference between a final state and the initial state—that is, final state minus initial state. In an exothermic reaction, the final state has less enthalpy than the initial state; hence the value for ΔH is negative. With endothermic reactions, the final state has more enthalpy than the initial state, and the value for ΔH is positive.

An example of an endothermic reaction is

$$CaCO_3(s) \rightarrow CaO(s) + CO_2(g) \qquad \Delta H = +42.5\ \text{kcal}$$

or

$$42.5\ \text{kcal} + CaCO_3(s) \rightarrow CaO(s) + CO_2(g)$$

Heat must be supplied continuously to decompose calcium carbonate to calcium oxide and carbon dioxide.

Note that these are molar ΔH values and that they apply to the balanced equation as written. For example, in the first reaction above, one mole of CH_4 reacts with 2 moles of O_2 to give the indicated products plus 213 kcal of heat. Two moles of CH_4 plus 4 moles of O_2 would give 2(213) kcal of heat.

Example 9–16: Thermochemical equations *Given the thermochemical equation*

$2H_2(g) + O_2(g) \rightarrow 2H_2O(l) + 136.6\ kcal,$

calculate the amount of heat liberated when 10.0 g of hydrogen completely reacts with oxygen to form water.

The problem can be solved by the same techniques we have been using. First make sure the equation is balanced.

Balanced equation:	$2H_2(g)$ +	$O_2(g)$ →	$2H_2O(l)$ +	136.6 kcal
Mole data:	2 moles	1 mole	2 moles	136.6 kcal

We see that 2 moles of hydrogen will liberate 136.6 kcal of heat when completely burned in oxygen.

The mole factor is

$$\frac{137\ \text{kcal}}{2.00\ \text{moles}\ H_2}$$

The only conversion factor needed is

$$\frac{1.00\ \text{mole}\ H_2}{2.00\ \text{g}\ H_2}$$

Therefore,

$$10.0\ \cancel{\text{g}}\ \cancel{H_2}\left(\frac{1.00\ \cancel{\text{mole}}\ \cancel{H_2}}{2.00\ \cancel{\text{g}}\ \cancel{H_2}}\right)\left(\frac{137\ \text{kcal}}{2.00\ \cancel{\text{moles}}\ \cancel{H_2}}\right) = 342\ \text{kcal}$$

Lavoisier and Laplace are credited with the discovery (about 1780) of one of the basic laws of thermochemistry—the amount of heat liberated when a given amount of a compound decomposes to its products equals the amount of heat needed to form the compound from those products. In other words, thermochemical equations can be reversed without destroying their quantitative meaning. For example, when two moles of hydrogen react with one mole of oxygen, 136.6 kcal of heat is liberated.

$$2H_2(g) + O_2(g) \rightarrow 2H_2O(l) + 136.6\ \text{kcal}$$

Precisely the same amount of heat is required for the reverse reaction—decomposing two moles of water to hydrogen and oxygen.

$$2H_2O(l) + 136.6\ \text{kcal} \rightarrow 2H_2(g) + O_2(g)$$

Every reaction that is exothermic in one direction is endothermic in the opposite direction, and vice versa.

PROGRAMMED SEQUENCE:

Turn now to p. 323 and work Set 7 of the Programmed Sequences. These problems deal with stoichiometry calculations. As with any of the quantitative aspects of chemistry, the key to understanding is working enough problems to become familiar with the procedures and the mathematics. Use these sequences as a review and as a self-test of your understanding of the chapter.

NEW TERMS

Catalyst: a substance that changes the rate of a chemical reaction but does not itself appear in the reaction products.

Contact process: a process for the commercial preparation of sulfuric acid via the catalytic oxidation of sulfur dioxide.

Dehydrating agent: a substance with a strong affinity for water.

Dihydrate: a hydrate that contains two H_2O units per molecule of compound.

Enthalpy: the heat change of a reaction measured at constant pressure.

Limiting reagent: a reactant not present in excess.

Mole factor: number of moles of desired substance divided by the number of moles of starting substance.

Natural gas: a natural petroleum product containing approximately 80 percent methane, CH_4.

Percent yield: actual yield divided by the theoretical yield times 100.

Stoichiometry: calculations based on the various relationships implied by chemical equations.

Thermochemical equations: equations that show the heat change accompanying chemical reactions.

PROBLEM SET A

1. Hydrochloric acid is made commercially by reacting salt (sodium chloride) with sulfuric acid.
   $2NaCl + H_2SO_4 \rightarrow Na_2SO_4 + 2HCl$
   How many moles of HCl can be produced from 500 g of sulfuric acid? How many grams?
2. How many moles of each reactant does it take to produce 300 g of Fe_2O_3 via the reaction
   $4Fe + 3O_2 \rightarrow 2Fe_2O_3$?
3. What mass of coke, C, is needed to produce 20.0 kg of zinc metal from pure ZnO ore?
   $2ZnO(s) + C(s) \rightarrow Zn(s) + CO_2(g)$
4. What mass of pure sulfuric acid is needed to completely dehydrate 2.00 moles of table sugar?
5. How many moles are contained in 10.0 g of sulfuric acid dihydrate?
6. How much bismuth nitrate pentahydrate is formed when 40.0 g of bismuth react with nitric acid?
   $Bi + 4HNO_3 + 3H_2O \rightarrow Bi(NO_3)_3 \cdot 5H_2O + NO$

7. How many grams of oxygen are needed to burn 1.00 kg of butane, C_4H_{10}?
   $2C_4H_{10} + 13O_2 \rightarrow 8CO_2 + 10H_2O$
   What mass of CO_2 is produced in the reaction?
8. Oxygen is liberated when the compound potassium chlorate is heated. How many moles of oxygen can be obtained from 200 g $KClO_3$?
   $2KClO_3 \rightarrow 2KCl + 3O_2$
9. Write balanced equations for the preparation of sulfuric acid from the ore iron pyrites, FeS_2. (Hint: first react FeS_2 with O_2 to get SO_2.)
10. How many kg of sulfuric acid can be obtained from 100 kg of 90 percent pure FeS_2 ore?
11. Given the following balanced equation:
    $3Cu + 8HNO_3 \rightarrow 3Cu(NO_3)_2 + 2NO + 4H_2O$
    Calculate:
    a. Moles of NO formed from 5.00 moles of Cu.
    b. Kilograms of H_2O formed from 2.00 moles of HNO_3.
    c. Mass of Cu needed to prepare 300 g $Cu(NO_3)_2$.
    d. Moles of HNO_3 needed to prepare 0.500 g of $Cu(NO_3)_2$.
12. A solution of sulfuric acid reacts with a solution of sodium hydroxide to form sodium sulfate solution and water. Write and balance the equation for this reaction and calculate the number of moles of sulfuric acid needed to react completely with a solution containing 200 g of sodium hydroxide.
13. How many grams of sodium sulfate are formed when 100 g of sulfuric acid react with 100 g of sodium hydroxide?
14. Carbon tetrachloride was prepared by reacting 500 g of carbon disulfide with chlorine.
    $CS_2 + 3Cl_2 \rightarrow CCl_4 + S_2Cl_2$
    If 800 g of CCl_4 were obtained, what was the percentage yield?
15. Methane gas reacts with oxygen to form carbon dioxide and water.
    $CH_4(g) + 2O_2(g) \rightarrow CO_2(g) + 2H_2O(g)$
    a. How many liters of oxygen measured at STP are needed to react with 86.0 g of methane?
    b. How many moles of methane are needed to produce 10.0 liters of carbon dioxide measured at STP?
    c. How many grams of water are formed from the reaction of 20.0 g of methane with 20.0 g of oxygen?
16. What volume of hydrogen, collected in the laboratory at 23°C and 755 torr, is formed when 1.20 moles of zinc completely react with sulfuric acid?
17. Rewrite the following equation to include heat as a term in the equation:
    $N_2(g) + 3H_2(g) \rightarrow 2NH_3(g) \qquad \Delta H = -22$ kcal
18. How much energy is liberated from the combustion of 30.0 kg of carbon according to the reaction
    $C(s) + O_2(g) \rightarrow CO_2(g) \qquad \Delta H = -94.1$ kcal
19. How much energy is needed to reduce 100 kg of iron(III) oxide to metallic iron?
    $3C(s) + 2Fe_2O_3(s) + 110.8 \text{ kcal} \rightarrow 4Fe(s) + 3CO_2(g)$

PROBLEM SET B

1. Elemental phosphorus is prepared by the following reaction:
   $2Ca_3(PO_4)_2 + 6SiO_2 + 10C \rightarrow P_4 + 6CaSiO_3 + 10CO$
   How many moles of P_4 can be produced from complete reaction of 750 grams of calcium phosphate? How many grams of P_4?

2. How many moles of each product are formed by the decomposition of 1,000 grams of nitric acid?
$4HNO_3 \rightarrow 4NO_2 + 2H_2O + O_2$

3. What mass of coke, C, is needed to produce 25.0 kg of cadmium metal from 79% pure CdO ore? Assume that CdO is the only component of the ore to react with coke.

4. Compute the mass of nitric acid necessary to produce 250 kg of the explosive nitroglycerine $C_3H_5(NO_3)_3$.
$C_3H_5(OH)_3 + 3HNO_3 \rightarrow C_3H_5(NO_3)_3 + 3H_2O$

5. Given 10.0 grams of iodine pentoxide and 25.0 grams of carbon monoxide, calculate the number of grams of I_2 produced by the following reaction:
$I_2O_5 + 5CO \rightarrow I_2 + 5CO_2$

6. How many grams of antimony(III) sulfide are produced by the complete reaction of 25.0 grams of antimony(III) chloride?
$3H_2S + 2SbCl_3 \rightarrow 6HCl + Sb_2S_3$

7. How many grams of oxygen gas are needed for the complete combustion of 75.0 grams of octane (C_8H_{18})?
$2C_8H_{18} + 25O_2 \rightarrow 16CO_2 + 18H_2O$

8. Given a test tube containing a mixture of potassium chloride and potassium chlorate. The mixture has a mass of 9.65 grams and after heating for a period of time achieves a constant mass of 9.06 grams. From these data, calculate the percentage of potassium chlorate in the original mixture. (Assume that potassium chloride does not decompose.)
$2KClO_3 \rightarrow 2KCl + 3O_2$

9. Calculate the number of liters of carbon dioxide gas formed by the combustion of 40.0 grams of acetylene. Assume STP.
$2C_2H_2 + 5O_2 \rightarrow 4CO_2 + 2H_2O$

10. Given an ore sample that contains 86.0% chromite ($FeCr_2O_4$). If the process of producing chromium metal is 90.0% efficient, how many grams of chromium metal would be produced from a 500 gram sample of the ore?

11. Given the following balanced chemical equation:
$4Zn + 10HNO_3 \rightarrow 4Zn(NO_3)_2 + NH_4NO_3 + 3H_2O$
    a. How many moles of ammonium nitrate are produced by the complete reaction of 15.0 grams of zinc?
    b. How many liters of gaseous water at a temperature of 110°C and 1 atm pressure are formed by the complete reaction of 3.00 moles of nitric acid?
    c. How many grams of zinc are needed to produce 100 grams of ammonium nitrate?

12. Given the following balanced chemical equation:
$CH_4(g) + 2O_2(g) \rightarrow CO_2(g) + 2H_2O(g)$
If 10.0 liters of methane and 10.0 liters of oxygen gas mix and react, how many liters of each gas remain after the reaction has stopped? Assume the same conditions of temperature and pressure before and after the reaction.

13. How much energy is needed to decompose 75.0 grams of sodium hydrogen carbonate by the following reaction?
$2NaHCO_3 + 31\text{kcal} \rightarrow Na_2CO_3 + CO_2 + H_2O$

14. Rewrite the following equations including the heat as a term in the equation:
    a. $2H_2(g) + O_2(g) \rightarrow 2H_2O(g) \quad \Delta H = -116\text{kcal}$
    b. $C(s) + 2S(s) \rightarrow CS_2(g) \quad \Delta H = +28.7\text{kcal}$

15. Calculate the kcal of energy released from the combustion of 250 grams of propane (C_3H_8).
$C_3H_8(g) + 5O_2(g) \rightarrow 3CO_2(g) + 4H_2O(g) \quad \Delta H = -531\text{kcal}$

10

subatomic particles and the nucleus

THIS WILL SAVE YOU TIME:

Refer to this list as you read through the chapter. In answering the questions, you will understand the important concepts you are most likely to be asked about later on.

After reading this chapter, you should be able to:

1. *Tell how radioactivity was discovered.*
2. *Explain what is meant by positively and negatively charged particles.*
3. *Explain how electric-discharge tubes can be used to obtain a value for the ratio of charge to mass of an electron.*
4. *Describe an experiment with an electric-discharge tube that shows the existence of protons.*
5. *Describe an experiment demonstrating that three different kinds of subatomic particles come from a natural radioactive source.*
6. *Give the nuclear designations for electrons, protons, alpha particles, neutrons, and gamma particles.*
7. *Describe Rutherford's experiment and explain the results and conclusions derived from it.*
8. *Tell what is meant by atomic numbers and give the exact definition of an element.*
9. *Calculate the number of protons and neutrons in an element given its atomic number and mass number.*
10. *Define "isotope" and give examples.*

Atomic structure, the topic of this chapter and the next, is the key to our understanding of modern chemistry. How do atoms of unlike elements differ? Why do certain atoms unite in a given manner, while others do not? What holds atoms together in molecules? These questions can be answered only in terms of atomic structure. In this chapter we first consider the three important fundamental particles: the electron, proton, and neutron. We then discuss the structure and composition of atomic nuclei.

The study of atomic structure began with the discovery of radioactivity. This research showed that matter contains particles smaller than atoms and that these particles are the basic structural units of atoms themselves.

THE DISCOVERY OF RADIOACTIVITY AND SUBATOMIC PARTICLES

Wilhelm Konrad Roentgen (1845–1923, Nobel prize in physics 1901) stumbled onto radioactivity in the midst of research on ultraviolet light. He used electric-discharge tubes like the one in Figure 10–3 as a source, and he detected the light released from the tubes by letting it hit the compound $BaPt(CN)_4$, which fluoresces when exposed to ultraviolet radiation. One day in 1895, Roentgen noticed this fluorescence even when the ultraviolet source was some distance from the $BaPt(CN)_4$. Wrapping the source in black paper, which stops ultraviolet light, did not stop the fluorescence, and Roentgen knew that the explanation for the fluorescence must lay with something other than ultraviolet light. His search for an answer led to the discovery of **x-rays**—high-energy light of short wavelengths. The following year, Antoine-Henri Becquerel (1852–1908) discovered radioactivity in a similar manner while testing ores for their ability to fluoresce. He found that if unexposed photographic plates were well wrapped and placed in contact with uranium-containing ores, halos appeared on the plates when they were developed. He correctly attributed this phenomenon to the emanation of rays (subatomic particles) from the atoms, particles that could pass through the wrapping paper. In 1898, Marie Sklodowska Curie (1867–1934) and her husband Pierre (1859–1906) identified two new elements—polonium, Po, and radium, Ra—that continuously emitted subatomic particles. The Curies and Becquerel shared the 1903 Nobel prize in physics in recognition of this work.

These experiments at the close of the nineteenth century started an avalanche of important discoveries that led to our present understanding of atomic structure.

ELECTRICAL PHENOMENA

It has been known since ancient times that a piece of amber rubbed with a woolen cloth acquires the ability to attract small, lightweight objects. Materials that exhibit this phenomenon are said to be electrically charged. If a charged amber rod is touched to the top of an electroscope, Figure 10–1, the thin pieces of gold foil separate. Since both pieces of gold foil contain the same kind of charge, it seems reasonable that like charges must repel each other. Similar results are obtained if a glass rod and silk are substituted for the amber and wool.

If a suspended pith ball is touched with a charged glass rod (Figure 10–2) and another pith ball with a charged amber rod, the two balls will attract each other. We

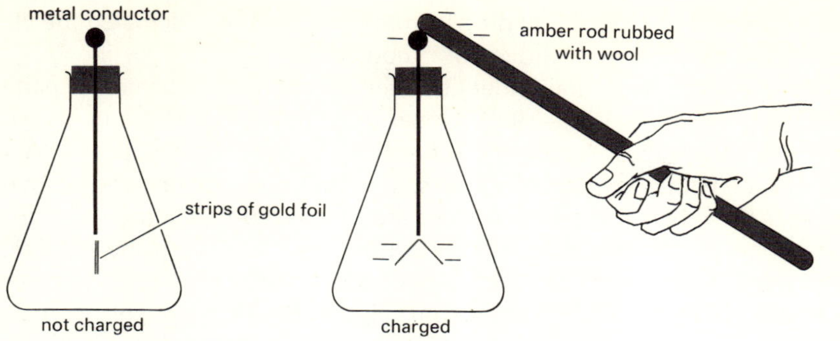

Figure 10–1 The electroscope.

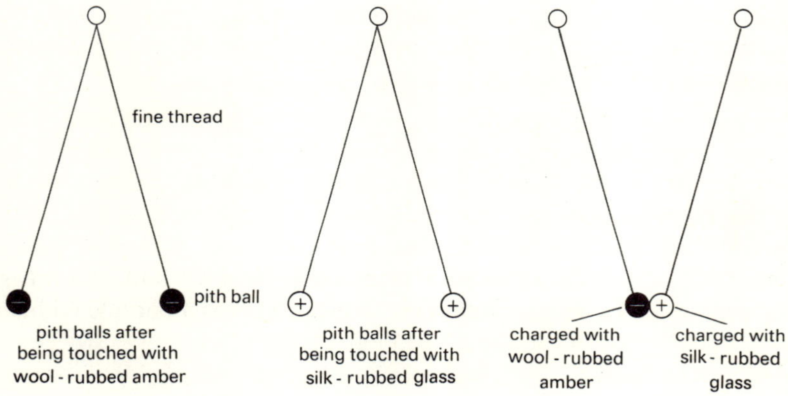

Figure 10–2 Pith-ball experiment showing charges of static electricity.

already know that like charges repel. Apparently, the balls have unlike charges, and unlike charges attract. Benjamin Franklin (1706–1790) called these different charges *positive* (glass rubbed with silk) and *negative* (amber rubbed with wool) electricity. Experiments with these charges led to the conclusion that a positively charged body contains an excess of positive electricity, a negative body an excess of negative electricity, and an uncharged body equal amounts of positive and negative electricity. These ideas were mistaken. We now know that electricity consists of electrons and that electrons themselves are negative. A negatively charged object contains an excess of electrons, while a positively charged object is one from which electrons have been removed. The flow of electrons from negative charge to positive charge is called an electric current.

ELECTRONS

Electrons are *subatomic particles that have negative charges.* Obviously the amount of charge on the electron is important. Robert Andrews Millikan (1868–1953, Nobel prize in physics 1923) gathered data on electron charge in an experiment centered on the movement of charged oil drops. From these data he calculated the charge on an individual electron as -1.60×10^{-19} coulomb.*

The scientist who actually identified the electron as a subatomic particle was Joseph John Thomson (1856–1940, Nobel prize in physics 1906) in his work with electric-discharge tubes. A typical discharge tube, shown in Figure 10–3**A,** consists of a glass tube with two metal electrodes, a positive anode at one end and a negative cathode at the other. If these electrodes are connected to a source of high voltage and the tube is partially evacuated, electrons will flow from the cathode to the anode. When a screen coated with a fluorescent compound such as zinc sulfide, ZnS, is placed in the path of the beam (Figure 10–3**B**), a bright, glowing streak appears. If a magnetic field is placed near the discharge tube, the electrons are deflected in a direction that shows that they have a negative charge (Figure 10–3**C**). By knowing the strength of the magnetic field and by then measuring the deflection of the electrons, Thomson was able to determine the ratio of the charge of an electron to its mass, that is, *e*/*m*. The value he obtained was -1.76×10^{8} coulomb/g. Dividing this value for *e*/*m* into Millikan's value for the electronic charge gives a value for the mass of an electron:

$$\frac{e}{e/m} = \frac{-1.60 \times 10^{-19}\ \text{coulomb/electron}}{-1.76 \times 10^{8}\ \text{coulomb/g}}$$

$$m = 9.11 \times 10^{-28}\ \text{g/electron}$$

Putting Millikan's calculation of charge together with Thomson's measurement of mass provides the basic description of the electron: It is a particle with a negative

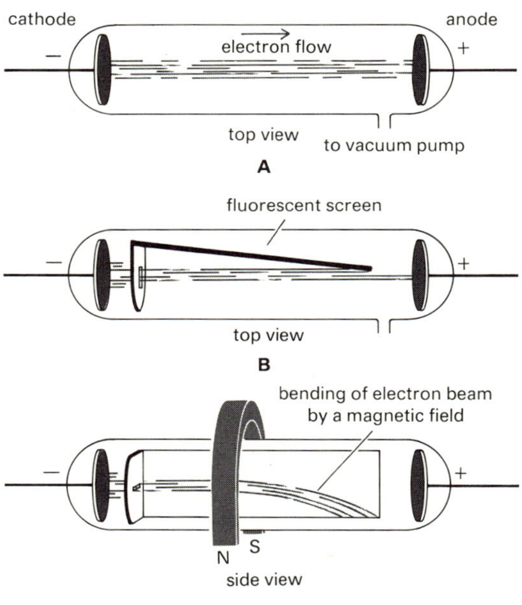

Figure 10–3 Electric discharge tubes.

*A coulomb is an electrical unit named for Charles Augustin Coulomb (1736–1806). One mole of electrons is equivalent to 96,487 coulombs.

charge of -1.60×10^{-19} coulomb and a mass of 9.11×10^{-28} g. Because the electron has the smallest known electrical charge, it is used as the reference for all other charged particles. For convenience, we say that the charge on an electron is -1. When we say, for example, that a particle has a $+3$ charge, it means a charge of $+3(1.60 \times 10^{-19})$ coulomb. The $+$ indicates that it is opposite in sign to the electronic charge, and the 3 means it is 3 times as large.

The mass of an electron on the atomic mass scale (carbon-12 isotope = 12. exact) is 0.00054874 amu. This figure is so small that, unless high precision is required, the approximate value of zero is usually used. The electron is represented by the symbol

$$\text{electron} = {}_{-1}^{0}\text{e} = {}_{\text{charge}}^{\text{mass}}\text{particle symbol}$$

THE PROTON

Figure 10–4 shows an electric-discharge tube somewhat different from the electric-discharge tubes shown above: the difference is a hole in the cathode. If hydrogen gas is pumped inside and the tube is then slowly evacuated, at a certain pressure a beam of particles that are not electrons can be seen streaming through the hole in the cathode. Since these particles travel in a direction opposite to that of the electrons, they must have a positive charge.

Where does the positive beam arise? The electron beam bombarding the gas in the tube knocks additional electrons off the hydrogen gas molecules (H_2). Although this reaction can be written in a variety of ways, the following equations describe what happens:

$$H_2 + {}_{-1}^{0}e \rightarrow H_2^+ + 2\,{}_{-1}^{0}e$$

$$H_2 + {}_{-1}^{0}e \rightarrow H^+ + H + 2\,{}_{-1}^{0}e$$

Stripping gas molecules of their electrons produces positively charged particles, or ions. In this experiment, the stream consists of hydrogen ions.

It is possible to measure the charge and mass of these positively charged particles in the same manner as electrons. In the case where hydrogen is the bombarded gas, the resulting H^+ ion has been given the name **proton** and is designated by the symbols

${}_{1}^{1}H$ or ${}_{1}^{1}p$

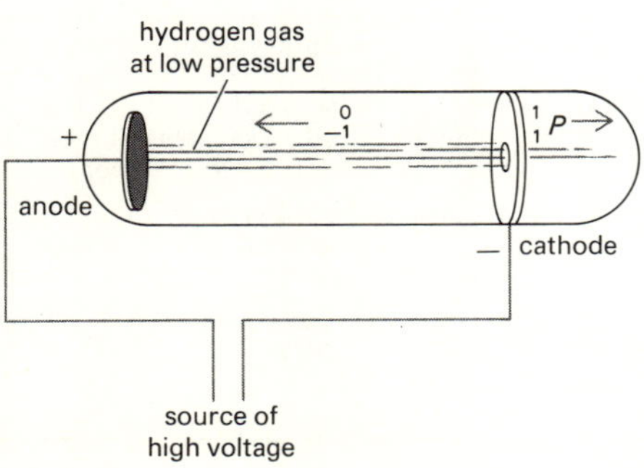

Figure 10–4 Electric-discharge tube used to show the presence of positively charged particles.

As the notation shows, the charge of the proton is equal to the charge of the electron in magnitude but opposite in sign; that is, $+1.60 \times 10^{-19}$ coulomb. The mass determined by e/m experiments amounts to 1.67×10^{-24} g, equal to a value of 1.007595 amu.

THE NATURE OF RADIOACTIVITY

If a piece of radioactive matter is placed in a lead container such as that shown in Figure 10–5, the beam of subatomic particles can be detected with a fluorescent screen. When a magnetic or electric field is placed near the path of the particles, the beam is separated into three distinct parts. One beam is deflected toward the positive electrode, indicating that it consists of negative particles; these are electrons, or **beta particles,** ${}_{-1}^{0}\beta$. The beam deflected toward the negative electrode is composed of doubly charged helium ions, called **alpha particles,** ${}_{2}^{4}\alpha$. The neutral or undeflected beam consists of light radiation of very short wavelengths that have more energy than x rays; these are called **gamma rays,** or **gamma particles** and are

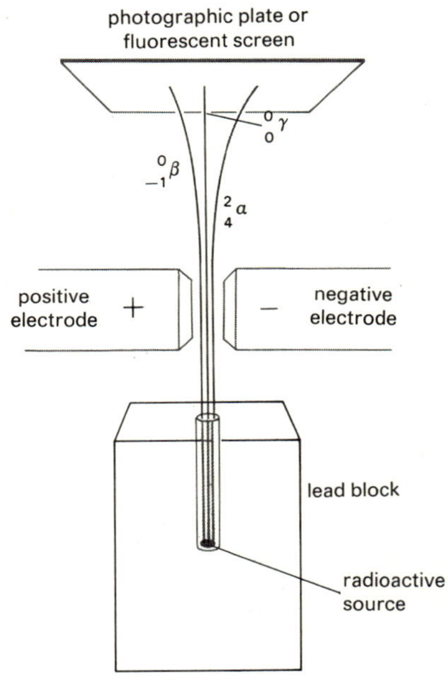

Figure 10–5 Separation of α, β, and γ particles coming from a radioactive source.

TABLE 10–1 PROPERTIES OF SUBATOMIC PARTICLES

Particles	Designation	Mass	Charge	Speed	Penetration power*
Alpha	${}_{2}^{4}He$, ${}_{2}^{4}\alpha$	4	2	9000 to 15,000 miles/s	Stopped by all but the thinnest sheets of metal or a piece of paper
Beta	${}_{-1}^{0}\beta$, ${}_{-1}^{0}e$	0	−1	Up to 100,000 miles/s	Greater than α particles; stopped by thin sheets of metal
Gamma	${}_{0}^{0}\gamma$	0	0	186,000 miles/s (speed of light)	Can penetrate 8 to 10 in. of Pb

* The relative penetrating power of the particles α, β, and γ is in the approximate ratio 1:100:10 000.

designated by the notation ${}_{0}^{0}\gamma$, indicating no charge and essentially no mass. Table 10–1 summarizes the properties of these three particles. Note the differences in penetration power and average velocity.

THE NUCLEUS: RUTHERFORD'S EXPERIMENT

If helium is used in place of hydrogen in an electric-discharge-tube experiment, the mechanism generates positive rays consisting of helium ions:

$$He + {}_{-1}^{0}e \rightarrow He^{+} + 2\,{}_{-1}^{0}e$$

$$He^{+} + {}_{-1}^{0}e \rightarrow He^{2+} + 2\,{}_{-1}^{0}e$$

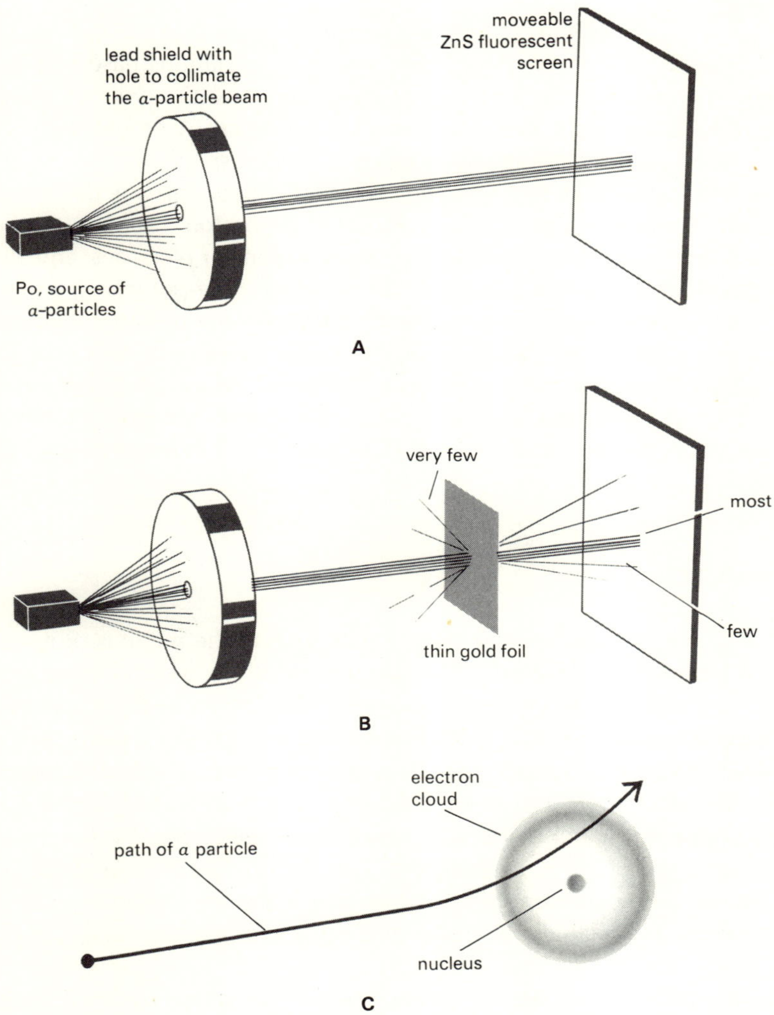

Figure 10–6 The historic experiment that proved the existence of the nucleus. **A** At the start of the experiment a stream of α particles produces flashes of light at one spot on the ZnS fluorescent screen. **B** Thin strips of gold foil cause the deflection of some particles; some are even deflected backwards. **C** Deflection of α particles by a nucleus.

Note that more than one electron can be removed from an atom or molecule by electron bombardment. The doubly charged helium ions produced in this manner are identical in properties to the alpha (α) particles emitted from radioactive materials. These particles have a mass of 4 amu and are designated ^4_2He or $^4_2\alpha$.

Rutherford's colleagues, Hans Geiger (1882–1945) and E. Marsden, performed a series of experiments in which they bombarded various materials with α particles from a radioactive source. The details of one such experiment are shown in Figure 10–6. Alpha particles from a sample of polonium, Po, were allowed to hit a thin piece of gold foil, Figure 10–6**B.** A fluorescent zinc sulfide screen placed behind the gold foil flashed when struck with α particles. This allowed the experimenters to see what path the α particles followed after passing through the gold. The data produced by this experiment led to two important deductions about atomic structure.

First of all, most of the α particles passed straight through the gold foil and produced a bright spot on the fluorescent screen. Calculations based on the size of gold atoms showed that the foil was approximately 1000 layers of atoms thick. Since the α particles penetrated a seeming wall of atoms without hitting anything, Rutherford concluded that atoms must be mostly open space.

However, approximately 1 in every 100,000 α particles was deflected from a straight path. Collision with a simple electron or proton could not cause such deflection. Alpha particles are much larger than protons or electrons, and a collision between them would be like a billiard ball striking a ping-pong ball. Since most α particles passed through and since something massive enough to deflect an α particle did exist, it appeared that most of the atom's mass must be concentrated in one place. This concentrated mass repulsed the deflected α particles, indicating that this concentrated mass had positive charge. Rutherford correctly concluded that atoms have positively charged nuclei that are very small in comparison to the size of the atom.

ATOMIC NUMBER

The next step was to measure the electrical charge on the nucleus. Rutherford observed that the size of the positive charge depended on what element was being observed. The critical experiment, however, was performed by Henry Gwyn-Jeffreys Moseley (1887–1915). He studied the wavelengths of x rays given off from electric-discharge tubes (cathode-ray tubes) in which various elements were used as targets for the electron beam. A typical x-ray tube is shown in Figure 10–7. The x rays are produced by bombarding the anode with electrons. Moseley discovered that the wavelength of the x rays depended on the nuclear charge of the particular element under bombardment. He found that different elements had different nuclear charges and that all atoms of the same element had the same nuclear charge.

With this information, we can define precisely what we mean by an element: *all atoms that have the same nuclear charge are the same* **element.** All atoms with a nuclear charge of 29 are copper, all with a nuclear charge of 1 are hydrogen, all with charge 2 are helium, and so on. *The positive charge on the nucleus of an atom is called its* **atomic number.** The atomic number of copper is 29. This number defines the element, for only copper atoms can have a +29 nuclear charge. The

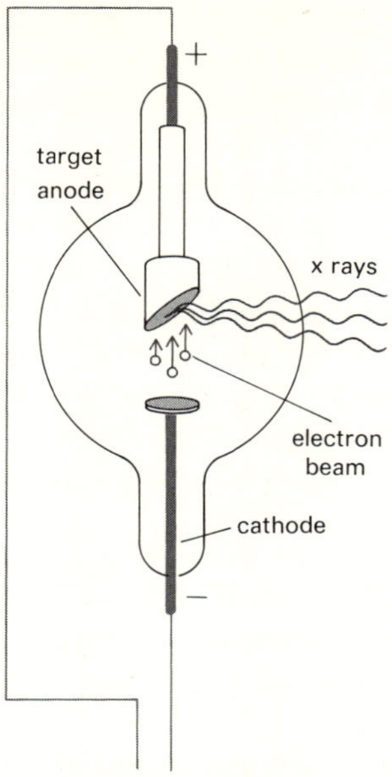

Figure 10–7 X-ray tube. Moseley discovered a relationship between the wavelength of the emitted x rays and the nuclear charge of the element used to make the target anode of the tube.

values of the atomic numbers range from the atomic number of 1 for hydrogen to 106 for the newly discovered element (unnamed) which, by definition, must have a +106 nuclear charge.

NUCLEAR STRUCTURE

We can now stop and review the data gathered thus far about the structure of atoms. It is known that all atoms in a given mass of an element are electrically neutral—that is, the sum total of all the charges of the subatomic particles present is zero. By certain experimental techniques, however, electrons can be removed from atoms to leave positively charged particles behind. These positive charges are in the nuclei, which are very small in size compared to the whole atoms. Nevertheless, the nuclei have most of the atomic mass. The magnitude of the nuclear charge defines the element. The total nuclear charge is called the atomic number. To complete the picture, we must now explain the mass and charge of nuclei in terms of subatomic particles.

The positive charges of nuclei result from the presence of protons. The element fluorine, for example, has an atomic number of 9; this defines the element and tells us that the nucleus has a positive charge of 9. If this positive charge is attributed to 9 protons, then an atomic mass of 9 amu is accounted for. The actual atomic mass, however, is 19. The difference of 10 amu must result from the presence of something else in the nucleus, something without an electrical charge—that is, a neutral subatomic particle.

By 1920, a number of scientists had postulated such neutral particles, but it wasn't until 1932 that James Chadwick (1891–   , Nobel prize in physics 1935) isolated and identified them. The reaction he studied was

$${}^{9}_{4}\mathrm{Be} + {}^{4}_{2}\mathrm{He} \rightarrow {}^{12}_{6}\mathrm{C} + {}^{1}_{0}n$$

(the subscripts refer to the nuclear charge or atomic number, and the superscripts to the mass number). Chadwick allowed helium ions (α particles from a radioactive source) to hit a sample of the element beryllium, Be. The reaction produced the expected neutral particles, called neutrons.

The **neutron** is defined as *an uncharged subatomic particle* having a mass of 1.008 665 44 amu and is designated by the notation ${}_{0}^{1}n$.

Because the mass of a neutron is approximately 1 amu, the fluorine nucleus must consist of 9 protons and 10 neutrons. *The sum of the number of protons and neutrons in a nucleus is called the* **mass number.** Fluorine has an atomic number of 9 and a mass number of 19. The number of neutrons in a nucleus is always equal to the mass number minus the atomic number.

We can now explain the existence of **isotopes.** Atoms having the same nuclear charge but different mass numbers are called isotopes.

Example 10–1: Isotopes *The two most abundant isotopes of tin have mass numbers of 120 and 118. Discuss the structure of their nuclei.*

From the table on back cover pages we find that tin is element number 50—that is, tin has an atomic number of 50, its nuclear charge is +50, and there are 50 protons in the nucleus of each atom. The isotope with a mass number of 120 must have 70 neutrons in each nucleus:

$$\begin{aligned}\text{Number of neutrons} &= \text{mass number} - \text{atomic number}\\ &= 120 - 50 = 70 \text{ neutrons.}\end{aligned}$$

The 118 isotope has 50 protons and 68 neutrons.

We cannot go into great detail about the structure of the nucleus, not only because such a discussion is beyond the scope of this text, but also because the nucleus is still poorly understood. What holds the nucleus together continues to be a mystery. Since the protons are of like charge, they should repel each other, but instead they stay together. A number of alternative theories have been proposed to explain this phenomenon and other nuclear unknowns. However, while study of the nucleus is a major focus of modern physics, chemistry pays less attention to the structure of the nucleus than to the arrangement of electrons around it. Chemists are primarily interested in ordinary—nonnuclear—reactions. In these reactions, the nuclei of the atoms remain unchanged; only the electrons are responsible for chemical bonding and chemical reactions.

NEW TERMS

Alpha particle: a subatomic particle that has a mass of 4 and a charge of +2, a doubly charged helium ion. It is designated by the notations ${}_{2}^{4}\alpha$, ${}_{2}^{4}\text{He}$, or He^{2+}.

Atomic number: the number of protons in the nucleus of an atom. It equals the positive charge on the nucleus.

Cathode ray: the electron beam produced in gas-discharge tubes.

Coulomb: the unit of electrical charge.

Electron: a subatomic particle having a charge of −1 and a mass of approximately zero. Notations are ${}_{-1}^{0}e$ and ${}_{-1}^{0}\beta$.

Element: all atoms having the same nuclear charge are the same element.

Fluorescence: light given off by a substance when it is struck by a beam of subatomic particles, or radiation.

Isotopes: atoms having the same nuclear charge but different numbers of neutrons in their nuclei.

Mass number: the sum of the number of protons and neutrons in the nucleus.

Neutron: an uncharged subatomic particle having a mass of 1; the notation is ${}_0^1n$.

Nucleus: the positively charged part of an atom that contains most of the mass.

Proton: a subatomic particle having a charge of +1 and a mass of 1; the designations are ${}_1^1H$, ${}_1^1p$, or H^+.

X rays: radiation of short wavelength.

PROBLEM SET A

1. Explain the meaning of the notation ${}^{131}_{54}Xe$ by describing the nucleus in terms of its atomic number, mass number, number of protons, and number of neutrons.
2. This table describes the nuclear makeup of three pairs of isotopes. Fill in the missing data.

Nucleus	${}_1^1H$	${}_1^2H$	${}^{12}_6C$	${}^{14}_6C$	${}^{235}_{92}U$	${}^{238}_{92}U$
number of protons	___	___	___	___	___	___
number of neutrons	___	___	___	___	___	___
mass number	___	___	___	___	___	___

3. How many electrons have to be added to each of the nuclei in Problem 2 to make each atom neutral?
4. Complete the table below by filling in the missing numbers and giving the nuclear notation of each species:

number of protons	___	2	17	___	___	___
number of neutrons	2	2	___	20	22	20
mass number	3	___	35	37	40	39
nuclear notation symbol	___	___	___	___	___	___

5. Carbon has three isotopes of mass 12, 13, and 14. Oxygen has three isotopes of mass 16, 17, and 18. How many different molecules of carbon monoxide are possible? (Different, that is, in molecular mass but not in chemical properties.)
6. If an atom of argon is hit by an electron in such a way that one of its electrons is released, according to the equation
$Ar + {}_{-1}^{0}e \rightarrow 2\,{}_{-1}^{0}e +$ ____

what new species is formed? (Hint: the sum of the charges and masses of the reactants must equal the sum of the charges and masses of the products.)

7. The radius of a gold atom is approximately 1.50×10^{-8}cm. How many gold atoms does an α particle pass through when it penetrates gold foil 0.000 100 in. thick? (Assume that the atoms are spherical and that they touch each other.)

PROBLEM SET B

1. Complete this table:

Nucleus	$^{19}_{9}F$	___	$^{56}_{26}Fe$	___	___	___
number of protons	___	___	___	26	___	99
number of neutrons	___	20	___	___	31	153
mass number	___	40	___	54	57	___

2. How many electrons have to be added to each of the nuclei in Problem 1 to make each atom neutral?
3. Given the information in Problem 5 of Set A, how many different molecules of carbon dioxide are possible?
4. If the argon ion, Ar^+, is hit by an electron in such a way that one of its remaining electrons is released, according to the equation
$Ar^+ + {}_{-1}^{0}e \rightarrow 2\,{}_{-1}^{0}e + \_\_\_$
what new species is formed?

11

the electronic structure of atoms

THIS WILL SAVE YOU TIME:

Refer to this list as you read through the chapter. In answering the questions, you will understand the important concepts you are most likely to be asked about later on.

After reading this chapter, you should be able to:

1. *List the various types of radiation in order of increasing energy.*
2. *Describe the construction and operation of a spectrograph.*
3. *State the basic postulates of the Bohr theory of atomic structure.*
4. *Explain what is meant by an energy-level diagram.*
5. *Tell what is meant by the terms quanta, photons, and quantization.*
6. *Explain the origin of spectral lines in terms of the Bohr theory.*
7. *Discuss the physical significance of the four quantum numbers.*
8. *Tell what is meant by the wave-mechanical model of atoms.*
9. *Explain what is meant by the principal quantum number.*
10. *Explain what is meant by the azimuthal, or orbital, quantum number.*
11. *List the types of orbitals in the first four energy levels.*
12. *Explain what is meant by the magnetic quantum number.*
13. *List the number of orbitals possible for each sublevel.*
14. *Explain the meaning of the spin quantum number.*
15. *Tell what is meant by the Pauli exclusion principle.*
16. *State the maximum number of electrons in any main energy level.*
17. *Write the electron assignments for any element.*
18. *List the order of filling for electron orbitals.*
19. *Illustrate the shapes of s and p orbitals and tell why these spatial arrangements are important.*
20. *Illustrate the application of Hund's rule to the assignment of electrons in atoms.*

To understand chemistry in the most essential way, we need to determine how electrons are arranged within the atom. All the multitudinous combinations and reactions of atoms and molecules are determined mainly by the precise arrangement, or structure, of electrons around nuclei. To know the how and why of bonding and reaction, we must know the how and why of electronic structure.

The discussion of electronic structure begins with electromagnetic radiation. This topic is basic to the science of spectroscopy, which has been the source of most of our knowledge of atomic structure.

ELECTROMAGNETIC RADIATION

Electromagnetic radiation is *a form of energy—radiant energy, energy transmitted through space without the apparent transmission of matter.* Visible light is one form of electromagnetic energy; some others (Table 11–1) are radio, radar, infrared, ultraviolet, x rays, γ rays, and cosmic rays. Each kind of radiation has a different energy. Radio waves have very low energies, while gamma (γ) radiation has high energy.

TABLE 11–1 THE ELECTROMAGNETIC SPECTRUM

Type of radiation	Approximate wavelength, meters	Energy per photon, calories
	2×10^{4}	2.4×10^{-30}
Radio, TV		
	4×10^{1}	1.2×10^{-27}
Microwaves Radar		
	3.3×10^{-4}	1.4×10^{-22}
Infrared heat		
	7×10^{-7}	6.9×10^{-20}
Visible light		
	4×10^{-7}	1.2×10^{-19}
Ultraviolet		
	1.2×10^{-8}	3.8×10^{-19}
Short UV		
	5×10^{-9}	9.6×10^{-18}
X rays		
	1.2×10^{-11}	3.8×10^{-15}
Gamma rays		
	1×10^{-14}	4.6×10^{-12}
Cosmic rays		
	1×10^{-15}	4.6×10^{-11}

Visible	Color	Wavelength ($\times 10^{-10}$ m)
	Red	6300–7600
	Orange	5900–6300
	Yellow	5600–5900
	Green	4900–5600
	Blue	4500–4900
	Violet	3800–4500

(10^{-10} m = 1 Ångstrom unit, Å)

Because of the pioneering work of Max Planck (1858–1947, Nobel prize in physics 1918), we now know that all forms of radiation can be thought of either as waves or as particles. If radiation is thought of as a wave, it can be described in terms of wavelength. Radio waves have a wavelength of approximately 2×10^4 m, and γ rays a wavelength of 1×10^{-14} m: the shorter the wavelength, the greater the energy.

Radiation can also be thought of as a stream of particles of extremely small mass. These particles are called **photons,** or **quanta.** The larger the number of quanta, the greater the total energy.

SPECTROSCOPY

A **spectroscope** is *an instrument used to separate light coming from a radiation source into its component parts and to display the parts in order of increasing wavelength.* The displayed wavelength distribution is called a **spectrum.** If the spectrum is displayed in graphical form, the instrument is called a **spectrograph.** Passing light from an emitting source through a glass prism results in an **emission spectrum.** As shown in Figure 11–1**A,** sunlight broken into its components by a prism produces a rainbow of colors.

If a spectroscope is illuminated with light from an electric-discharge tube containing hydrogen or with light from a mercury arc, a number of bright lines appear on the emission spectrum. This is called a **line spectrum.** Each line records light of a different wavelength. Figures 11–1**B** and 11–1**C** illustrate the different lines in the visible portions of the spectra obtained for mercury and hydrogen.

Each element produces a unique and characteristic line spectrum, no matter what form the element is in. The same lines appear for sodium whether the emitting source is sodium metal, sodium chloride, or some complex organic molecule containing sodium. This fact makes the spectrograph a valuable analytical tool for identifying the various elements.

Because each element has its own characteristic spectrum and because the thing that distinguishes one element from another is atomic structure, any explanation of spectra must also explain the structure of atoms. In 1890, Johannes Rydberg (1854–1919) discovered by trial and error a mathematical formula that he used to predict accurately the wavelengths at which the lines of a spectrum would appear. It fell to Niels Bohr (1885–1962, Nobel prize in physics 1922) to give Rydberg's calculations theoretical significance. He explained the equation in terms of a model of electrons arranged around the nucleus. That model is the Bohr theory of the atom, one of the steppingstones to modern chemistry and physics.

THE BOHR THEORY OF THE ATOM

Bohr's atomic model hinges on four basic postulates:

1. The electrons in atoms are located in orbits, or energy levels, around the nucleus.
2. The electrons in orbits closest to the nucleus are of lower energy than those in orbits farther away from the nucleus.
3. Every electron in an atom may have only certain allowed energies. This energy determines what orbit the electron occupies.

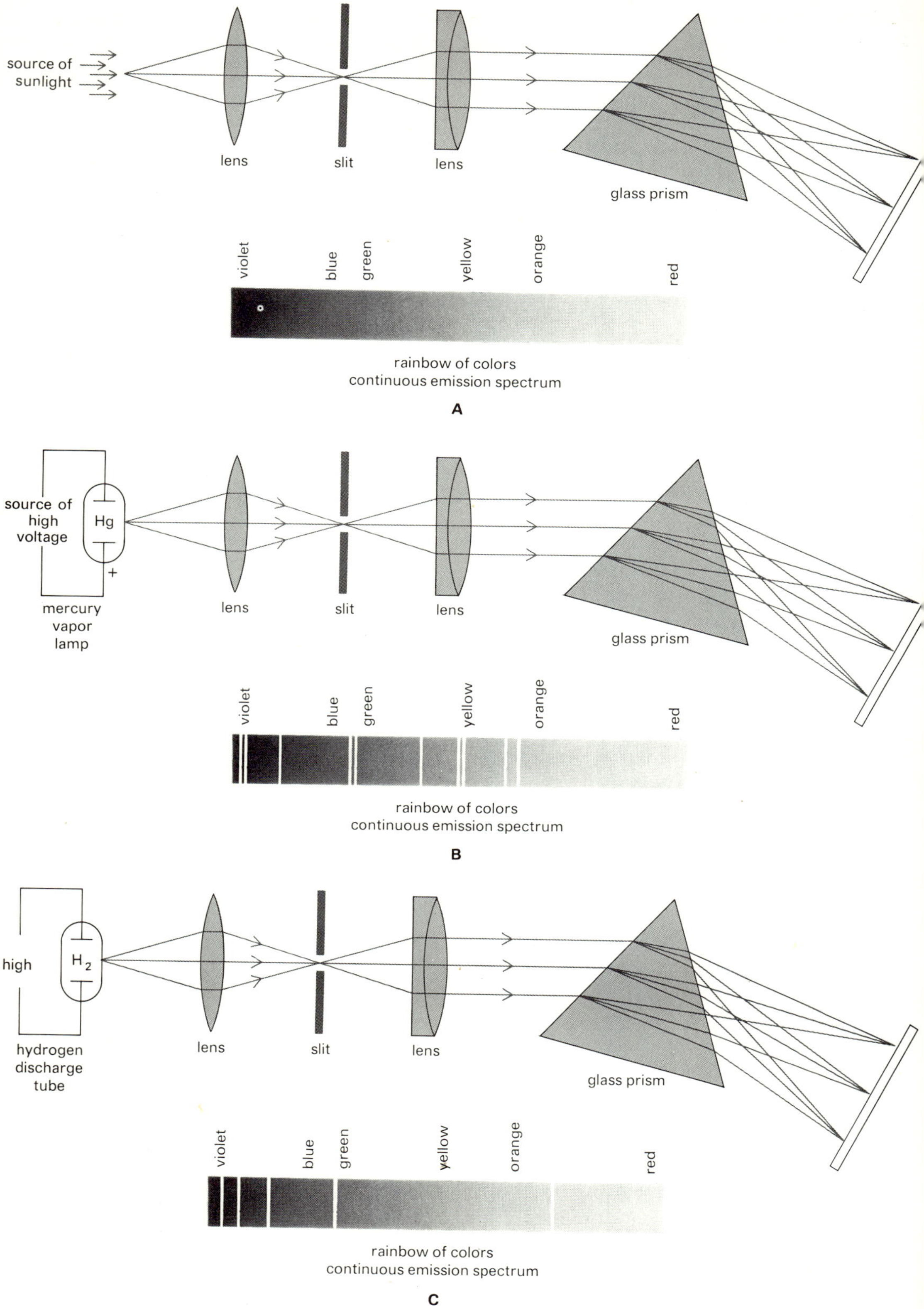

Figure 11–1 Spectrograph. **A** Spectrum of sunlight. **B** Mercury spectrum. **C** Visible region of hydrogen spectrum.

4. Electrons can move from one orbit to another. To do this, an electron must gain or lose a certain exact amount of energy—a **quantum** of energy.

Bohr assumed that the electrons moved about the nucleus in much the same way that the earth orbits the sun. He said that the force of attraction drawing the negative electrons toward the positive nucleus equals exactly the tangential force pulling the electrons away. As Bohr saw it, an electron behaves something like a ball on a string whirled in circles around your head. The string exerts a force drawing the ball toward your face; the tangential force, the one that will send the ball flying across the room if you release the string, pulls the ball away from your face. The two forces balance each other and keep the ball in a precise circular path. The electron orbit likewise results from a balance of attractive and tangential forces. Figure 11–2 gives two-dimensional representations of the hydrogen and helium atoms in terms of Bohr's model. Keep it in mind that the scale of the pictures is very misleading: were the nucleus the size of a pinhead, the electron would be 40 feet away.

We now know that the Bohr theory is not entirely correct, but the model acts as an important tool for grasping the contemporary concept of atomic structure.

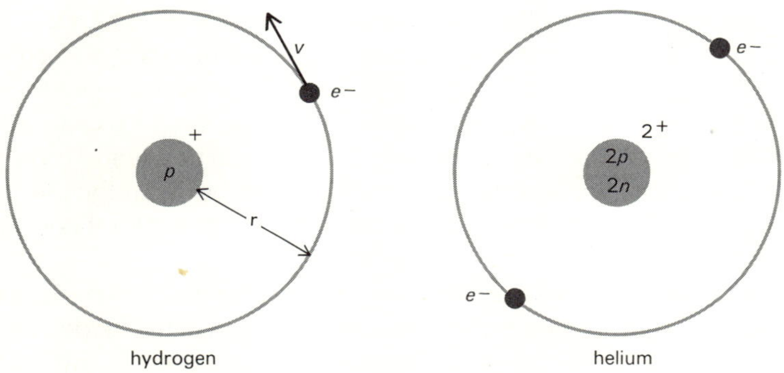

Figure 11–2 Bohr representations of the hydrogen and helium atoms. The Bohr radius is indicated by r and equals 5.29×10^{-11} m. The electron velocity v is 2.188×10^{6} m/s.

HOW BOHR'S THEORY EXPLAINS ATOMIC SPECTRA

Light from a hydrogen-containing electric-discharge tube passed through a prism produces the line spectrum shown in Figure 11–1**C.** Each line represents light of an exact wavelength, or, in terms of particles, light quanta of a precise energy value. These light quanta originate from the movement of electrons in the hydrogen atoms.

The lone electron in a hydrogen atom can occupy any one of a number of orbits. These orbits are known as **energy levels,** and each one has a number, starting with the lowest level—called the **ground state**—the one closest to the nucleus. Figure 11–3 shows the lower energy levels of the Bohr hydrogen atom. The numbers assigned to them are called **principal quantum numbers** and are designated by the letter n; thus, when we say $n = 5$, we are referring to principal quantum number five, or the fifth energy level out from the nucleus of an atom.

For an electron to move from $n = 1$, the ground state, to a higher energy level,

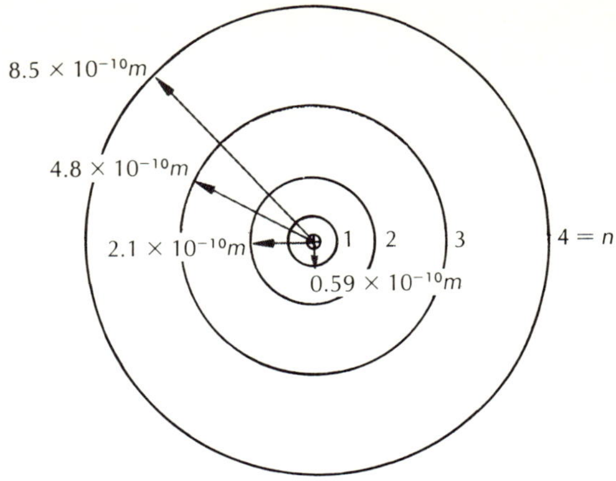

Figure 11–3 Some principal energy levels for the Bohr hydrogen atom. Bohr radii are shown by arrows.

it must gain *a precise amount of energy*—a **quantum of energy.** This energy may come from any one of many sources. In electric-discharge tubes, for example, the energy is supplied by high-voltage electricity. To go from $n = 1$ to $n = 2$ requires an exact amount of energy. This energy can be calculated from the mathematics of the Bohr theory; it amounts to 3.90×10^{-19} cal/atom. To move the electron in a hydrogen atom from $n = 1$ to $n = 3$ requires a larger amount of energy, exactly 4.63×10^{-17} cal/atom. The electron must absorb these exact amounts of energy: it cannot accept an intermediate amount and come to rest somewhere between the $n = 1$, $n = 2$, or $n = 3$ levels. This rule is the basic postulate of Bohr's quantum theory: the energy of the electrons is quantized.

The movement of an electron through its energy levels can be likened to climbing a ladder. Your feet represent electrons, and the rungs of the ladder are the energy levels. You start out on the ground—the ground state. As you climb, moving your electron-feet up the ladder, you gain altitude one rung—or energy level—at a time. You can move only from one rung to another; there is no place to put your feet between rungs—an illustration of quantization. If you lose energy, you fall to earth, returning your electron-feet to the ground state. The one false note to this analogy is that the rungs of a ladder are spaced at regular intervals. This is not true of the electron's energy levels. It takes 3.90×10^{-19} cal/atom to go from ground state to $n = 2$, but only an additional 0.73×10^{-19} cal/atom to move from $n = 2$ to $n = 3$.

The information in Figure 11–4 can serve as a useful key to the Bohr theory. At the bottom of the illustration, the electron in the hydrogen atom is pictured first in the ground state, where the principal quantum number has a value of 1. If this electron receives energy from some outside source, it can accept a certain exact amount of the energy and move to an **excited state**—*an energy level where the value of n is greater than 1.* If the quantum of energy absorbed by the electron happens to be equal to 4.88×10^{-19} cal/atom the electron moves to the fourth energy level, $n = 4$, as indicated by the gray arrow. This excited electron will get rid of the extra energy it has acquired. It can do this by dropping back to any orbit of lower energy. If the change is from energy level $n = 4$ to $n = 3$, a given amount of energy is liberated—a quantum of energy, or a photon. This change is marked with a bold line. In this particular case, the energy corresponds to a spectral line in the infrared—a photon of infrared light. The transition from $n = 4$ to $n = 3$ is also shown as a bold arrow in

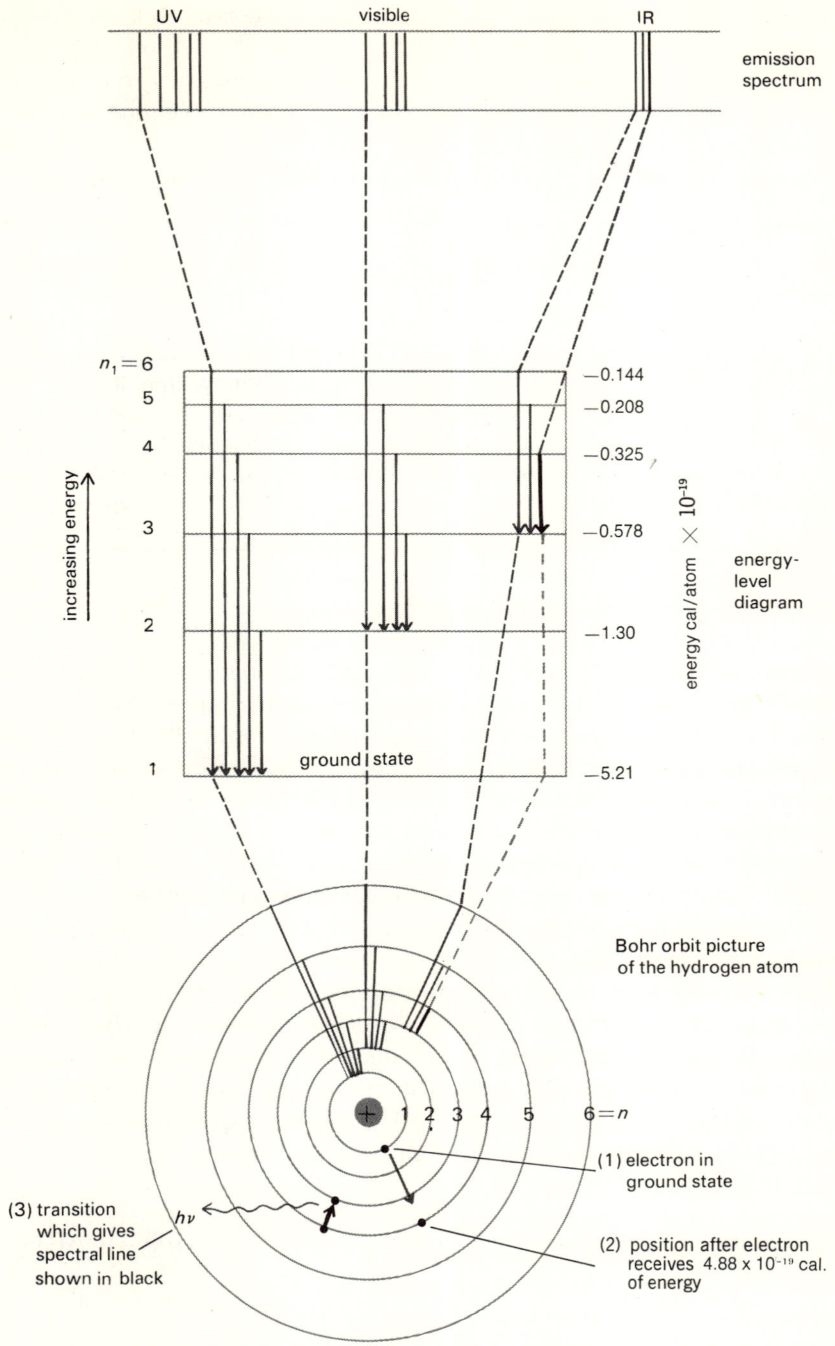

Figure 11–4 Hydrogen emission spectrum, energy-level diagram, and Bohr representation for the hydrogen atom.

the energy-level diagram in the center of Figure 11–4 and as a black line in the emission spectrum at the top of the figure.

All the lines in the hydrogen spectrum can be accounted for in a similar manner. When electrons move from an upper to a lower energy level, the loss of energy appears as light quanta of fixed wavelengths. Figure 11–4 relates all the lines on the hydrogen emission spectrum to the transitions in the energy-level diagram.

Example 11–1: Explaining spectral lines *How are the many lines in the hydrogen spectrum accounted for?*

When the atoms in hydrogen gas receive energy from some source such as an electric discharge, the electrons in millions of atoms absorb energy and move from the $n = 1$ ground state to higher energy levels. As shown in Figure 11–4, all the electrons that give up energy and return directly to the ground state give off energy corresponding to light quanta in the ultraviolet portion of the spectrum. Transitions from higher levels to the $n = 2$ level result in visible light, and transitions from higher levels to $n = 3$ account for infrared light emission.

ORIGIN OF THE FOUR QUANTUM NUMBERS

As spectroscopy techniques improved, questions that could not be answered by the Bohr model cropped up. High-resolution spectrographs showed that many of the lines on a typical emission spectrum were not single lines at all but closely spaced groups of lines. To account for the new lines, the electron orbit was seen as an ellipse, similar to the orbit the earth follows around the sun rather than the circular orbits assumed by Bohr. Since it takes two numbers to describe an ellipse, a new number was added to the principal quantum number. The new number, designated by the letter l, is called the **orbital,** or **azimuthal, quantum number.**

Problems with the Bohr model did not end with the elliptical orbit. Further experimentation showed that putting the light-emitting source in a strong magnetic field divided the lines of a spectrum into still more lines. A third value, called the **magnetic quantum number** and designated m_l, was introduced to explain the magnetic effect. Finally, a fourth quantum number was introduced to take care of the spinning of the electron. The hypothesis states that an electron can spin either clockwise or counterclockwise about an axis, and each type of spin gives a characteristic spectral line. This number is called the **spin quantum number,** m_s.

THE WAVE-THEORY APPROACH TO THE ELECTRONIC STRUCTURE OF ATOMS

In 1924, Louis de Broglie (1892–, Nobel prize in physics 1929) hypothesized that electrons behave like waves. Two years later, Edwin Shrödinger (1887–1961, Nobel prize in physics 1933) used Broglie's idea to devise an equation that is the basis for the science of wave mechanics. The equation is built around four constants analogous to the four quantum numbers of the Bohr theory. Solving the Schrödinger equation yields numerical values for the energy of the electrons and predicts the electrons' most likely locations about the nucleus. The amazing thing about the Schrödinger equation is that it remains a postulate with no theoretical derivation. The mathematical expressions in the equation are so complex that they can be solved only for the very simplest atoms. However, the Schrödinger model remains a valuable scientific tool in understanding the essential nature of the atom.

WAVE-MECHANICAL MODEL OF THE ATOM

According to the wave-mechanical picture of the atom, electrons are located in energy levels around the nucleus, and their energies are quantized—just as they were in the Bohr theory of atomic structure. These energy levels, however, are described in quite a different manner. Each principal energy level consists of one or more sublevels, or subshells; these subshells, in turn, comprise one or more orbitals. **Orbitals** are defined *as three-dimensional regions in the space around the nucleus where there is the greatest probability of locating a particular electron.* It is not possible to specify a particular path for an electron; one can only talk about the probability of the electron's presence in a certain region of space around the nucleus.

The electrons can also be described in terms of electron clouds—regions of electronic charge distribution around the nucleus. The shapes and densities of these clouds indicate the probability of finding an electron in that particular part of the atom. The most accurate picture we can now draw of the atom is one of a nucleus surrounded by electron clouds. The shape of the atom depends on the shapes of the electron clouds.

A DESCRIPTION OF THE QUANTUM NUMBERS

The wave-mechanical model describes each electron in terms of four quantum numbers. These numbers permit us to calculate the electron's energy and to predict the area around the nucleus where the electron may be found. In greater detail, the quantum numbers are:

n = **principal quantum number.** The principal quantum number can have any positive integral value, but for the presently known elements, the values range from $n = 1$ to $n = 7$. The principal quantum number determines the main energy level, or shell, in which the electron is located. The $n = 1$ shell is closest to the nucleus and has the lowest energy.

l = **orbital,** or **azimuthal, quantum number.** This quantum number designates the sublevel, or subshell, within the main energy level. It tells the shape of the electron cloud, or orbital, around the nucleus. Since numbers are used for the principal quantum numbers, the orbital quantum numbers are designated by the letters *s, p, d,* or *f.* l can have values of 0, 1, 2, 3, . . . , $n - 1$. When an electron has an l value of 0, it is said to be an *s* orbital. When $l = 1$, the electron is in a *p* orbital; for $l = 2$, a *d* orbital; for $l = 3$, an *f* orbital. For example, the second main energy level, $n = 2$, consists of two sublevels which we call the *s* and *p* orbitals. The types of orbitals for the first four energy levels are:

Energy level	Orbital type
$n = 1$	*s*
$n = 2$	*s* and *p*
$n = 3$	*s, p,* and *d*
$n = 4$	*s, p, d,* and *f*

m_l = **magnetic quantum number.** The magnetic quantum number designates the orientation of the electron orbitals in space. Each sublevel consists of one or more electron orbitals, and the magnetic quantum number describes the number of orbitals of a given kind in each main energy level. The magnetic quantum number can have values from $+l$ to $-l$, including 0. For *p* orbitals where $l = 1$, m_l can have three values, 1, 0, and −1: Thus there are three *p* orbitals. The number of orbitals possible in each sublevel is:

Sublevel	Number of orbitals possible
s	one *s* orbital
p	three *p* orbitals
d	five *d* orbitals
f	seven *f* orbitals

m_s = **spin quantum number.** It describes the spin orientation of the electron. It can have values of $+\frac{1}{2}$ or $-\frac{1}{2}$. The two orientations usually are designated by arrows pointing up and down, ↑ and ↓.

THE EXCLUSION PRINCIPLE

A law of nature known as the **Pauli exclusion principle** dictates the number of electrons possible in any principal energy level. The law is named after Wolfgang Pauli (1900–1958, Nobel prize in physics 1945) who found that each electron must have its own set of quantum numbers and that no two electrons in a given atom can share the same quantum numbers. On the basis of the Pauli principle, the total possible electrons in any one shell is represented by the expression $2n^2$. Thus 2 electrons are permitted in the first shell, 8 in the second, and 18 in the third.

Example 11–2: Electrons in energy levels *What is the maximum number of electrons permitted in the n = 4 energy level?*

The total possible is $2n^2$. Hence, for the fourth level, $2(4)^2 = 32$ is the maximum number possible.

ELECTRON ASSIGNMENTS IN ATOMS

Since the electrons in the outermost energy levels are responsible for such important chemical phenomena as ion formation, bonding, and reactivity, an obvious key to understanding the chemistry of the elements is knowing how electrons are distributed within the atom. We can begin to build electronic structure by starting with what we have already learned.

The electrons in atoms are located in energy levels around the nucleus. Each of the main energy levels is designated by a principal quantum number. The first main energy level is closest to the nucleus and contains the electrons of lowest energy. It consists of one *s* orbital capable of holding two electrons. The second en-

TABLE 11–2 ELECTRON ORBITALS FOR THE FIRST FOUR ENERGY LEVELS

Energy level	Sublevels	Number of orbitals in each sublevel	Number of electrons
$n = 1$	*s*	1	2
$n = 2$	*s*	1	2
	p	3	6
$n = 3$	*s*	1	2
	p	3	6
	d	5	10
$n = 4$	*s*	1	2
	p	3	6
	d	5	10
	f	7	14

ergy level, $n = 2$, has two sublevels. One of these is an *s* orbital that can hold 2 electrons. The other sublevel consists of three *p* orbitals, each of which can hold two electrons for a total of 6 electrons. These data are summarized in Table 11–2.

Now our problem is to assign all the electrons in a given atom to the proper energy levels and orbitals. For convenience, the number and locations of electrons in atoms are specified by the following symbols:

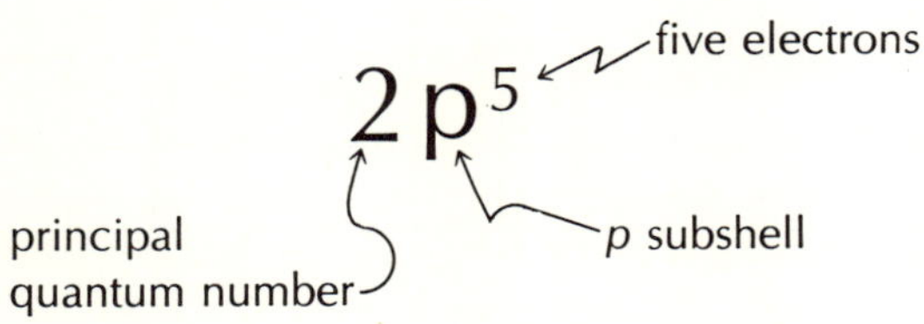

Example 11–3: Electronic notation *Explain the notation $4d^5$.*

The notation denotes 5 electrons, which are in the fourth energy level, $n = 4$, and the *d* subshell where $l = 2$.

Hydrogen contains only one electron. When the atom is in its ground state—the condition in which the electron has its lowest energy—the electron is in the first energy level. When $n = 1$, there is only one *s* orbital available; hence the electron assignment for the hydrogen atom is $1s^1$. The next element, helium, atomic number 2, has two electrons. Since the 1*s* orbital has the lowest energy and since it can hold two electrons, the electron assignment for helium is $1s^2$. For element 3, lithium, the third electron cannot go into the 1*s* orbital since it is already filled. This electron has to go into the $n = 2$ energy level. We now have to decide whether this electron will be assigned to the 2*s* orbital or to one of the three 2*p* orbitals. Electrons fill the lowest energy orbitals first and only then fill higher energy orbitals. To determine which orbital the remaining lithium electron will fill, we first have to list the increasing energies for each sublevel. The following list details the order in which electrons fill the orbitals and the number of electrons each of these sublevels will hold:

$$1s^2\ 2s^2\ 2p^6\ 3s^2\ 3p^6\ 4s^2\ 3d^{10}\ 4p^6\ 5s^2\ 4d^{10}\ 5p^6\ 6s^2\ 4f^{14}\ 5d^{10}\ 6p^6\ 7s^2\ 5f^{14}\ 6d^{10}$$

Given this order, the proper assignment for lithium is $1s^22s^1$.

Example 11–4: Electron assignments *What are the electron assignments for the element chlorine?*

Chlorine is element number 17; hence there are 17 electrons. The assignments follow the order of increasing energies. The electronic structure of chlorine is $1s^22s^22p^63s^23p^5$. Note that the last level is not completely filled; the $3p$ level can hold a total of 6 electrons; but chlorine has only 5 available for it.

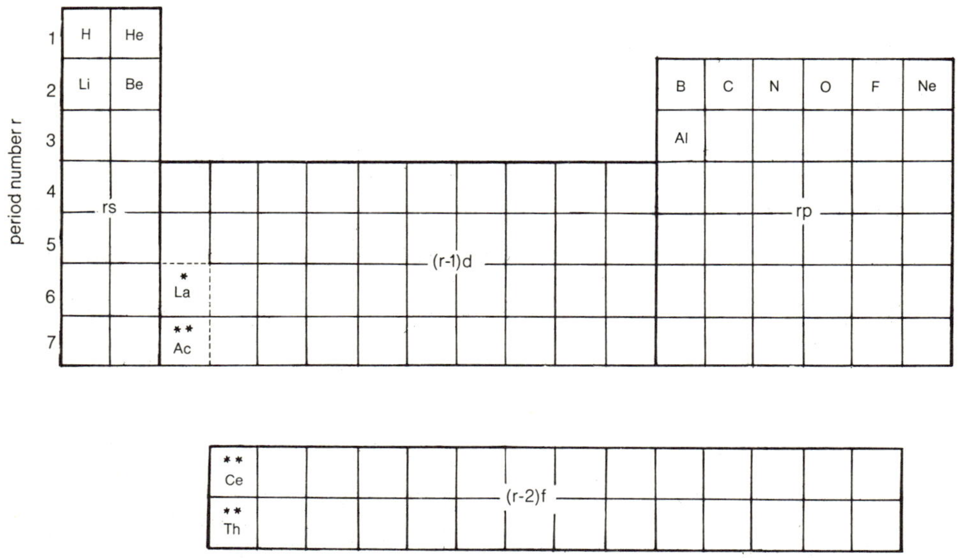

Figure 11–5 Periodic table indicating the type of orbital being filled.

The order of increasing energies for the atomic orbitals can be deduced from the periodic table. Figure 11–5 shows a periodic table divided into four areas that designate the type of orbital being filled. Note that the element helium was moved next to hydrogen to simplify the scheme. Each horizontal row is called a period and is given a number r from 1 to 7. In the first block on the left labeled rs, the s orbitals are being filled. The r, corresponding to the period or horizontal row number, tells us the principal quantum number. In the second large block, labeled $(r-1)d$, the d orbitals are being filled, and the principal quantum number for these d orbitals is equal to $r-1$. The block to the right contains the elements whose p orbitals are being filled, and the principal quantum number is equal to r. For the lanthanides and actinides, the $(r-2)f$ orbitals are being filled. Although shown as a separate block at the bottom of the table, the lanthanides belong in the square in period 6 marked La, and the actinides belong in the square labeled Ac in period 7.

You can obtain the electronic configuration of any given element by starting at period 1 and proceeding period by period from left to right, adding electrons in the order indicated by the table, until you get to the element of interest. Take gallium, atomic number 31, as an example. Starting at the left of period 1, we see that the 1*s* orbital fills first with 2 electrons. Proceeding to period 2, the 2*s* level then fills, followed by the 2*p* level, for a total of 8 electrons. Moving to period 3, the filling order is 3*s*, then 3*p*, using another 8 electrons. The element gallium is in period 4. Here the order of filling is 4*s*, then ($r-1$)*d* or 3*d*. Finally, at element 31, the last electron goes into the *rp*, or 4*p*, orbital. This procedure can be summarized in this way:

Period	Electron assignments	Number of electrons
1	$1s^2$	2
2	$2s^22p^6$	8
3	$3s^23p^6$	8
4	$4s^23d^{10}4p^1$	13

The complete assignment is therefore: $1s^22s^22p^63s^23p^64s^23d^{10}4p^1$.

Example 11–5: Electron assignments *What is the predicted electron structure for the element cerium, atomic number 58?*

Using the periodic-table technique, we obtain the following filling order:

Period	Electron assignments	Number of electrons
1	$1s^2$	2
2	$2s^22p^6$	8
3	$3s^23p^6$	8
4	$4s^23d^{10}4p^6$	18
5	$5s^24d^{10}5p^6$	18
6	$6s^24f^2$	4
		58

The complete assignment is: $1s^22s^22p^63s^23p^64s^23d^{10}4p^65s^24d^{10}5p^66s^24f^2$

Figure 11–6 shows an easily constructed chart that serves as an alternative method for determining the increasing energy of atomic orbitals.

The electronic assignments for a number of elements, such as $_{24}$Cr, $_{29}$Cu, and $_{41}$Nb, show some deviations from the general rules just discussed. Chromium, for example, has the configuration

$1s^22s^22p^63s^23p^64s^13d^5$

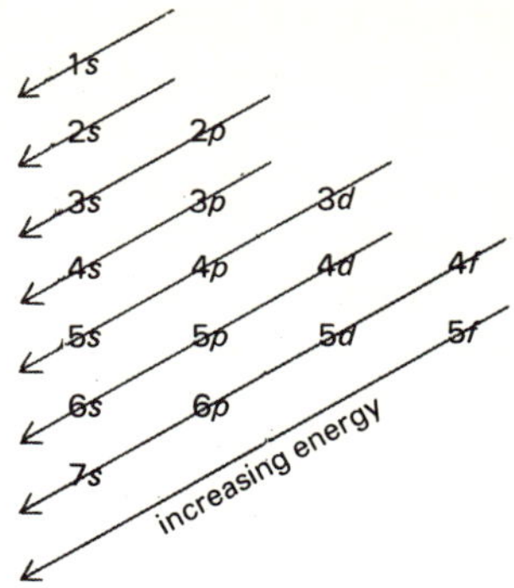

Figure 11–6 Order of increasing energy for atomic orbitals.

rather than the predicted arrangement

$1s^22s^22p^63s^23p^64s^23d^4$

Once the orbitals of an atom have been filled, the order of orbital energies is not necessarily the same as the filling order. The actual order of orbital energies in the nickel atom, for example, is $1s^22s^22p^63s^23p^63d^84s^2$ rather than the filling order $1s^22s^22p^63s^23p^64s^23d^8$. As the elements gain size, nuclear charge increases, changing the electron environment and the energies of the orbitals.

A PHYSICAL PICTURE OF THE ATOM

What does an atom really look like? This question can be answered only in terms of a rather vague, shadowy picture. Modern theory tells us where an electron can be found most of the time—in other words, the most probable location of the electron. The graph shown in Figure 11–7**B** plots the probability of finding a 1*s* electron at given distances from the nucleus. The peak of the curve gives the electron's most probable location; this distance coincides with the fixed radius of the old Bohr theory. Note that sometimes the electron is close to the nucleus and at other times it is quite far away. The region of space covered by the electron is an orbital, and an electron can be at any place in the orbital at any time. Figure 11–7**A** is a two-dimensional representation of the 1*s* orbital of the hydrogen atom. The shading becomes denser as the probability of finding the electron in that area increases. The hydrogen atom, as we now understand it, can be pictured as a nucleus surrounded by a spherically symmetrical electron cloud.

Only the *s* orbitals are spherically symmetrical about the nucleus; the other orbitals have certain preferred directions in space. Because only two electrons can occupy a given orbital, the six electrons in the *p* levels are divided into three distinct orbitals. These electron clouds occupy a somewhat ellipsoidal region on each side of the nucleus, as illustrated in Figure 11–8. The three *p* orbitals have the same dumbbell shape and are oriented in three different directions, corresponding to the *x*, *y*, and *z* axes of the usual coordinate system. They are referred to as p_x, p_y, and p_z orbitals.

At first glance, the perpendicularity of a p_x orbital to a p_z orbital may seem like the most trivial of details, but in truth it is a significant piece of information. The spatial orientation of atomic orbitals gives a molecule its shape, and shape plays

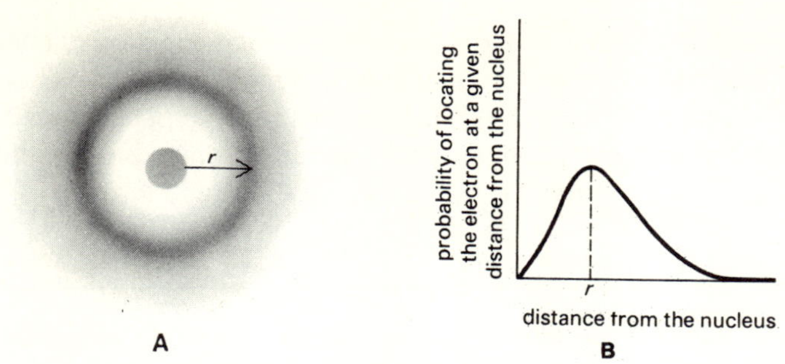

Figure 11–7 The hydrogen atom. **A** Spherically symmetrical electronic cloud around the hydrogen nucleus. The most probable distance is designated by *r*. **B** Graph showing the probability of locating a 1*s* electron at any given distance from the nucleus.

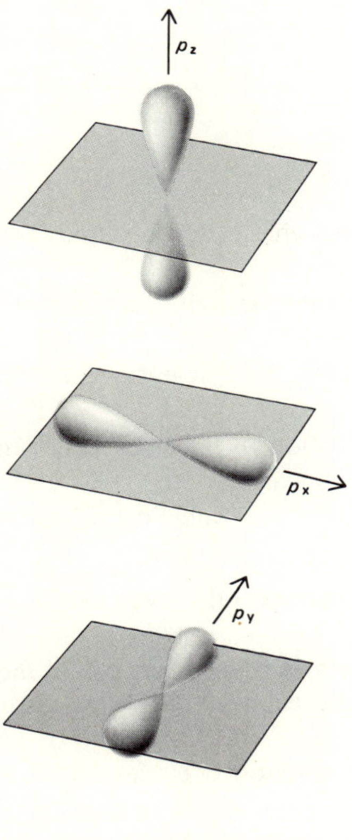

Figure 11–8 Spatial orientations of the *p* orbitals.

an important role in determining the molecule's chemical and physical properties. For example, the curious characteristics of water—its existence as a liquid rather than a gas at room temperature, its solvent properties—are due in large part to the bent shape of the water molecule. The geometrical arrangement of atoms in molecules can best be explained and understood in terms of electron orbitals. The spatial orientations of the *d* and *f* orbitals will not be needed for the present discussion.

One other rule will be needed, however, for our discussion of chemical bonding —the so called maximum multiplicity **rule of Hund.** It states: When electrons are assigned to atoms and several orbitals of the same type are available, a single electron is placed into each orbital before any electrons are permitted to pair. This rule is illustrated by Example 11–6.

Example 11–6: Hund's rule *Illustrate the application of Hund's rule to the assignment of electrons to the orbitals of the element vanadium, atomic number 23.*

The method of Example 11–4 gives us the assignment

$1s^2 2s^2 2p^6 3s^2 3p^6 4s^2 3d^3$

Because the $3d$ level can hold 10 electrons (5 orbitals) and only 3 electrons are available, 1 electron will go into each of 3 orbitals and 2 will contain no electrons. This distribution can be illustrated by writing the $3d^3$ of the assignment scheme as

$3d^1 3d^1 3d^1 3d^0 3d^0$.

NEW TERMS

Azimuthal quantum number: denotes the subshell that an electron occupies and specifies the shape of the electron cloud.

Electromagnetic radiation: a form of energy transmitted through space without the apparent transmission of matter.

Electron cloud: an orbital, the region in space around the nucleus where the probability of finding an electron is greatest.

Emission spectrum: the displayed wavelength distribution of light coming from an emitting source.

Energy level: one of a set of energy states that an electron in an atom may assume.

Excited state: an energy level of greater energy than the ground state.

Ground state: the state in which electrons are in the lowest energy levels available to them—the lowest energy state of an atom.

Hund's rule: when electrons are assigned to atoms and several orbitals of the same type are available, a single electron is placed into each orbital before electrons are permitted to pair.

Magnetic quantum number: denotes the orbital an electron occupies within a subshell.

Orbit: another name for the principal energy levels occupied by electrons in atoms.

Orbital: region in space around the nucleus where the probability of finding an electron is greatest. Each orbital can hold a maximum of two electrons.

Orbital quantum number: same as azimuthal quantum number.

Pauli exclusion principle: no two electrons in a given atom can have the same set of quantum numbers.

Photons: light quanta.

Principal quantum number: the number used to designate the principal energy levels that an electron may occupy in an atom.

Quanta: particles of light.

Shell: another designation for the principal energy levels.

Spectroscope: an instrument used to separate and display, in the order of increasing wavelengths, the component parts of light coming from a radiation source.

Spin quantum number: describes the spin orientation of the electron.

Wave equation: an equation discovered by Schrödinger that is the basis for the science of wave mechanics.

Wavelength: the distance between crests of a wave.

PROBLEM SET A

1. Why would you expect gamma radiation to be more dangerous than x rays?
2. Which has greater energy, light of short or long wavelengths?
3. Explain how Bohr used the principles of the quantum theory to explain atomic structure.
4. Is more energy liberated when an electron in a hydrogen atoms falls from $n = 4$ to $n = 2$ or when the jump is from $n = 5$ to $n = 3$? Refer to Figure 11–4.
5. Explain what is meant by the statement, "The Pauli exclusion principle is a law of nature."
6. Give electron assignments for the elements bromine, mercury, xenon, and barium.
7. Prepare a table listing the quantum-number assignments for all the electrons in the first three energy levels.
8. How many *d* and *f* electrons are possible in the $n = 5$ energy level?
9. Identify the atom that has the electronic structure
   $1s^2 2s^2 2p^6 3s^2 3p^6 4s^2 3d^{10} 4p^6 5s^2 4d^{10} 5p^6 6s^2 5d^1$
10. Use the element manganese, atomic number 25, to illustrate the application of Hund's rule to the assignment of the 3*d* electrons.

PROBLEM SET B

1. How can a spectroscope be used to determine which elements are found in a metal alloy?
2. When two electrons are located in the same orbital, what can you say about the values of their four quantum numbers? 3 # are same, other opposite in charge
3. Associate each of these electronic "jumps" in the hydrogen atom—6 to 2, 5 to 2, 4 to 2, and 3 to 2—with a color in the visible spectrum. Refer to Table 11–1 and Figure 11–4.
4. If each atomic orbital were limited to a maximum of three electrons rather than two, which elements would represent a just-filled first energy level and a just-filled second energy level?
5. Give the electron assignments for the elements silicon, iron, argon, and potassium.
6. Prepare a table listing the quantum number assignments for all the electrons in the $n = 4$ energy level.
7. Identify the atom that has the electronic structure:
   $1s^2 2s^2 2p^6 3s^2 3p^6 4s^2 3d^4$.

8. Use the element phosphorus, atomic number 15, to illustrate the application of Hund's rule.
9. State the number of unpaired electrons in each of the following species: He, C, S, As.
10. Many substances fluoresce (give off visible light) when exposed to ultraviolet light. How would you explain this in terms of the electronic structure of atoms?

12

the periodic table

THIS WILL SAVE YOU TIME:

Refer to this list as you read through the chapter. In answering the questions, you will understand the important concepts you are most likely to be asked about later on.

After reading this chapter, you should be able to:

1. *State the periodic law and explain the scientific basis for the periodic table.*
2. *Explain what is meant by transition elements and by the difference between the A and B subgroups of elements.*
3. *Draw a blank periodic table and locate the following groups of elements: alkali metals, nonmetals, metals, noble gases, lanthanides, transition elements, inner transition elements, actinides, and A and B subgroups.*
4. *Explain what is meant by inner transition elements, and explain why they are located at the bottom of the periodic table.*
5. *Give examples of transuranium elements.*
6. *Name the most recently discovered elements.*
7. *Name the elements that are gases and liquids at normal atmospheric conditions.*
8. *Explain what is meant by ionization potentials.*
9. *Illustrate the periodicity of some elements using one chemical and one physical property.*
10. *Explain how it is possible to predict the properties of a new element from the periodic table.*
11. *Name and discuss the physical and chemical properties of the noble gases.*
12. *Name and discuss the physical and chemical properties of the alkali metals.*
13. *Name and discuss the physical and chemical properties of the halogens.*

Why do certain elements have similar properties? With modern atomic theory, we can answer this question in terms of electronic structures. Different elements whose atoms have similar electronic structures in their outer shells have many properties in common. This idea relating similarity in structures to similarity in properties is the basis for the periodic law.

In this chapter, we shall discuss the theory underlying the periodic law and use the periodic table as the basis for studying the chemistry of several groups of elements.

THE PERIODIC LAW

As we saw earlier in Chapter 3, Mendeleev and Meyer discovered the periodicity in properties revealed by arranging the elements in order of increasing atomic weights (masses). Several unexplained errors, however, cropped up in such an arrangement, but with the discovery of isotopes, the **periodic law** could be cast in its correct form: *The properties of the elements are a periodic function of their atomic number.* Using the periodic law, Mendeleev predicted very accurately the chemical and physical properties of several undiscovered elements. In honor of his achievement, element 101 was named after him—mendelevium.

THE BASIS FOR THE PERIODIC ARRANGEMENT

A modern periodic table is shown in Table 12–1 and inside the front cover. The properties of the elements are a function of the electronic structure of their atoms, or, more specifically, of the arrangement of electrons in the outermost electronic energy levels. Elements with a similar arrangement of electrons in the outer orbitals are grouped together in vertical columns. These **periodic groups** are labeled from I to VIII and are divided into A and B subgroups. Each **period,** or horizontal row, contains the elements with the same maximum principal quantum number for the ground-state structure of the atom.

In Table 12–1, a formula under the group number indicates the electronic arrangements. All the elements in group 1A have an outer orbital designated by ns^1, where n is the principal quantum number of the outermost shell. The electronic structures of the group 1A elements are shown in Table 12–2.

The elements in group VIIA all have outer orbitals designated by ns^2np^5. Iodine is in the fifth period of the table; therefore, $n = 5$, and the outer electronic sublevels are $5s^25p^5$.

Example 12–1: Relating electron assignments and periodic table *The elements selenium, tellurium, and polonium are all in the same periodic group. What are the assignments for the electrons in their outermost orbitals?*

These elements are in group VIA. From Table 12–1 we see that they all have outer orbitals designated by the formula ns^2np^4. Hence the assignments are

Se	in period 4	$4s^24p^4$
Te	in period 5	$5s^25p^4$
Po	in period 6	$6s^26p^4$

BLE 12–1 PERIODIC TABLE

IA	IIA	IIIB	IVB	VB	VIB	VIIB	VIII	VIII	VIII	IB	IIB	IIIA	IVA	VA	VIA	VIIA	Noble gases
ns^1	ns^2	$(n-1)d^1ns^2$	$(n-1)d^2ns^2$	$(n-1)d^3ns^2$	$(n-1)d^4ns^2$	$(n-1)d^5ns^2$	$(n-1)d^6ns^2$	$(n-1)d^7ns^2$	$(n-1)d^8ns^2$	$(n-1)d^{10}ns^1$	$(n-1)d^{10}ns^2$	ns^2np^1	ns^2np^2	ns^2np^3	ns^2np^4	ns^2np^5	ns^2np^6
H 1																	He 2
Li 3	Be 4											B 5	C 6	N 7	O 8	F 9	Ne 10
Na 11	Mg 12											Al 13	Si 14	P 15	S 16	Cl 17	Ar 18
K 19	Ca 20	Sc 21	Ti 22	V 23	Cr 24	Mn 25	Fe 26	Co 27	Ni 28	Cu 29	Zn 30	Ga 31	Ge 32	As 33	Se 34	Br 35	Kr 36
Rb 37	Sr 38	Y 39	Zr 40	Nb 41	Mo 42	Tc 43	Ru 44	Rh 45	Pd 46	Ag 47	Cd 48	In 49	Sn 50	Sb 51	Te 52	I 53	Xe 54
Cs 55	Ba 56	La 57 (58–71)	Hf 72	Ta 73	W 74	Re 75	Os 76	Ir 77	Pt 78	Au 79	Hg 80	Tl 81	Pb 82	Bi 83	Po 84	At 85	Rn 86
Fr 87	Ra 88	Ac 89 (90–103)	(Ku or Kh) Rf 104	(Ha) 105	106	107	108	109	110	111	112	113	114	115	116	117	118
119	120	121 (122–1)															

Lanthanide series	Ce 58	Pr 59	Nd 60	Pm 61	Sm 62	Eu 63	Gd 64	Tb 65	Dy 66	Ho 67	Er 68	Tm 69	Yb 70	Lu 71	$4f^{0-14}5d^{0-1}\ 6s^2$
Actinide series	Th 90	Pa 91	U 92	Np 93	Pu 94	Am 95	Cm 96	Bk 97	Cf 98	Es 99	Fm 100	Md 101	No 102	Lw 103	$5f^{0-14}6p^{0-2}\ 7s^2$
	122	123	124	125	126										

TABLE 12–1 PERIODIC TABLE

TABLE 12–2 ELECTRONIC CONFIGURATION OF GROUP 1A ELEMENTS

Element	Symbol	In period	Electronic structure
Lithium	Li	2	$1s^2\underline{2s^1}$
Sodium	Na	3	$1s^22s^22p^6\underline{3s^1}$
Potassium	K	4	$1s^22s^22p^63s^23p^6\underline{4s^1}$
Rubidium	Rb	5	$1s^22s^22p^63s^23p^64s^23d^{10}4p^6\underline{5s^1}$
Cesium	Cs	6	$1s^22s^22p^63s^23p^64s^23d^{10}4p^65s^24d^{10}5p^6\underline{6s^1}$

TRANSITION ELEMENTS

The **transition elements** are those in the B subgroups and the subgroup VIII. The properties of these elements can be attributed to the delayed filling of the next-to-outermost electronic energy levels. As an example, consider the elements in the period containing elements 21 to 30, scandium to zinc. The 4*s* level of all these elements is filled, and the 3*d* level gains from 1 to 10 electrons as the atomic number increases. In general, one can say that the transition elements have next-to-outermost electronic energy levels that are not completely filled; these energy levels are responsible for the properties of the transition elements.

The **inner transition elements** comprise two series that are placed separately at the bottom of the periodic table for convenience. The first series, the **lanthanides** (formerly called the rare earths), contains the elements from 57 through 71; these elements have a structure indicated by the electronic configuration $4f^{0-14}5d^{0-1}6s^2$—three incomplete shells. As the atomic numbers of these elements increase in going from cerium to lutetium, the 4*f* orbitals become more populated with electrons, even though the $n = 5$ and $n = 6$ energy levels have only a few electrons. These elements have similar chemical and physical properties; they are commonly found together in nature and formerly were very difficult to separate from one another.

The **actinide series,** elements 89 to 103, thorium to lawrencium, are also inner transition elements with similar properties; their outer electronic configuration is $5f^{0-14}6d^{0-2}7s^2$—three incomplete shells.

TRANSURANIUM ELEMENTS

Elements with atomic numbers greater than uranium are called **transuranium elements.** All these elements are produced artificially by nuclear bombardment reactions. They could have existed naturally on the earth shortly after its creation but would have disappeared relatively rapidly because of radioactive disintegration. Traces of plutonium are found in nature, however, as a result of some neutron absorption by uranium. The transuranium elements are tabulated in Table 12–3.

Element 102, nobelium, will serve as an example of these elements. Nobelium was first produced at the Lawrence Radiation Laboratory in 1958 by reacting the

TABLE 12–3 THE TRANSURANIUM ELEMENTS

Atomic number	Name	Symbol	Mass number*	Half-life for α emission
93	Neptunium	Np	237	2.25×10^6 y
94	Plutonium	Pu	239	2.43×10^3 y
95	Americium	Am	241	500 y
96	Curium	Cm	247	4×10^7 y
97	Berkelium	Bk	247	7×10^3 y
98	Californium	Cf	244	45 min
99	Einsteinium	Es	254	480 d
100	Fermium	Fm	252	25 h
101	Mendelevium	Md	256	1 h
102	Nobelium	No	254	3 s
103	Lawrencium	Lr	256	45 s
104	(Rutherfordium)	(Rf)	257	3 s
105	(Hahnium)	(Ha)	260	1.6 s
106	(not named)			

* Only one isotope of each element is shown.

isotope of the element curium, $^{246}_{96}Cm$, with high-energy carbon ions accelerated by the heavy ion linear accelerator (HILAC):

$$^{246}_{96}Cm + ^{12}_{6}C \rightarrow ^{254}_{102}No + 4^{1}_{0}n$$

Russian scientists at the Research Institute in Dubna have claimed discovery of element 104. They bombarded plutonium-242 with neon nuclei:

$$^{242}_{94}Pu + ^{22}_{10}Ne \rightarrow ^{260}_{104}Ku + 4^{1}_{0}n$$

In 1969, Albert Ghiorso (1915–, codiscoverer of elements 95 through 103) and his colleagues at the Lawrence Radiation Laboratory bombarded 60 micrograms (two-millionths of an ounce) of californium-249 with carbon-12 nuclei and produced element 104, of mass 257. They have not been able to duplicate the Russian experiments and believe that the Russian claim is questionable. The name for element 104 varies with nationality: kurchatovium (in honor of Igor Kurchatov, a nuclear scientist), Ku or Kh, for the Russians, and rutherfordium, Rf, for the Americans. Neither name has been officially accepted.

Element 105 was discovered in 1970 by Ghiorso and his research group. They bombarded californium-249 with a beam of nitrogen-15 nuclei:

$$^{249}_{98}Cf + ^{15}_{7}N \rightarrow ^{260}_{105}Ha + 4^{1}_{0}n$$

The name hahnium, symbol Ha, has been proposed for element 105 to honor Otto Hahn, the discoverer of the fission reaction.

Many exciting experiments still remain to be performed. Theoretical predictions indicate that element 114 with a mass number of 298 will be a relatively stable isotope. A probable route for its synthesis is bombarding uranium-238 nuclei with uranium-238 nuclei:

$$^{238}_{92}U + ^{238}_{92}U \rightarrow ^{476}_{184}(184) \rightarrow ^{298}_{114}(114) + ^{170}_{70}Yb + 8^{1}_{0}n$$

IONIZATION POTENTIALS

Ionization potential is a *measure of the energy necessary to remove an electron from a neutral atom*. You can measure this energy for gaseous elements by finding the voltage that must be applied to an electric-discharge tube to prompt a sudden large increase in the flow of electrons. For hydrogen atoms, the reaction

$$H \rightarrow H^{+} + ^{0}_{-1}e$$

occurs at an applied potential of 13.6 volts. By convention, we say that the ionization potential amounts to 13.6 electron volts per atom, which equals 313 kcal/mole. A tabulation of these ionization potentials, Table 12–4, shows that they follow a periodic relationship, although many deviations appear in the transition elements. In general, the ionization potentials decrease as the atomic numbers of the elements increase in a given periodic group.

Ionization potentials are related to the metallic character of the elements. The lower the ionization potential, the greater the metallic character. From Table 12–4,

TABLE 12–4 FIRST IONIZATION POTENTIALS OF THE ELEMENTS

IA	IIA	IIIB	IVB	VB	VIB	VIIB	VIII	VIII	VIII	IB	IIB	IIIA	IVA	VA	VIA	VIIA	Noble gases
H 1 313																	He 2 567
Li 3 124	Be 4 215											B 5 191	C 6 260	N 7 336	O 8 314	F 9 402	Ne 10 497
Na 11 119	Mg 12 176											Al 13 138	Si 14 188	P 15 254	S 16 239	Cl 17 300	Ar 18 363
K 19 100	Ca 20 141	Sc 21 151	Ti 22 158	V 23 156	Cr 24 156	Mn 25 171	Fe 26 182	Co 27 181	Ni 28 176	Cu 29 178	Zn 30 216	Ga 31 138	Ge 32 187	As 33 231	Se 34 225	Br 35 273	Kr 36 323
Rb 37 96	Sr 38 131	Y 39 152	Zr 40 160	Nb 41 156	Mo 42 166	Tc 43 167	Ru 44 173	Rh 45 178	Pd 46 192	Ag 47 175	Cd 48 207	In 49 133	Sn 50 169	Sb 51 199	Te 52 203	I 53 241	Xe 54 280
Cs 55 90	Ba 56 120	La 57 129 (58–71)	Hf 72 127	Ta 73 138	W 74 184	Re 75 182	Os 76 201	Ir 77 212	Pt 78 207	Au 79 213	Hg 80 241	Tl 81 141	Pb 82 171	Bi 83 185	Po 84 —	At 85 —	Rn 86 248
Fr 87 —	Ra 88 —	Ac 89 — (90–103)	(Rf) —	Ha 105													

Lanthanide series	Ce 58 159	Pr 59 133	Nd 60 145	Pm 61 133	Sm 62 129	Eu 63 131	Gd 64 142	Tb 65 155	Dy 66 157	Ho 67 —	Er 68 —	Tm 69 —	Yb 70 143	Lu 71 115
Actinide series	Th 90 —	Pa 91 —	U 92 —	Np 93 —	Pu 94 —	Am 95 —	Cm 96 —	Bk 97 —	Cf 98 —	Es 99 —	Fe 100 —	Md 101 —	No 102 —	Lr 103 —

you can see that on the basis of this definition, cesium is one of the most metallic elements, and fluorine one of the least.

Note the high ionization potentials of the noble gases; this property helps to account for their chemical inertness.

The ionization potentials from Table 12–4 measure the energy needed to remove the first electron from the ground state of the atom. The **second ionization potential** is always much greater; *it is a measure of the energy needed to remove a second electron from a singly-charged atom*. For example,

$He \rightarrow He^{+} + {}_{-1}^{0}e$ 567 kcal/mole   first ionization potential

$He^{+} \rightarrow He^{2+} + {}_{-1}^{0}e$ 1254 kcal/mole   second ionization potential

PERIODIC RELATIONSHIPS IN OTHER PHYSICAL PROPERTIES

The melting and boiling points of the elements display the periodic relationship shown in Table 12–5. As an example, note the melting and boiling points of the three elements gallium, indium, and thallium in group IIIA. The element gallium, melting point 29.8°C, will turn to a liquid when held in the palm of the hand; its

TABLE 12–5 MELTING AND BOILING POINTS OF THE ELEMENTS

(in °C)

Key:

31 Ga (atomic number, symbol)
2237 ← boiling point
29.8 ← melting point

IA	IIA	IIIB	IVB	VB	VIB	VIIB	VIII	VIII	VIII	IB	IIB	IIIA	IVA	VA	VIA	VIIA	Noble gases
1 H -252.7 -259.2																	2 He -268.9 -269.7
3 Li 1330 108.5	4 Be 2770 1277											5 B - (2030)	6 C 4830 3727	7 N -195.8 -210	8 O -183 -218.8	9 F -188.2 -219.6	10 Ne -246 -248.6
11 Na 892 97.8	12 Mg 1107 650											13 Al 2450 660	14 Si 2680 1410	15 P 280 44.2	16 S 444.6 119.0	17 Cl -34.7 -101.0	18 Ar -185.8 -189.4
19 K 760 63.7	20 Ca 1440 838	21 Sc 2730 1539	22 Ti 3260 1668	23 V 3450 1900	24 Cr 2665 1875	25 Mn 2150 1245	26 Fe 3000 1536	27 Co 2900 1495	28 Ni 2730 1453	29 Cu 2595 1083	30 Zn 906 419.5	31 Ga 2237 29.8	32 Ge 2830 937.4	33 As 613 817	34 Se 685 217	35 Br 58 -7.2	36 Kr -152 -157.3
37 Rb 688 38.9	38 Sr 1380 768	39 Y 2927 1509	40 Zr 3580 1852	41 Nb 3300 2415	42 Mo 5560 2610	43 Tc — 2200	44 Ru 4900 2500	45 Rh 4500 1966	46 Pd 3890 1552	47 Ag 2210 960.8	48 Cd 765 320.9	49 In 2000 156.2	50 Sn 2270 231.9	51 Sb 1380 630.5	52 Te 989.8 449.5	53 I 183 113.7	54 Xe -108.0 -111.9
55 Cs 690 28.7	56 Ba 1640 714	57 La 3470 920	72 Hf 5400 2222	73 Ta 5425 2996	74 W 5930 3410	75 Re 5900 3180	76 Os 5500 2700	77 Ir 5300 2454	78 Pt 4530 1769	79 Au 2970 1063	80 Hg 357 -38.4	81 Tl 1457 303	82 Pb 1725 327.4	83 Bi 1560 271.3	84 Po — 254	85 At — (302)	86 Rn (-62) (-71)
87 Fr — (27)	88 Ra — 700	89 Ac — 1050	104 Rf (Ku or Kh)	105Ha													

Lanthanide series	58 Ce 3468 795	59 Pr 3127 935	60 Nd 3027 1024	61 Pm — (1027)	62 Sm 1900 1072	63 Eu 1439 826	64 Gd 3000 1312	65 Tb 2800 1356	66 Dy 2600 1407	67 Ho 2600 1461	68 Er 2900 1497	69 Tm 1727 1549	70 Yb 1427 824	71 Lu 3327 1652
Actinide series	90 Th 3850 1750	91 Pa — (1230)	92 U 3818 1132	93 Np — 637	94 Pu 3235 640	95 Am — —	96 Cm — —	97 Bk — —	98 Cf — —	99 Es — —	100 Fm — —	101 Md — —	102 No — —	103 Lr — —

boiling point, however, is 2237°C. The melting points of indium and thallium are 156°C and 303°C, respectively.

Atomic size is another physical property with a periodic relationship. This property has been shown earlier, in Figure 3–3.

Elements in the same periodic group also exhibit regularity in the chemical and physical properties of their compounds. The solubility of the group IIA sulfates in water is a good example. Calcium sulfate, $CaSO_4$, is moderately soluble in water; strontium sulfate, $SrSO_4$, is less soluble; and barium sulfate, $BaSO_4$, is even less soluble than strontium sulfate. On the basis of these solubility data, the Curies predicted that the sulfate of the element radium should be practically insoluble, and they used this information to isolate the new element.

We have already discussed the periodicity of the oxidation numbers of elements. Other physical properties that show periodicity are densities, heats of vaporization, and heats of fusion.

PREDICTING THE PROPERTIES OF NEW ELEMENTS

The periodic table played a major role in the discovery of the elements, for it showed workers where undiscovered elements were to be expected and gave them a way of predicting their chemical and physical properties. The periodic table can still be used for this purpose. Scientists at the Oak Ridge National Laboratory have recently predicted some of the properties for elements 113 and 114; these two elements are expected to be quite stable. Element 113 is located directly under thallium in group IIIA and element 114 under lead in group IVA. We can thus predict that the chemical and physical properties of these two elements will resemble those of thallium and lead, respectively. Some of the predicted properties are shown in Table 12–6.

TABLE 12–6 PREDICTED PROPERTIES FOR ELEMENTS 113 AND 114

	Element 113	Element 114
Periodic group	IIIA	IVA
Atomic mass	297	298
Most stable oxidation state	+1	+2
First ionization potential, eV*	7.4	8.5
Metallic radius, 10^{-10} m	1.75	1.85
Ionic radius, 10^{-10} m	1.48	1.31
Atomic volume, ml/mole	18	21
Density, g/ml	16	14
Melting point, °K	700	340
Boiling point, °K	1400	420
Heat of vaporization, kcal/mole	31	9

* eV = electron volt = 23.0 kcal/mole.

To illustrate how the data of Table 12–6 were obtained, let us consider the predicted melting point of element 113. Plotting the melting points of the known elements gallium, indium, and thallium by the period or row in which these elements are located in the periodic table yields a straight line. Extrapolating this straight line to the seventh periodic row permits us to predict 700°K as the melting point for element 113 (see Figure 12–1).

When elements 113 and 114 are discovered, possibly within the next decade, it will be interesting to compare the data of Table 12–6 with the actual properties of the elements.

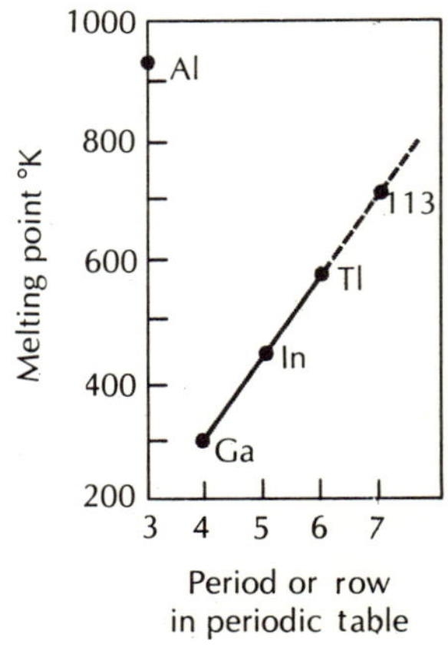

Figure 12–1 Predicting the melting point of element 113.

NOBLE GASES

The **noble gases** are those elements found in the extreme righthand column of the table. Their outer shells contain eight electrons. Until Neil Bartlett discovered the stable solid compound $XePtF_6$ in 1962, it was thought that these elements formed no chemical compounds, and they were called inert gases. We now know that they can form chemical bonds; however, only compounds of xenon, krypton, and radon, primarily with fluorine and oxygen, are known at this time.

Dry air contains approximately 0.93 percent argon, 0.002 percent neon, and 0.0005 percent helium; only traces of krypton and xenon are present. The heaviest noble gas, radon, is radioactive and does not occur in the atmosphere in detectable amounts. Argon and neon are recovered from air by low-temperature fractional distillation. The primary source of helium is natural gas wells in the Texas-Kansas area, where it occurs in concentrations from 1 to 7.5 percent. The United States has a virtual monopoly on the world's supply of helium and regulates its production carefully.

Helium is used extensively in laboratories as an inert gas for many chemical procedures. It is in great demand for the atomic energy and space programs. One important commercial use for helium is as a replacement for nitrogen in the air breathed by deep-sea divers. At the high pressures a diver experiences under water, nitrogen dissolves in the blood. Should the diver surface too fast, lowering pressure quickly, the dissolved nitrogen forms bubbles in the blood, causing painful and sometimes fatal bends. If the nitrogen in the air is replaced with helium, however, the mixture can be breathed safely at great depths because of the very low solubility of helium in the blood. Another important use for helium is as a coolant for very low temperature operations; liquid helium has a temperature of −269°C.

Argon is used extensively as an inert filler for ordinary light bulbs; argon does not react with incandescent tungsten filaments as do nitrogen and oxygen. Electric-discharge tubes filled with neon give the characteristic red color of the neon emission spectrum, hence its use in neon light bulbs.

There are few commercial uses for xenon and krypton because of their high cost; xenon costs about $90/liter. Radon, however, has found some use in medical radiation therapy.

ALKALI METALS

The group IA elements from lithium to francium are called the **alkali metals.** The word alkali comes from the Arabic word *al-qili,* meaning plant ashes, because the ashes of plants contain considerable amounts of the alkali compounds sodium and potassium carbonate. The alkali metals do not occur in the free state in nature, only as compounds, primarily as silicates and as salts like sodium chloride. The element francium is radioactive and is extremely rare. All the alkali metals are soft and have a silvery luster when cut with a sharp knife blade. They conduct an electric current, are malleable (can be hammered into sheets), and are ductile (can be drawn into wires).

The melting points decrease from 180.5°C to 28.7°C and the densities increase from 0.53 g/ml to 1.90 g/ml as you proceed from lithium to cesium.

The pure metals can be prepared by **electrolysis,** a *process in which an electric current is passed through a molten salt to give a deposit of pure metal on the nega-*

tive electrode. Once prepared, however, the metals must be kept under oil because they are so reactive.

The element sodium finds extensive use as a cooling agent in nuclear reactors. The metal's excellent heat-conductivity properties compensate for the difficulty of handling such a reactive substance.

Cesium has the lowest ionization potential of all the elements measured so far; the outer 6*s* electron can be readily removed by a beam of light. This property of cesium is used in **photocells**—*devices for converting light signals into electrical signals*. A simple photocell consists of an evacuated electric-discharge tube containing two electrodes, one of which is coated with cesium metal. When the tube is connected to a high-voltage source, no electrical current will flow. When light hits the cesium-coated electrode, however, electrons are given off, and these electrons move to the positive electrode, thus creating an electric current that can be used to regulate any electrically controlled device. These photocells find extensive use in television pickup devices, remote controls such as automatic garage-door openers, and for producing sound from the sound tracks of movie films.

The physical properties of the alkali metals are summarized in Table 12–7.

TABLE 12–7 PHYSICAL PROPERTIES OF ALKALI METALS

Property	Li	Na	K	Rb	Cs
Melting point, °C	180	98	63.4	38.8	28.7
Boiling point, °C	1336	889	757	679	690
Density, g/ml at 20°C	0.535	0.971	0.862	1.53	1.90
Electronegativity	1.0	0.9	0.8	0.8	0.7
Atomic radius, 10^{-10} m	1.52	1.86	2.31	2.44	2.62
Ionic radius, 10^{-10} m	0.60	0.95	1.33	1.48	1.69

The alkali metals are very reactive chemically. With water they give hydroxides and hydrogen gas:

$$2Li(s) + 2H_2O \rightarrow 2LiOH(aq) + H_2(g) + \text{energy}$$
$$2Na(s) + 2H_2O \rightarrow 2NaOH(aq) + H_2(g) + \text{energy}$$
$$2K(s) + 2H_2O \rightarrow 2KOH(aq) + H_2(g) + \text{energy}$$

In a more general form,

$$2M(s) + 2H_2O \rightarrow 2MOH(aq) + H_2(g) + \text{energy}$$

where M denotes any of the alkali metals. These reactions are so vigorous that the heat given off is enough to cause an explosion when the hydrogen mixes with the oxygen of the air.

All the salts of the alkali metals, with a few exceptions, are soluble in water; all form +1 ions.

Some other important reactions of the alkali metals include the following:

$2M + X_2 \rightarrow 2MX$ (X denotes a halogen, a group VIIA element)
$2M + S \rightarrow M_2S$ (the same reaction for Se and Te)

$$2M + 2NH_3 \xrightarrow{\text{catalyst}} 2MNH_2 + H_2$$
$$2M + H_2 \rightarrow 2MH$$

HALOGENS

The elements fluorine, chlorine, bromine, iodine, and astatine, located in group VIIA of the periodic table, are called **halogens.** The term halogen is derived from the Greek words *halos,* meaning salt, and genes, meaning born; hence the term means salt maker. All these elements exist as diatomic molecules in the free state. Fluorine, F_2, and chlorine, Cl_2, are gases; bromine, Br_2, is a liquid; and iodine, I_2, is a solid at room temperature. The element astatine is radioactive; it has not been available in quantity sufficient to determine its properties accurately.

TABLE 12–8 PHYSICAL PROPERTIES OF THE HALOGENS

Property	Fluorine	Chlorine	Bromine	Iodine
Appearance at room temperature	pale-yellow gas	greenish-yellow gas	brown-red liquid	gray-black solid
Formula	F_2	Cl_2	Br_2	I_2
Melting point, °C	−223	−102	−7	113
Boiling point, °C	−187	−35	59	184
Atomic radius, 10^{-10} m	0.64	0.99	1.14	1.33
Ionic radius, 10^{-10} m	1.36	1.81	1.95	2.16
Electronegativity	4.0	3.0	2.8	2.5
Density, g/ml at 20°	1.11 at −187°	1.57 at −35°	3.14	4.94

Some of the important physical properties of the halogens are listed in Table 12–8.

Because the halogens are very reactive, none are found free in nature. Fluorine occurs in the common minerals fluorite, CaF_2, cryolite, Na_3AlF_6, and apatite, $CaF_2 \cdot 3Ca_3(PO_4)_2$. Chlorine is found in combination with many metals as chlorides. Sea water contains approximately 3 percent NaCl; this compound occurs in large underground deposits and is widely distributed. Bromine occurs in natural salt brine as bromides and is extracted in commercial quantities from sea water. The main source of iodine is the compound sodium iodate, $NaIO_3$, which is found in Chile along with saltpeter, $NaNO_3$. There is some iodine in sea water, and it can be recovered from kelp, a type of seaweed.

Fluorine gas is extremely reactive and corrosive, burns the skin seriously on contact, and is very poisonous. It is prepared commercially by electrolysing HF, which is produced by the action of sulfuric acid on CaF_2:

$$CaF_2(s) + H_2SO_4(aq) \rightarrow 2HF(g) + CaSO_4(s)$$

$$2HF \xrightarrow[\text{molten KF} \cdot 2\text{HF}]{\text{electrolysis in}} H_2(g) + F_2(g)$$

Fluorine gas can be stored in metal containers because the gas reacts with the container to form a protective layer of the metal fluoride. Fluorine was widely used after World War II to separate uranium isotopes; today it is used to prepare many commercially important compounds.

Chlorine gas is prepared by electrolyzing aqueous sodium chloride solutions:

$$2NaCl(aq) + 2H_2O \xrightarrow[\text{current}]{\text{electric}} 2NaOH(aq) + H_2(g) + Cl_2(g)$$

Chlorine has a very irritating odor and is poisonous; it was used as a chemical warfare weapon in World War I. The gas is easily liquefied and can be stored in steel cylinders. One important use for chlorine is as a bleaching agent; the actual bleaching, however, is done by hypochlorous acid formed when chlorine reacts with water.

$$Cl_2(g) + H_2O \rightarrow HCl(aq) + \underset{\text{hypochlorous acid}}{HClO(aq)}$$

Large amounts of chlorine are used to bleach wood pulp in the manufacture of paper and for treating drinking water to kill harmful bacteria. Chlorine also plays a role in the industrial processes for making dyes, drugs, hydrochloric acid, and a host of other important compounds.

Bromine is recovered from the NaBr(aq) in sea water by treatment with chlorine gas:

$$2NaBr(aq) + Cl_2(g) \rightarrow Br_2(l) + 2NaCl(aq)$$

Bromine is a deep-red, volatile liquid with a pungent odor; it is poisonous and burns the skin badly on contact. The major industrial use of the element is the manufacture of two compounds: silver bromide, AgBr, and dibromoethane, CH_2BrCH_2Br. Silver bromide is the light-sensitive material of photographic films. Dibromoethane is consumed in large quantities as a gasoline additive; 50,000 tons of bromine a year go to its production. Ethyl gasoline contains the antiknock compound tetraethyl lead, $Pb(C_2H_5)_4$, and the lead builds up in potentially damaging deposits in the engine. The problem is solved by adding dibromoethane to the gasoline; the lead reacts with bromine to form lead bromide, which is exhausted to the atmosphere. One can truthfully say that 90 percent of the bromine produced in the United States goes to produce smog and harmful atmospheric contaminants.

Iodine is obtained from sodium iodate by treatment with sodium hydrogen sulfite.

$$2NaIO_3(aq) + 5NaHSO_3(aq) \rightarrow I_2(s) + 2Na_2SO_4(aq) + H_2O + 3NaHSO_4(aq)$$

Iodine is a somewhat metallic-looking gray solid; when heated, it sublimates to a violet gas. In the recovery of iodine from iodates, the mixture is heated to dryness to sublimate the iodine. The gas is then condensed to a solid on a cool surface.

Iodine is used for the production of numerous compounds, including dyes, medicines, and silver iodide; the latter is used in photographic films. A solution of iodine in ethanol (ethyl alcohol), called tincture of iodine, is sometimes used as an antiseptic.

Lack of iodine in the human diet can lead to goiter, a disease of the thyroid gland. Seafoods, beans, spinach, and butter contain sufficient iodine compounds, but people usually supplement their iodine intake with iodized salt—table salt containing added sodium iodide.

Some reactions of the halogens are these:

1. *With metals.* (M = most metals. Br_2 and I_2 will not react with gold or platinum.)

   $M(s) + X_2 \rightarrow MX(s)$

2. *With water.* Bromine reacts in the same manner as does chlorine.

   $$\underset{\text{hypobromous acid}}{Br_2 + H_2O \rightarrow HBrO(aq)} + \underset{\text{hydrobromic acid}}{HBr(aq)}$$

   Fluorine reacts to give hydrofluoric acid.

   $2F_2(g) + 2H_2O \rightarrow 4HF(aq) + O_2(g)$

3. *With nonmetals* (Z = certain nonmetals, such as B, P, As, and Sb).

   $2Z + 3X_2 \rightarrow 2ZX_3$

4. *With hydrocarbons* (compounds containing only hydrogen and carbon).

   $$X_2 + \underset{\text{methane}}{CH_4} \rightarrow CH_3X + HX$$

5. *With halides.* A more active halogen will displace a less active halogen from its **halide**—*a binary compound of a halogen and another element.* The order of activity is F > Cl > Br > I.

   $F_2(g) + 2NaBr(aq) \rightarrow Br_2 + 2NaF(aq)$
   $Br_2 + CaI_2(aq) \rightarrow I_2 + CaBr_2(aq)$
   $Cl_2(g) + NaF(aq) \rightarrow$ no reaction

NEW TERMS

Actinides: the series of elements from 89 to 103, Ac to Lr. These are inner transition elements with 3 incomplete shells.

Alkali metals: the elements in group IA, excluding hydrogen.

Electrolysis: chemical decomposition by the action of an electric current.

Group: the term given to all the elements in a vertical column of the periodic table.

Halide: a binary compound containing a halogen and another element.

Halogen: the elements in group VIIA.

Inner transition elements: elements in the lanthanide or actinide series.

Ionization potential: the energy necessary to remove an electron from a neutral atom.

Lanthanides: the series of elements from 57 to 71, La to Lu.

Mendelevium: element 101.

Noble gases: the elements helium, neon, argon, krypton, xenon, and radon.

Period: horizontal row of elements in the periodic table.

Periodic law: the properties of the elements are a periodic function of their atomic numbers.

Periodic table: a chart in which the elements are arranged in groups according to their properties.

Photocell: a device for converting light signals to electrical signals.

Sublimation: the process of changing from the solid to the gas state without the appearance of the liquid state.

Transition element: all elements in the B subgroups and group VIII. They have more than one incomplete energy level.

Transuranium elements: all elements with atomic numbers greater than 92.

PROBLEM SET A

1. By reference to Table 12–1, construct a table like the one below for elements number 4 through 20 and also for elements number 34, 52, and 84. (See Example 12–1 for help with the last three.)

Element symbol	Name	Atomic number	Group number	Total number of outer orbital electrons
H	hydrogen	1	IA	1
He	helium	2	(noble gas)	2
Li	lithium	3	IA	1

2. From the table completed for Problem 1, list those elements that are members of
   a. the alkali metal family
   b. the halogen family
   c. the noble gases
3. In the periodic table, arsenic, element 33, adjoins four other elements: 15, 32, 34, and 51. Which of these have properties resembling arsenic's?
4. Without consulting the periodic table, select the elements from the following list that are located in the same periodic group and those that are in the same period: $_{20}$Ca, $_{16}$S, $_{19}$K, $_{56}$Ba, $_{30}$Zn, $_{34}$Se, $_{4}$Be.
5. Why would you expect the elements zinc, cadmium, and mercury to be in the same periodic group?
6. Name two elements most like tellurium (atomic number 52).
7. Predict some chemical and physical properties for the element francium (atomic number 87).
8. Complete and balance the following reactions (name reactants and products):
   a. $Rb + H_2O \rightarrow$
   b. $Na + S \rightarrow$
   c. $Li + H_2 \rightarrow$
   d. $Cl_2 + Cs \rightarrow$
   e. $H_2SO_4 + CaF_2 \rightarrow$
   f. $KCl + Br_2 \rightarrow$
   g. $KIO_3 + KHSO_3 \rightarrow$
   h. $F_2 + KBr \rightarrow$
   i. $As + Br_2 \rightarrow$
   j. $F_2 + H_2O \rightarrow$
   k. $Br_2 + Pt \rightarrow$
   l. $Zn + Cl_2 \rightarrow$

PROBLEM SET B

1. Prepare a table as in Problem 1, Set A, including elements number 31 through 38 and 49 through 56.

2. Which of the elements just listed in Problem 1 are members of
   a. the alkali metal family
   b. the halogen family
   c. the noble gases
3. Compare the electronic assignments for the elements Nd, Pm, Sm, and Eu and explain why they are inner transition elements.
4. The outer electrons in the elements Cu, Ag, and Au have the designation $(n-1)d^{10}ns^1$. Explain what this symbolism means.
5. Why are the noble gases chemically inert?
6. Explain why the second ionization potential is always larger than the first ionization potential.
7. List some elements that might be expected to have oxides corresponding to the formulas M_2O_3, M_2O, and MO_2.
8. Complete and balance the following reactions. Name the reactants and products, and if there is no reaction, so state:
   a. $K + H_2 \rightarrow$
   b. $Cs + H_2O \rightarrow$
   c. $Sb + Cl_2 \rightarrow$
   d. $Au + I_2 \rightarrow$
   e. $Na + Se \rightarrow$
   f. $Cl_2 + KI \rightarrow$
   g. $I_2 + KCl \rightarrow$
   h. electrolysis of NaCl

13
chemical bonds

THIS WILL SAVE YOU TIME:

Refer to this list as you read through the chapter. In answering the questions, you will understand the important concepts you are most likely to be asked about later on.

After reading this chapter, you should be able to:

1. *Explain the octet rule, and tell what is meant by a noble-gas structure.*
2. *Compare the properties of sodium ions with those of elemental sodium, and explain why they are so different.*
3. *List some common ions with an electronic structure similar to the noble gas xenon.*
4. *Explain the difference between a covalent and an electrovalent bond.*
5. *Give experimental evidence for the statement, "Electrons can have opposite spins."*
6. *Draw the Lewis structures for simple covalent compounds.*
7. *Give an example of a compound with a double bond and one with a triple bond. Show their Lewis structures.*
8. *Explain the meaning of the terms resonance and resonance hybrid.*
9. *Illustrate the resonance forms for a compound or a polyatomic ion using Lewis structures.*
10. *Tell what is meant by electronegativities, and show how they can be used to predict the ionic or covalent character of chemical bonds.*
11. *Explain the meaning of polar molecules and give examples.*
12. *Describe the molecular structure of water.*
13. *Explain what is meant by hydrogen bonding, and draw an illustration of hydrogen bonding in water.*
14. *List four types of solids based on the nature of the bonding forces. Describe these bonding forces, list some typical physical properties associated with them, and give an example for each type of solid.*

When two atoms approach one another, various forces are exerted between them. Some of these forces tend to draw the atoms together, others tend to push them apart. In most atoms, with the major exception of the noble gases, attractive forces outweigh repulsive, and the atoms are drawn together, forming a **bond.** The magnetic and gravitational forces of atoms are too weak to account for this bonding, and we say that the forces are mainly electrostatic in character.

In this chapter, we will look in detail at bonding, one of the most important phenomena in all chemical behavior.

THE OCTET RULE

The theory of chemical bonding followed close on the heels of Bohr's atomic theory. In 1916, Walther Kossel (1888–1956) and Gilbert N. Lewis (1875–1946) proposed quite similar schemes for explaining atomic bonding. Their mutual discovery was the **octet rule,** which says that *atoms interact to change the number of electrons in their outer electronic shells in an attempt to get an electronic structure similar to that of a noble gas.* The noble-gas structure consists of eight electrons in the outermost shell for all elements except hydrogen and helium, where a complete shell consists of only two electrons. Thus, according to the simplest form of the octet rule, the outermost electron shells of most elements tend to gain or lose electrons until the total number of electrons equals eight.

Lewis and Kossel recognized that the noble gases all have full quotas of electrons in their orbitals, and they concluded that this structure accounted for the chemical inertness of these elements. They also concluded that the tendency to achieve electronic structures similar to the noble gases explains chemical bonding in all compounds. There are many exceptions to the octet rule, but the exceptions can be explained by using the general principles of electron interactions, thus buttressing the theoretical ideas behind the rule.

The sodium atom is a good example to use to illustrate the octet rule. Sodium has a low ionization potential and can readily lose its 3*s* electron:

$$\underset{\substack{\text{sodium atom} \\ Na^0}}{1s^2 2s^2 2p^6 3s^1} \rightarrow \underset{\substack{\text{sodium ion} \\ Na^+}}{1s^2 2s^2 2p^6} + {}_{-1}^{0}e$$

The electron structure of the resultant sodium ion is exactly the same as that of the noble gas neon. This ion is a very stable species.

Other elements gain electrons to fill the outer shell to the stable configuration of eight electrons. Chlorine is an example:

$${}_{-1}^{0}e + \underset{\substack{\text{chlorine atom} \\ Cl^0}}{1s^2 2s^2 2p^6 3s^2 3p^5} \rightarrow \underset{\substack{\text{chloride ion} \\ Cl^-}}{1s^2 2s^2 2p^6 3s^2 3p^6}$$

In the case of chlorine, it might appear that there are two ways for the atom to acquire a noble gas structure: gaining one electron or losing seven. The second possibility, however, is not very likely. The ionization potential increases very rapidly with the removal of each additional electron, making the removal of seven electrons energetically unfeasible.

$1s^22s^22p^63s^23p^5 \rightarrow 1s^22s^22p^6 + 7\,{}_{-1}^{\;0}e$

This reaction does not occur.

Table 13–1 lists the noble-gas structure of some common ions.

TABLE 13–1 NOBLE GAS STRUCTURE OF VARIOUS IONS

Electronic arrangement	Noble gas	Ion
$1s^22s^22p^6$	Neon, Ne	Fluoride, F^- Sodium, Na^+ Magnesium, Mg^{2+} Aluminum, Al^{3+}
$1s^22s^22p^63s^23p^6$	Argon, Ar	Sulfide, S^{2-} Chloride, Cl^- Potassium, K^+ Calcium, Ca^{2+}
$1s^22s^22p^63s^23p^64s^23d^{10}4p^6$	Krypton, Kr	Selenide, Se^{2-} Bromide, Br^- Rubidium, Rb^+ Strontium, Sr^{2+}

ELECTROVALENT BONDS

The electrostatic attraction that one charged ionic species has for another results in the type of chemical bond called an **ionic,** or **electrovalent, bond.** Sodium and chlorine atoms unite to form sodium chloride in the following manner:

$1s^22s^22p^63s^1$ +	$1s^22s^22p^63s^23p^5$ →	$1s^22s^22p^6$ +	$1s^22s^22p^63s^23p^6$
sodium atom	chlorine atom	sodium ion	chloride ion
Na	Cl	Na^+	Cl^-

No single molecule of sodium chloride exists; the substance Na^+Cl^- is actually a combination of many sodium ions bound to many chloride ions. In the solid, they are packed so that six sodium ions are around each chloride ion and six chloride ions surround each sodium ion. We saw this structure earlier, in Figure 8–2.

Ions are entirely different species from the parent elemental atoms. Elemental sodium is a very reactive, soft metal, and chlorine is a poisonous, greenish-yellow gas. The sodium and chloride ions, however, have completely different chemical and physical properties. These ions are not poisonous; in fact, they are essential in the human diet.

The bonds of binary compounds formed from the union of the metallic elements of groups I and II with the nonmetals located in the upper right section of the periodic table are primarily ionic in nature. Ionic compounds are also formed from the union of positive metal ions and negative polyatomic ions such as hydroxide, sulfate, nitrate, and phosphate.

COVALENT BONDS

Another way for atoms to acquire a noble-gas configuration is to share electrons. The gas hydrogen chloride will serve as an example (Figure 13–1). The atom of chlorine gains a full complement of electrons by sharing the hydrogen electron, and at the same time, the hydrogen atom attains its full quota of two by sharing one of the electrons from chlorine. This sharing gives both atoms the noble-gas configuration. Irving Langmuir (1881–1957, Nobel prize in chemistry 1932), who extended the earlier work of Kossel and Lewis, called this type of bonding **covalence.**

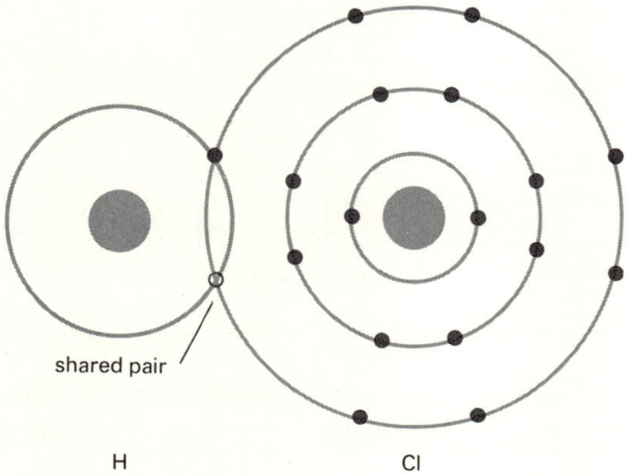

Figure 13–1 Bohr representation of covalent bonding in hydrogen chloride. The indicated pair of electrons, one from hydrogen and one from the chlorine atom, are shared by both atoms to form a covalent bond.

ELECTRON SPIN AND BONDING

There are only two values of the spin quantum number, corresponding to the two spin directions for electrons. In 1921, the famous experiment performed by Otto Stern (1888–1969, Nobel prize in physics 1943) and Walther Gerlach (1889–) provided direct experimental evidence for the existence of two spins. The experiment was based on the law of physics that a moving charged particle creates a magnetic field. The magnetic field of a charged particle (electron) spinning in one direction has north and south poles oriented in a direction opposite to that observed for a similar particle spinning in the opposite direction. Applying this principle, Stern and Gerlach passed silver atoms (as a gas) through a strong magnetic field and were able to observe two species of silver atoms distinguished by the direction of electron spin. All the energy levels of silver atoms are filled except for those containing the outer $5s^1$ electrons. These electrons are unpaired and can have either spin assignment. Both spin species were observed experimentally.

Only electrons of opposite spin can pair. This statement follows from the Pauli exclusion principle: no two electrons in an atom can have the same four quantum numbers.

LEWIS STRUCTURES

The outer-orbital electrons in an atom can be indicated by drawing dots—one for each electron—around the symbol for the element. These representations are called **Lewis structures,** and they serve as convenient tools for illustrating chemical

TABLE 13–2 LEWIS STRUCTURES FOR A NUMBER OF COMPOUNDS
The dots represent the outer-shell electrons, and a pair of dots between atoms indicates the location of a covalent bond. A dash (bond) placed between atomic symbols also indicates a shared pair of electrons.

Compound	Lewis structure	Dash notation
Hydrogen chloride		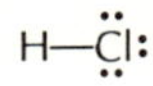
Ammonia	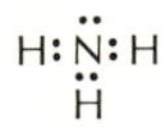	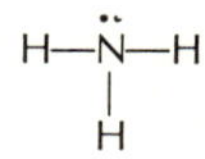
Methane	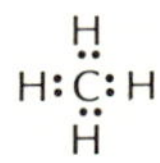	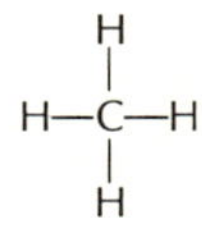
Sodium chloride	Na^+ 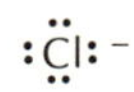	
Ammonium chloride	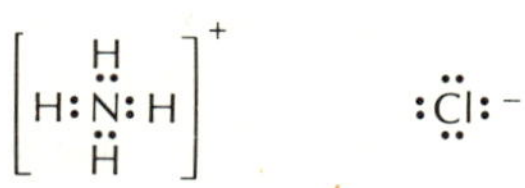	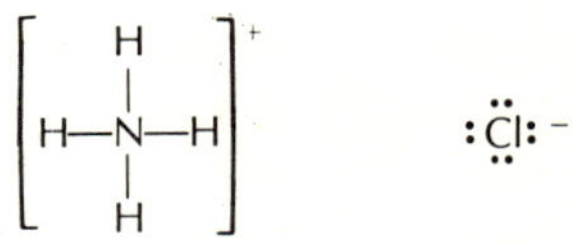

bonding. Another common practice is to indicate a shared pair of electrons by drawing a dash between the symbols of the two atoms forming a covalent bond. Both schemes are shown in Table 13–2.

Example 13–1: Lewis structures *Show the Lewis structure for the covalent compound carbon tetrachloride.*

Carbon has four electrons in its outer shell available for bonding, ·C·, and each chlorine has seven, :Cl:. Grouping the four chlorine atoms around the carbon and assigning a shared electron pair to each C—Cl bond gives the correct representation:

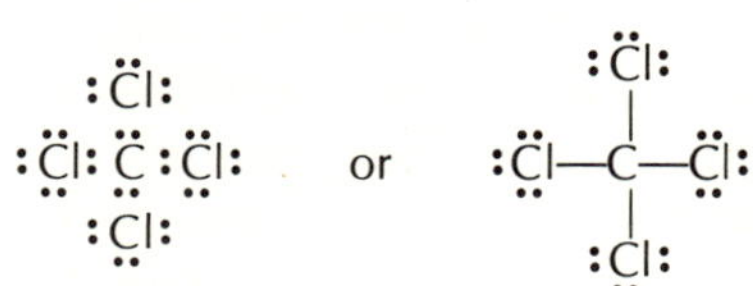

Example 13–2: Lewis structures *Give the notation for the electronic arrangement in the covalent compound dibromoethane (common name, ethylene dibromide),* CH_2BrCH_2Br.

The electrons associated with each atom are: ·C·, ·H, and :Br:. The Lewis structure for the molecule is

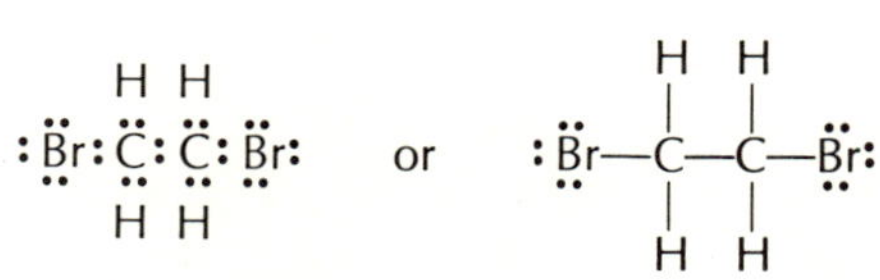

Example 13–3: Lewis structures *Show the Lewis structure for the compound magnesium chloride.*

Magnesium chloride is an ionic compound; magnesium exists as Mg^{2+} ions, and chlorine as chloride ions, Cl^-. The structure can be represented as

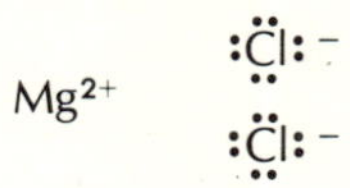

THE CARBON ATOM AND THE STRUCTURE OF METHANE

The carbon-hydrogen and the carbon-carbon bonds are found in practically all organic molecules. As the most common and important types of chemical bonds, they deserve careful examination.

The carbon atom has a total of six electrons, which, according to our rules, would be assigned as:

$1s^2 2s^2 2p^2$

You will recall that Hund's rule orders available electrons assigned to every one of several orbitals of the same type before two electrons are assigned to any one orbital. If we apply Hund's rule, the assignment is

↑ ↓	↑ ↓	↑	↑	
$1s^2$	$2s^2$	$2p_x$	$2p_y$	$2p_z$

In this notation, each box represents an orbital, and the arrows indicate the different spins on the electrons. This assignment, however, shows the presence of only two unpaired electrons. It cannot be correct for carbon atoms in compounds like methane, where four unpaired electrons are needed to form the four covalent bonds.

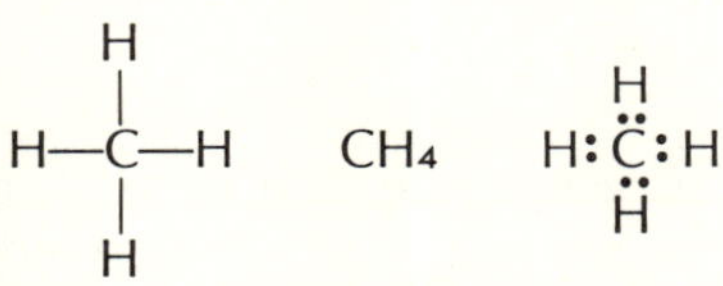

The solution to the apparent dilemma is to say that there must be **hybrid bonds.** The three *p*-orbital electrons and the 2*s* electron are assumed to be equivalent, as indicated by the notation

↑ ↓	↑	↑	↑	↑
$1s^2$	$2sp^3$			

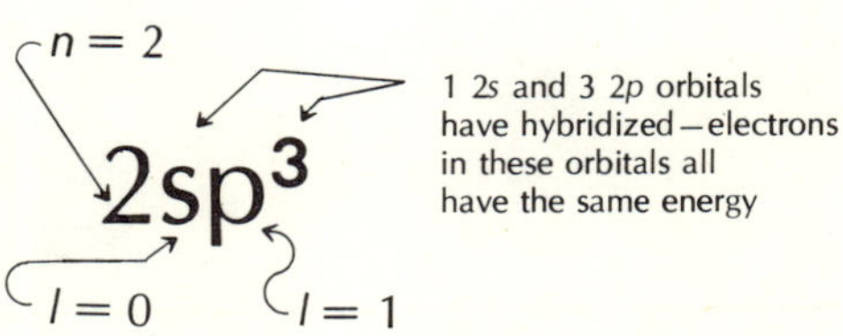

Theoretical calculations show that if the 2*s* and the three 2*p* electrons of carbon are assumed to be hybridized, the resultant sp^3 hybrid orbitals should look like Figure 13–2. The orbitals are directed toward the four corners of a regular tetrahedron and the angles made by the bonds are 109°28′. Each of the sp^3 orbitals overlaps with the *s* orbital of a hydrogen atom, an arrangement that accounts for the structure and geometry of methane. The structure of methane can be represented by the drawings shown in Figure 13–3.

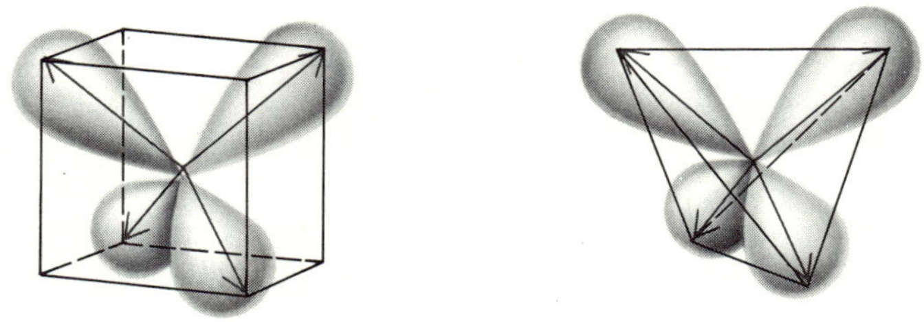

A
bond angles 109° 28′

B
bond angles 109° 28′

Figure 13–2 Spatial arrangement of sp^3 hybrid orbitals. The bonds are directed to the four corners of a regular tetrahedron, **B.** Figure **A** shows the positioning with respect to a perfect cube.

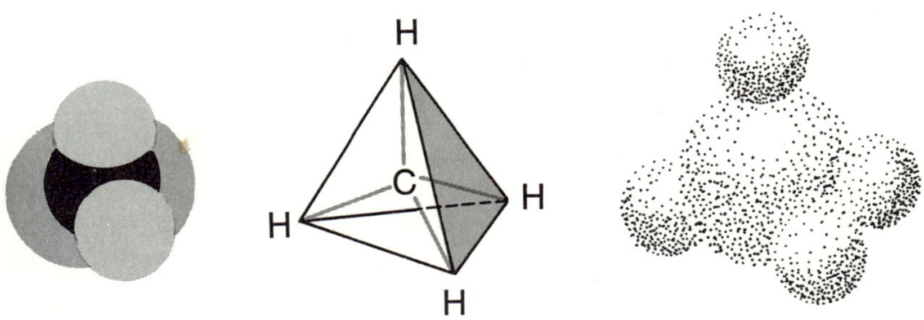

Figure 13–3 Some methods for representing the tetrahedral structure of CH_4.

Methane's tetrahedral structure can also be explained in terms of electron-pair repulsions. Each electron pair in a given system is repelled by the other electron pairs. When four electron pairs are located the same distance from a central nucleus, repulsion pushes each pair into the position that puts it the greatest possible distance from the other pairs. This arrangement is a regular tetrahedron.

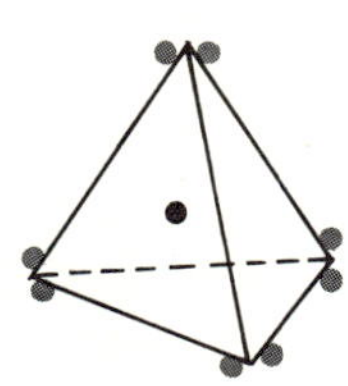

MULTIPLE BONDS

Many atoms share more than one pair of electrons to achieve a noble-gas configuration. When two pairs of electrons are shared, the bond is called a **double bond;** the bond resulting from the sharing of three pairs is called a **triple bond.**

Ethene (common name, ethylene) is an example of a compound with a double bond. To achieve an octet of electrons around each carbon atom in C_2H_4, we must draw the following structure:

```
H:C::C:H      or      H—C=C—H
  ¨  ¨                  |  |
  H  H                  H  H
```

Nitrogen gas is an example of a compound with a triple bond. The structure of each nitrogen atom is

$1s^2 2s^2 2p_x^1 2p_y^1 2p_z^1$

and the triple bond is formed by the pairing of the three p electrons of one nitrogen atom with the three p electrons of opposite spin from the second nitrogen atom.

	$1s^2$	$2s^2$	$2p_x$	$2p_y$	$2p_z$	
Nitrogen atom 1	↑↓	↑↓	↑	↑	↑	↖ these electrons
Nitrogen atom 2	↑↓	↑↓	↓	↓	↓	↙ pair
	$1s^2$	$2s^2$	$2p_x$	$2p_y$	$2p_z$	

The Lewis structure is

:N:::N: or   :N≡N:

Example 13–4: Multiple bonds *Draw the Lewis structure for the compound ethyne (common name, acetylene),* C_2H_2.

This compound has a triple bond. The structure is

H:C:::C:H   or   H—C≡C—H

RESONANCE

Figure 13–4 shows three possible ways of writing the Lewis structure of the nitrate ion. Because it is possible to position the double bond between the nitrogen and any one of the three oxygen atoms, the question arises as to whether there are actually three different species of nitrate ion. We can answer this question by studying the bond length, the distance between two atoms. The interatomic distance for the N=O bond is always shorter than the N—O bond length. A measurement of the bond lengths in the nitrate ion shows that none of the bonds is double

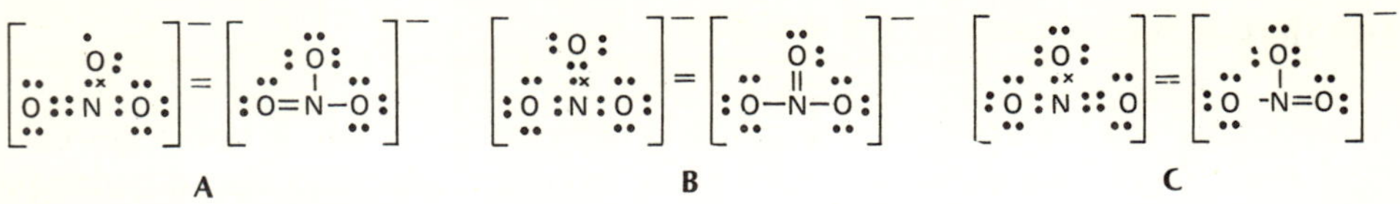

Figure 13–4 Resonance hybrid structures for the nitrate ion in pairs **A**, **B**, and **C**. The electrons originally associated with oxygen are indicated by black, those from nitrogen by gray, and the added electron that gives the negative charge by x. These assignments have no meaning, but they assist in the bookkeeping process.

and none is single. The actual bond length lies in between that for a single and that for a double bond. This observation is explained by a **resonance hybrid,** a *structure in which each of the illustrated forms contributes to the actual structure of the molecule.* The term **resonance** is used for *the situation where more than one Lewis structure can be drawn for a molecule.* When you are drawing these structures, however, remember that only electrons can be relocated: the nuclei of the atoms in each structure must remain in fixed positions. The situation in the nitrate ion might be likened to three cans of paint; a can of black paint representing the double bond, and two cans of white paint representing the single bonds. When the paint is mixed, the result is three cans of gray paint, the gray corresponding to the resonance hybrid bond.

In the Lewis structures for the nitrate ion, Figure 13–4, note that the same atom, nitrogen, contributes the shared pair of electrons in some of the bonds. Nevil Vincent Sidgwick (1873–1952) gave the name **coordinate covalence** to this form of bonding, *in which one atom apparently contributes both of the shared electrons.* We know, however, that none of the bonding electrons can be distinguished from one another and that it is incorrect to try to make assignments as in Figure 13–4. Nevertheless, these diagrams do show how the total number of electrons can be accounted for and that the octet rule is obeyed.

The resonance theory serves as a useful tool: the more Lewis structures one can write for a molecule, the more stable the structure appears to be. A certain amount of bonding energy associated with these resonance hybrids seems to give extra stability to the molecules.

ELECTRONEGATIVITY

Every element has been assigned a positive number, called the **electronegativity,** that *shows an atom's ability to attract and hold the bonding electrons.* Linus Pauling (1901–, Nobel prize in chemistry 1954, Nobel peace prize 1963) developed an electronegativity scale (Table 13–3). He assigned the number 4 to fluorine, the element with the greatest power to attract and hold electrons—that is, the most electronegative element. Cesium and francium, the elements that hold electrons most weakly, have the lowest value, 0.7. Pauling developed the scale from a study of the amount of energy needed to break a bond between pairs of elements. The cesium fluoride bond, for example, is one of the most difficult to break; in other

TABLE 13–3 ELECTRONEGATIVITIES OF THE ELEMENTS

H 2.1																	He
Li 1.0	Be 1.5											B 2.0	C 2.5	N 3.0	O 3.5	F 4.0	Ne
Na 0.9	Mg 1.2											Al 1.5	Si 1.8	P 2.1	S 2.5	Cl 3.0	Ar
K 0.8	Ca 1.0	Sc 1.3	Ti 1.5	V 1.6	Cr 1.6	Mn 1.5	Fe 1.8	Co 1.9	Ni 1.9	Cu 1.9	Zn 1.6	Ga 1.6	Ge 1.8	As 2.0	Se 2.4	Br 2.8	Kr
Rb 0.8	Sr 1.0	Y 1.2	Zr 1.4	Nb 1.6	Mo 1.8	Tc 1.9	Ru 2.2	Rh 2.2	Pd 2.2	Ag 1.9	Cd 1.7	In 1.7	Sn 1.8	Sb 1.9	Te 2.1	I 2.5	Xe
Cs 0.7	Ba 0.9	La-Lu 1.0-1.2	Hf 1.3	Ta 1.5	W 1.7	Re 1.9	Os 2.2	Ir 2.2	Pt 2.2	Au 2.4	Hg 1.9	Tl 1.8	Pb 1.9	Bi 1.9	Po 2.0	At 2.2	Rn
Fr 0.7	Ra 0.9	Ac-Lr 1.1-1.4	(Rf) (Ku or Kh)	(Ha)													

words, separating cesium ions from fluorine ions in the compound CsF requires more energy than that needed to separate other chemical bonds. As you might expect, electronegativities are related to first ionization potentials. Cesium has the lowest electronegativity and also the lowest ionization potential, whereas fluorine has both the highest electronegativity and the highest ionization potential (excluding the noble gases helium and neon).

One of the uses of the electronegativity scale is in predicting the type of chemical bonds formed between atoms. The greater the difference in the electronegativities of two elements, the greater the electrostatic attraction the atoms have for each other. Consider the compound KCl as an example. It is possible to write both an ionic and a covalent formula:

$$K^{+}:\ddot{\underset{..}{Cl}}:^{-} \qquad K:\ddot{\underset{..}{Cl}}:$$

Because the difference in electronegativities of these two elements is quite large, 2.2, a shared pair of electrons would be expected to be much closer to the chlorine than to the potassium. Chlorine's attraction for the electron pair is so great that the bond is primarily ionic, and the electron from potassium is almost completely associated with the chloride ion.

When the difference in electronegativities is less than 1.7, the bond is primarily covalent. The bond is always covalent in substances such as bromine, Br_2, where both atoms have exactly the same electronegativity.

The bond between most elements is neither completely ionic nor completely covalent, and it is correct to speak of many chemical bonds as partially ionic or partially covalent. One can even go so far as to assign a percent figure to the ionic and covalent character of a compound. Figure 13–5 gives a plot of the percent ionic character of a single chemical bond as a function of the difference in the electronegativities of the two bonded atoms. Remember, however, that the ionic and covalent classifications represent two absolute models for bonding. Real bonding in most molecules lies somewhere in between these absolute bounds.

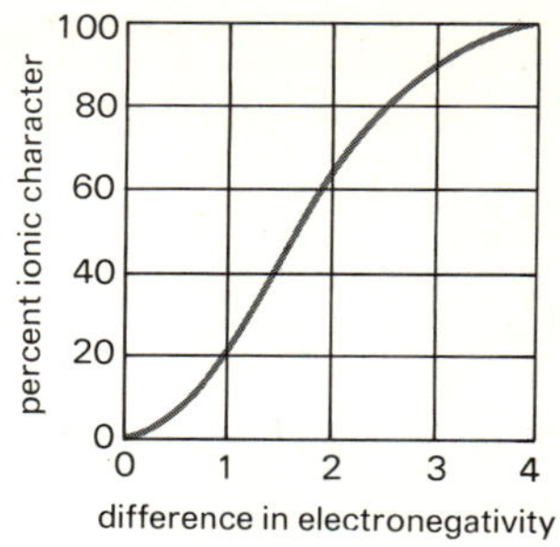

Figure 13–5 Percent ionic character of a single bond plotted as a function of the difference in electronegativities of the two bonded atoms.

Example 13–5: Electronegativities and type of chemical bond *What percentage of the bond in the compound potassium chloride is ionic, and what percentage is covalent?*

The difference in electronegativities of the two elements is 2.2. Reference to Figure 13–5, shows that this bond is expected to be about 70 percent ionic and 30 percent covalent. KCl is therefore considered an ionic compound.

The differences in electronegativities for the hydrogen halides are shown in Table 13–4. The ionic character of the bonds changes from the more ionic HF (45 percent ionic) to the covalent HI (4 percent ionic) as the electronegativity difference decreases from 1.9 to 0.4. Because the electronegativities are based on bond strengths, it is apparent that the H—I bond is easier to break than the H—F bond. These bond strengths are also listed in Table 13–4.

TABLE 13–4 ELECTRONEGATIVITY DIFFERENCES, IONIC AND COVALENT CHARACTER, AND BOND STRENGTHS FOR THE HYDROGEN-HALIDE BONDS

Bond	Electro-negativity difference	Percent ionic character	Percent covalent character	Bond strength (kcal/mole)
H—F	1.9	45	55	135
H—Cl	0.9	17	83	103
H—Br	0.7	12	88	87
H—I	0.4	4	96	72

POLAR MOLECULES

When a chemical bond forms between two like atoms, the bond is a **nonpolar bond.** The bonds formed between unlike atoms, where there is a difference in electronegativity, are called **polar bonds.** In polar bonds, there is a partial separation of the electronic charge. Hydrogen bromide will serve as an example. From Table 13–4 it is seen that the H—Br bond is 12 percent ionic and 88 percent covalent as a result of the 0.7 difference in electronegativities. Because bromine attracts electrons more strongly than hydrogen, the bonding electrons are most likely located closer to the bromine nucleus than to the hydrogen nucleus. This arrangement can be illustrated by the Lewis structure:

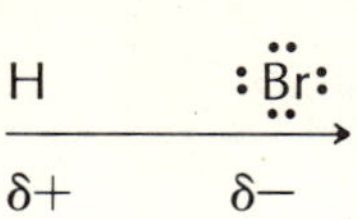

(the δ [delta] notation signifies a partial charge separation)

This compound is not ionic, yet the greater attraction that bromine has for the electrons creates an unbalanced electron distribution, making the bromine side of the molecule a little more negative than the hydrogen side. These electrically unbalanced molecules are called **polar molecules,** and the extent of the partial charge is spoken of in terms of the **dipole moment** of the molecule. Molecules with a dipole moment—polar molecules—have an electrical force of attraction between them analogous to the magnetic force between magnets. This phenomenon is most important in explaining the chemical and physical behavior of compounds. As we will see, water owes its remarkable properties to its large dipole moment; in other words, it is a very polar molecule.

Example 13–6: Polar molecules *Methanol (methyl alcohol), CH_3OH, molecules are polar. Explain.*

The Lewis structure for methanol is

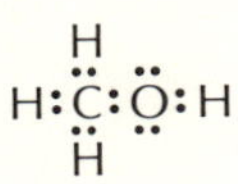

The electronegativity of carbon is 2.5 and that of oxygen is 3.5; hence the oxygen atom attracts the shared pair of electrons more strongly. This imbalance produces a polar molecule.

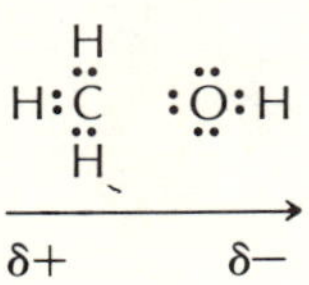

THE MOLECULAR STRUCTURE OF H_2O

The water molecule is not linear: the angle between the O–H bonds is not 180° and the atoms do not lie in a straight line. The actual angle between the bonds has been shown by x-ray diffraction methods to be 104.7° (Figure 13–6**A**). The oxygen atoms have a greater electronegativity than the hydrogen atoms, and as a result, the electrons are pulled toward the oxygen, as shown in Figure 13–6**B.** This imbalance of charge makes the oxygen atom a little more negative than the hydrogen atoms. The charge separation is designated by the delta plus and delta minus signs to indicate that the charge separation is only partial and that the molecule is not an ion. Water molecules have a permanent dipole moment (Figure 13–6**C**). Water's outstanding ability to dissolve other substances, particularly ionic salts, probably results from these dipoles.

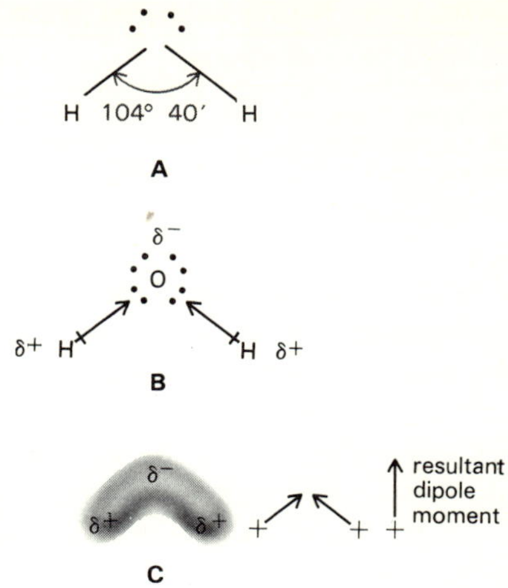

Figure 13-6 Diagrams illustrating the presence of a dipole moment in the water molecule.

HYDROGEN BONDING

Since not all of the electron pairs in the oxygen atom of a water molecule form covalent bonds, there is a force of attraction between these non-bonding electrons of the oxygen atom of one water molecule and the hydrogen atoms of other water molecules. This effect is illustrated in Figure 13–7. These attractive forces are much

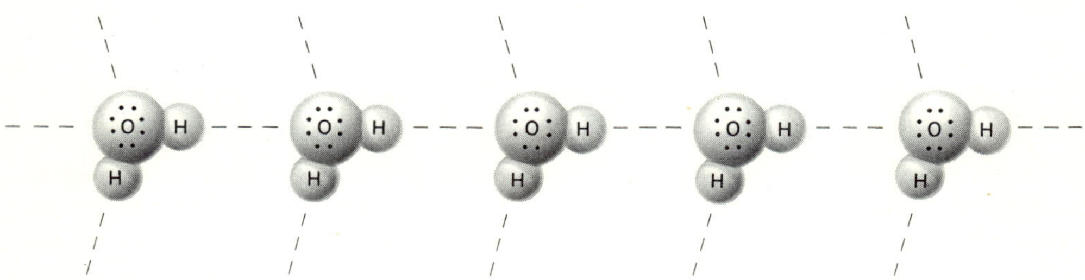

Figure 13–7 Hydrogen bonding between water molecules.

weaker than regular covalent or electrovalent bonds. This type of bond, called a **hydrogen bond,** is formed only between hydrogen and small electronegative atoms such as fluorine, oxygen, and nitrogen. The molecule HF is another good example of a compound with hydrogen bonding.

TYPES OF BONDING IN SOLIDS

Solids can be divided into four different classes based on the type of bonding forces that hold the structural units together. These types are summarized in Table 13–5.

TABLE 13-5 SOLIDS CLASSIFIED BY BONDING FORCE

	Ionic	Covalent	Molecular	Metallic
Structural units that occupy the lattice sites	Ions	Atoms	Molecules	Positive ions
Bonding force	Ionic or electrostatic	Sharing of electron pairs	Molecular interaction forces	Electrostatic attraction between positive ions and "sea of electrons"
Melting points	Relatively high	Very high	Very low	Moderate to high
Hardness	Quite hard and brittle	Very hard	Very soft	Variable
Electrical conduction	Very low	Very low	Very low	Good conductors
Examples	NaCl AgBr CaO	C (diamond) SiC (carborundum) SiO_2(quartz)	CO_2 N_2 CCl_4	Fe Cu Au

Ionic solids

Sodium chloride is an example of an ionic solid. The main forces holding the crystal together are ionic bonds between positive sodium and negative chloride ions. Other compounds with the same crystalline form as sodium chloride are the other halides of sodium, NaBr, NaF, NaI; the halides of lithium; the halides of potassium; also CaO, FeO, AgBr, MgO, and PbS.

All ionic solids are quite hard and brittle. They do not conduct an electric current; they all have relatively high melting points. Some examples are LiF, 870°C; KCl, 368°C; and NaBr, 755°C.

Covalent solids

The structural unit in covalent solids is the atom. These atoms are bound to other atoms by shared electron pair bonds, that is, by covalent bonds. Covalent solids are characterized by their very high melting points and extreme hardness. They do not conduct an electric current. Examples are quartz (SiO_2), diamond, and carborundum (SiC).

The cubic unit cell of the diamond lattice is illustrated in Figure 13-8. Each

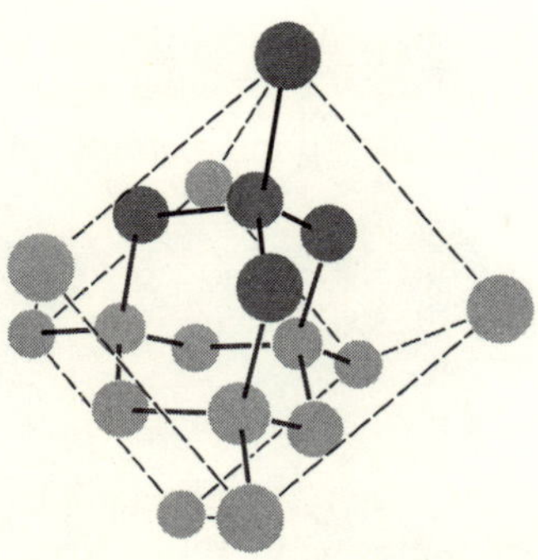

Figure 13-8 The diamond crystal.

carbon atom is bonded tetrahedrally to four other carbon atoms. The type of bonding in diamonds is identical to the single covalent bonds found between carbon atoms in organic compounds. Similar covalent structures are formed by other elements of the fourth periodic group: germanium, tin (grey form), and silicon. The compounds ZnS, AgI, and SiC also have tetrahedral structures.

Molecular solids

The structural units in molecular crystals are molecules. In these solids, the bonding forces between the atoms are much stronger than the weak interaction forces between the molecules. Examples of molecular crystals are solid nitrogen, sugar, carbon dioxide, and carbon tetrachloride. The binding forces in these molecular solids are the same as the intermolecular forces responsible for the deviation of gases from ideal behavior. One such force results from dipole-dipole interactions. Molecules with no permanent dipoles, such as those listed above, may have dipoles induced temporarily as a result of their nearness to other molecules. For our purposes, we need simply recognize that weak forces between the molecules of these molecular solids do exist.

The bonds of molecular solids are extremely weak, making the compounds very soft and their melting points low. They do not conduct an electric current.

Metallic solids

Still another type of chemical bond is needed to account for the structure and properties of metallic solids. One property of metals that distinguishes them from other types of solids is their ability to conduct an electric current. This property implies that electrons must be able to migrate freely throughout the structure. To explain this migration, we describe the metallic bond as positive ions bound together by a "sea of electrons." The electrons lost from the outer shells of the atoms are free to move throughout the structure. The resulting bonding force comes from the cementing effect of the mobile electron cloud, the attractive forces between the positive ions and the "sea of electrons." A two-dimensional representation of a metal composed of doubly charged ions is shown in Figure 13–9**A.**

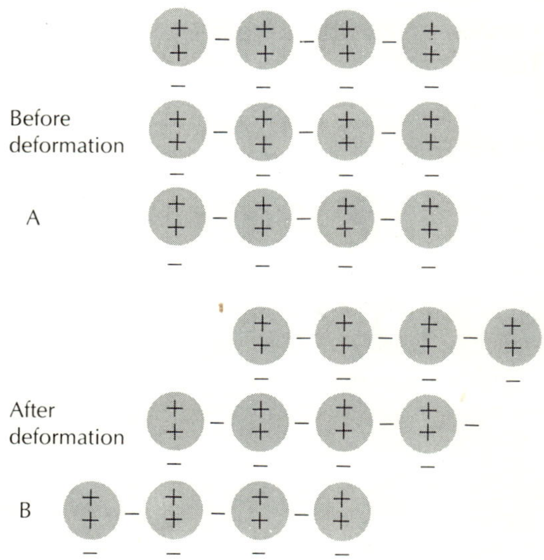

Figure 13–9 Two-dimensional representation of a metallic crystal. The movement of ions in the metallic solid produces no change in the nature of the bonding forces. This model explains the malleability and ductility of metals.

The mobile electron picture of metals also accounts for the malleability and ductility of metals. A metal sample can be hammered, rolled, drawn, and worked mechanically without destroying its integrity. In contrast, ionic crystals shatter when struck with a hammer. If the metal in Figure 13–9**A** is subjected to a stress, the metal can be deformed by the movement of ions to new positions. As shown in Figure 13–9**B,** the bonding forces in the new position after deformation are similar to those before.

NEW TERMS

Coordinate covalence: covalent bonds in which both shared electrons are furnished by the same atom.

Covalent bond: chemical bonds formed by the sharing of electron pairs.

Dipole moment: the measure of the extent of charge separation in a polar molecule.

Double bond: a covalent bond in which two pairs of electrons are shared.

Electronegativity: a measure of an atom's ability to attract and hold bonding electrons.

Electrovalent bond: an ionic bond, the type of chemical bond that results from the attraction of positive and negative ions.

Hydrogen bond: bonds formed between hydrogen atoms and fluorine, oxygen, or nitrogen atoms. They are weaker than ionic or covalent bonds.

Ionic bond: an electrovalent bond.

Lewis structure: representations showing the outer orbital electrons in atoms. They are convenient for illustrating chemical bonding.

Octet rule: the tendency for atoms to acquire eight electrons in their outer electronic shells when they combine chemically. (There are numerous exceptions.)

Polar bond: bonds between atoms with different electronegativities.

Polar molecule: a molecule in which the center of positive charge does not coincide with the center of negative charge.

Resonance hybrid: a molecule or ion to which no single electron formula can be assigned. The actual structure is a hybrid of two or more contributing structures.

Triple bond: a covalent bond in which two atoms share three pairs of electrons.

PROBLEM SET A

1. Use equations to illustrate the formation of a calcium ion and a fluoride ion from their respective atoms.
2. Name an element other than silver that might be suitable for a Stern-Gerlach experiment. Give its electron configuration.
3. Give Lewis structures for the following:
   CH_4, CO_2, $MgCl_2$, Cl_2, NH_3, NH_4^+, CF_4.
4. The difference in electronegativities between the elements in the compound MX is 1.2. Is this compound ionic or covalent? Explain.
5. Arrange the following compounds in the order of increasing ionic character of their bonds:
   NaCl, KBr, MgF_2, HI, CaS.

6. One resonance form for the benzene molecule, C_6H_6, is

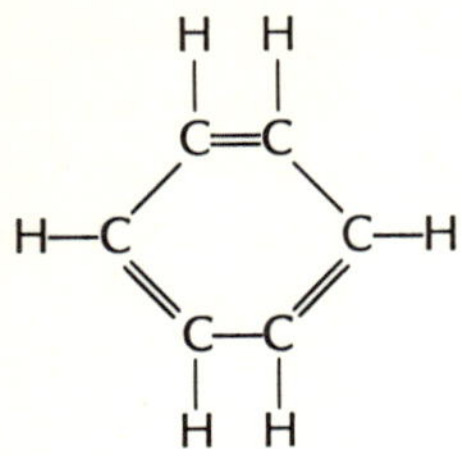

Draw a Lewis structure for another possible resonance structure.

7. Draw three resonance structures for the carbonate ion, CO_3^{2-}.
8. The molecule CCl_4 is not polar, but HOH is. Can you explain these facts?

PROBLEM SET B

1. Illustrate by equations the formation of a potassium ion and a sulfide ion from their respective atoms.
2. In general, would metals or nonmetals have higher first ionization potentials? Why?
3. Give Lewis structures for the following: C_2H_6, SO_4^{2-}, Na_2S, CO_2, and N_2.
4. The molecule CF_4 has polar bonds but the molecule itself is nonpolar. How can this be? (Hint: is the molecule symmetrical?)
5. Arrange the following compounds in the order of increasing ionic character of their bonds: $CaCl_2$, F_2, K_2S, RbI, and HCl.
6. Suggest why molecules with more than one acceptable Lewis structure are more stable than might be predicted.
7. Draw all of the resonance structures for the infamous pollutant molecule sulfur dioxide, SO_2.
8. In a certain experiment, acetic acid, CH_3CO_2H, was determined to have a molecular mass of approximately 120. How can this be explained?

14

solutions

THIS WILL SAVE YOU TIME:

Refer to this list as you read through the chapter. In answering the questions, you will understand the important concepts you are most likely to be asked about later on.

After reading this chapter, you should be able to:

1. *Distinguish among a mixture, a solution, and a pure compound.*
2. *Give an example of each of the various types of solutions based on the states (solid, liquid, or gas) of the mixture's components.*
3. *Explain the meaning of the term solution of electrolytes.*
4. *Describe what happens when a compound like sodium chloride dissolves in water.*
5. *Explain the meaning of the terms saturated solution, solvation, and miscible.*
6. *Distinguish between solubility and rate of solution.*
7. *List the factors that affect the rate of solution of a solute in a solvent.*
8. *Discuss the factors that affect the solubility of a solid in a liquid.*
9. *Explain how temperature affects the solubility of different solids in liquids.*
10. *Tell how pressure changes affect the solubility of gases in liquids.*
11. *Solve problems involving molarity of solutions.*
12. *Explain what is meant by the colligative properties of solutions.*
13. *Tell what changes occur in vapor pressure, freezing point, and boiling point when a completely soluble nonvolatile solute is added to water.*
14. *Calculate the molecular mass of a compound from freezing-point data.*

In Chapter 3, we looked briefly at the two basic types of mixtures. In heterogeneous mixtures like granite, the component parts can be seen by the naked eye alone, yet not even the most powerful microscope can detect the components of a homogeneous mixture like seawater. Seawater is indeed a mixture, however, for when it evaporates, a residue of various compounds remains behind. But how do we know that this residue is evidence of a mixture and not simply the leftovers of a decomposed compound? Seawater is homogeneous—as a compound must be—but the proportions of its components are not constant—as a compound's must be. A large portion of a sample of seawater may evaporate, and the sample will remain homogeneous, even though the proportion of water to residual matter has been altered greatly. Such *a homogeneous mixture, whose composition can vary continuously between certain limits,* is called a **solution.** In terms of chemical theory, the difference between a solution and a compound is that the law of definite composition is true of compounds, but not of solutions.

Even the purest substances are actually mixtures or solutions. In fact, no absolutely pure compound exists. A sample of iron, for example, might be 99.99999 percent pure, but it still contains 0.00001 percent impurities. In a 1-g sample of this iron, there are $6.02 \times 10^{23}/55.8$ or 1.08×10^{22} atoms. If 0.00001 percent of these are impurities, the number of foreign atoms amounts to 1.08×10^{15}—a very large number.

In this chapter, we shall discuss the various types of solutions, their properties, and the methods for expressing their concentrations.

TYPES OF SOLUTIONS

When components are mixed to form a solution, *the substance present in the greater amount* is termed the **solvent;** *the other components* are called the **solutes.** For example, soda pop is a solution of carbon dioxide, coloring agents, and flavoring materials (sugar) in water. The water is the solvent, and the carbon dioxide, sugar, and coloring agents are solutes. In 90-proof vodka (45 percent alcohol in water) the water can be considered the solvent and the alcohol the solute; in 100-proof vodka (50 percent alcohol), either component can be considered the solute or solvent.

In any solution, the state of the solvent is the state of the solution. Our soda pop example above is a mixture of a liquid (water), a solid (sugar), and a gas (carbon dioxide). Since the water is the solvent, the solution is a liquid as well.

The three states of matter, solid, liquid, and gas, can be combined in nine different ways to form binary (two-component) mixtures shown in Table 14–1. Since all gases are soluble in each other, air and all other gas mixtures are properly called solutions. Solids and liquids, however, cannot dissolve in gases. On evaporation and sublimation, the vapor molecules of liquids and solids form solutions with air molecules, but one does not say that solids and liquids as such are soluble in gases.

Many **alloys** are solutions of solids in solids. Brass, a solid solution of copper and zinc, is a good example. Gold with silver and cobalt with nickel are two other combinations of metals that mix in all proportions to form solid solutions.

The classic example of a solution of a liquid in a solid is mercury metal in gold or silver. These solutions are called **amalgams** and are extensively used in dental work for repairing the cavities left by tooth decay.

Examples of gases dissolving in solids are not common; usually a relatively small amount of gas is taken up by the surface of a solid, and the result is not con-

TABLE 14–1 TYPES OF SOLUTIONS

Components		
Solute phase	Solution	Examples
Gas	Gas	All mixtures of gases, air
	Liquid	Carbonated beverages
	Solid	Hydrogen in palladium
Liquid	Gas	–
	Liquid	Alcoholic beverages (alcohol and water)
	Solid	Amalgams, mercury in gold
Solid	Gas	–
	Liquid	Sugar in water
	Solid	Alloys, brass, silver in lead

sidered a solution. Metallic palladium, however, can attract and hold large amounts of hydrogen; when palladium is heated to a dull-red color, it can absorb about 1000 times its own volume in hydrogen. The hydrogen breaks down into hydrogen atoms, which fit into the spaces between the palladium crystal layers. You might ask whether this mixture is really a solution, because the hydrogen molecules are chemically broken down into hydrogen atoms. The hydrogen is easily recovered as molecular hydrogen, however, and the definition of solutions is flexible enough to cover this case. Palladium's ability to dissolve hydrogen can be used in the purification of hydrogen. If impure hydrogen is placed under pressure on one side of a heated palladium thimble, pure hydrogen will be given off from the palladium on the other side, where the pressure is lower. Palladium does not dissolve gases other than hydrogen.

ELECTROLYTES AND NONELECTROLYTES

To the chemist, liquid solutions are of the most interest, and aqueous—water-solvent—solutions are the most important of all. Aqueous solutions can be divided into two classes: those that conduct an electric current are called **ionic solutions,** or **solutions of electrolytes;** those that do not conduct an electric current are **nonelectrolytes.** We will take a closer look at ionic solutions in the next chapter.

WHY DO COMPOUNDS DISSOLVE?

Many solution processes are complex and difficult to explain. The essential phenomenon behind all solution processes, however, is that the mixture of different species gives rise to various attractive and repulsive forces whose net result is solution.

Figure 14–1 shows the solution of sodium chloride crystals in water. Positive sodium and negative chloride ions on the surface of the sodium chloride crystal attract the polar water molecules. The loose bonds formed between the water dipoles and the ions are strong enough to pull the attracted ions away from the surface of the crystal. These ions then go into solution as aggregates (clusters) of ions surrounded by a variable number of water molecules. The process is called **solvation,** and the ions are said to be **solvated.** Solvation is *the interaction of solvent molecules with solute*

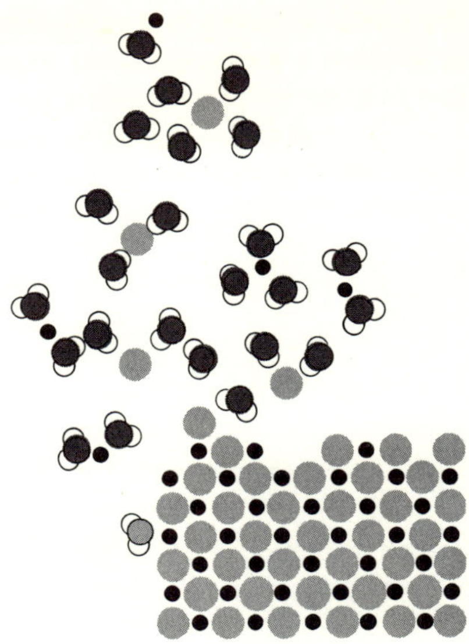

Figure 14–1 Solution process for sodium chloride in water. Polar water molecules attract ions and pull them away from the crystal to form clusters of solvated ions.

molecules, atoms, or ions to form aggregates in solution. Because of molecular motion, the solvated ions soon diffuse throughout the solution, and more and more of the crystal dissolves. If there is an excess of solid, an equilibrium is eventually reached where the rate at which molecules leave the surface equals the rate at which they return to the crystal, and we say that the solution is saturated. *A solution that contains undissolved solute in equilibrium with the solution at the given temperature and pressure* is a **saturated solution.**

Solvation also makes it possible for many nonionic compounds to be dissolved. Ordinary table sugar, $C_{12}H_{22}O_{11}$, consists of nonionic molecules that contain many —OH attachments. The bonds in these —OH groups are polar because of the same type of charge separation as in the water molecule. When a sugar crystal is placed in water, the polar water molecules orient themselves so that they form hydrogen bonds with the sugar molecules and they pull the large sugar molecules away from

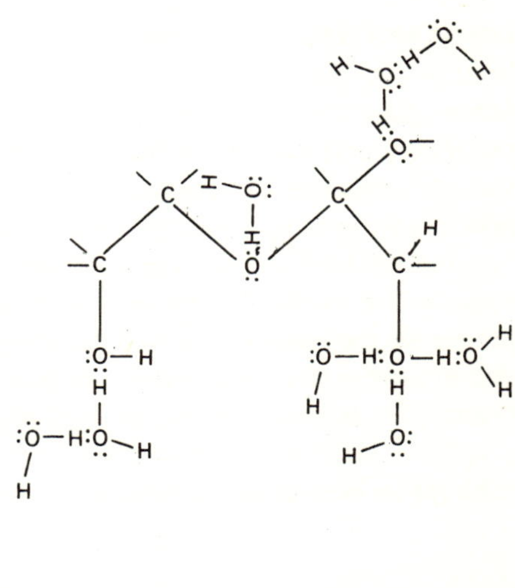

Figure 14–2 Diagram of a portion of a sucrose (table sugar) molecule, showing the formation of hydrogen bonds with water molecules.

the crystal to form solvated (hydrated) aggregates (Figure 14–2). Each sugar molecule in solution is surrounded by a large cluster of water molecules, and these aggregates can migrate freely throughout the solution.

Some gases are partially soluble in water; ammonia and carbon dioxide are examples. The aqueous solutions of these gases will conduct electric current; hence we know that they contain ions. Both gases react to some extent with water:

$$NH_3(g) + H_2O(l) \leftrightharpoons NH_3(aq) \leftrightharpoons NH_4^+ + OH^-$$
$$CO_2(g) + 2H_2O(l) \leftrightharpoons CO_2(aq) \leftrightharpoons [H_2CO_3(aq)]^* \leftrightharpoons H_3O^+ + HCO_3^-$$

The solutions contain many molecules of hydrated NH_3 and CO_2, but the chemical reactions account in part for the solubility. Both gases are recovered unchanged when these solutions are heated.

Metals are often said to dissolve in acids. It is more accurate to say that many metals react with acids; for example,

$$Zn(s) + \underset{\text{hydrochloric acid}}{2HCl(aq)} \rightarrow Zn^{2+} + 2Cl^- + H_2(g)$$

When the resulting solution evaporates, the original zinc is not recovered, but the ionic compound zinc chloride, $Zn^{2+}(Cl^-)_2$, is obtained.

SOLUBILITY AND THE RATE OF SOLUTION

Some solutes and solvents are completely soluble in one another; that is, they are **miscible** in all proportions. Ethyl alcohol and water are miscible in all proportions, and all gases are completely miscible in one another. Other solutes, however, may be able to dissolve only to a certain definite amount in a given solvent at a given temperature. When sugar is continuously added to coffee, a point is reached where no more sugar dissolves and the additional sugar accumulates as a solid at the bottom of the cup. We say that the coffee is **saturated** with sugar. At the boiling point of water, 487.2 g of sugar can dissolve in 100 g of water to give a solution that is 83.0 percent sugar by mass. For the sake of comparison, a saturated solution of sodium chloride can hold only 39.8 g of salt per 100 g of water at the boiling point, only 28.5 percent by mass.

The amount of solute a solvent is able to dissolve must not be confused with the rate at which the solute goes into solution. The amount of solute a given solvent can hold at a given temperature is a constant for that substance, but the rate at which a solute goes into solution varies with the experimental conditions. A solid lump of sugar dropped into the bottom of a cold-drink glass dissolves very slowly. The rate of solution depends on the rate at which the solvent molecules are able to diffuse to and from the solid surfaces. Mechanical stirring of the mixture speeds solution because it brings fresh solvent molecules to the surface of the solid more rapidly. The sugar will also dissolve more rapidly if it is added in the form of a fine powder which increases the area of the solid surfaces available to the molecules of the solvent. Warming the drink will hurry solution of the sugar; the greater the temperature, the higher the average kinetic energy of the solvent molecules, and the more rapidly

* This species exists in low concentrations.

the molecules can move to and from the surfaces of the solid. Thus temperature too affects the rate of solution.

FACTORS AFFECTING SOLUBILITY

The nature of the solvent and solute are very important in determining the solubilities of compounds. The old adage that like dissolves like applies to many substances. Gasoline, which is a mixture of many nonpolar hydrocarbons, is a good solvent for other nonpolar organic molecules. Similarly, the polarity of water molecules makes water an excellent solvent for other polar molecules and for ionic solids.

Water is not a solvent for gasoline, and vice versa. A two-layered system results when one is added to the other, and we say that they are **immiscible.** Actually, the gasoline is slightly soluble in the water, and the water is slightly soluble in the gasoline. The combination produces a saturated solution of water in gasoline at the top and a saturated solution of gasoline in water at the bottom.

Temperature is one of the most important factors affecting solubility. In general, solids are more soluble in hot than in cold solvents; however, with certain substances the reverse is true. Figure 14–3 plots the solubilities of several substances in water as a function of temperature.

An energy change always accompanies the solution of a crystalline solid. The structure of the crystal must first be broken down. This process, step I, always requires the addition of energy. The second energy change, step II, occurs when the solution process (solvation) takes place. These steps can be better described by equations:

Step I: crystalline solid + energy → crystal fragments
Example: NaCl(*s*) + energy → separate Na^+ and Cl^- ions
(endothermic process)
Step II: crystal fragments + solvent molecules → solution complexes + energy
Example: Na^+ + Cl^- + water dipoles → solvated ions + energy
(exothermic process)

The influence of temperature on the solubility of a salt in a solvent depends on the extent of the energy changes involved in these two processes. In the case of NaCl, the heat needed for step I is just about balanced by the energy produced in step II, and there is little variation in solubility with temperature.

If the overall solution process is exothermic,

solute + solvent → solution + heat

then the energy liberated in step II is more important. Adding heat, a product of the reaction, will usually make the reaction occur less readily, and the solubility will thus tend to decrease with an increase in temperature.

If the process is endothermic,

solute + solvent + heat → solution

the addition of heat increases the solubility. The solution of most solids in liquids follow this endothermic pattern.

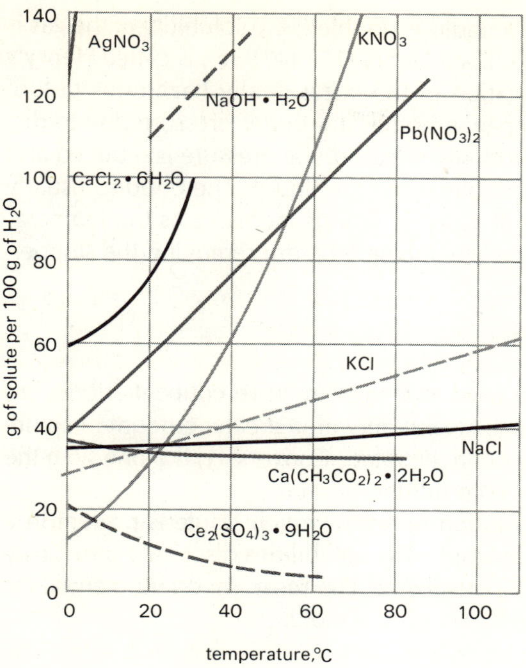

Figure 14–3 Solubilities of some ionic compounds in water. The indicated formulas are for the species that exist in equilibrium with the saturated solution at the indicated temperatures.

To summarize: the solubility of solids depends on the nature of the bonds being broken in the solid crystal and the molecular processes responsible for the solution phenomena. Table 14–2 summarizes the general rules for the solubility of a number of classes of compounds in water and some of the exceptions to the rules.

Pressure changes do not appreciably affect the solubility, except in the case of a gas dissolving in a liquid, when the solubility of the gas is directly proportional to the

TABLE 14–2 SOLUBILITIES OF SALTS IN WATER AT ROOM TEMPERATURE

Soluble Over 1 g/100 g H_2O	Very slightly soluble Less than 0.1 g/100 g H_2O
All nitrates	
All acetates	
All chlorates	
All sodium salts	
All potassium salts	
All ammonium salts	
Most chlorides	$AgCl$, $CuCl$, Hg_2Cl_2
Most bromides	$AgBr$, $CuBr$, Hg_2Br_2
Most iodides	AgI, CuI, Hg_2I_2, HgI_2
Most sulfates	$BaSO_4$, Hg_2SO_4, $PbSO_4$
Na_2CO_3, K_2CO_3, $(NH_4)_2CO_3$	Most carbonates
(Na_2O, K_2O hydrolyze)*	Most oxides
(Na_2S, K_2S, $(NH_4)_2S$, MgS, CaS, SrS, BaS all hydrolyze)	Most sulfides

* Reacts with water.

pressure. If the pressure of a gas above a liquid is doubled, the solubility of the gas is also doubled. This law, discovered by William Henry (1774–1836), is called **Henry's law.** Henry's law finds practical application in the occasionally hazardous task of opening a bottle of champagne. Champagne bottles are under pressure due to dissolved carbon dioxide. If the cork is suddenly released, the pressure is reduced and dissolved carbon dioxide escapes as bubbles from the liquid. The bubbles usually bring liquid with them. The solubility of gases in liquids increases as the temperature is lowered, so it is wise to chill your champagne before removing the stopper.

THE CONCENTRATION OF SOLUTIONS

The terms **dilute** and **concentrated** are used to express relative concentrations, but they have no exact quantitative meaning. A solution with a greater amount of solute than another solution is said to be more concentrated. Conversely, the one with the smaller amount of solute is said to be more dilute.

The absolute concentration of a solution tells how much solute is present in a given amount of solvent or in a given solution. The solubilities discussed previously were stated in terms of grams per 100 g of solvent. The most important method for expressing concentrations, however, is based on the mole.

MOLARITY

A **1 molar solution,** abbreviated 1 *M*, is defined as *one that contains one mole of solute per liter of solution:*

$$\text{molarity } (M) = \frac{\text{moles of solute}}{\text{volume of solution in liters}}$$

To determine the molarity of a solution, we need to know the amount of solute in moles and the total volume of the solution. Adding one mole of solute to one liter of solvent does not necessarily give a 1 *M* solution—the final volume may not be one liter. It is therefore important to note that the definition of a molar solution refers to moles of solute contained in each liter of the final solution, the one resulting from the mixture of solute and solvent.

Example 14–1: Molarity *A solution is prepared by dissolving 250 g of potassium hydroxide, KOH, in enough water to make 3.00 liters of solution. What is the molarity of the solution?*

To find the molarity of the solution, it is first necessary to find the number of moles of solute and then divide this value by the total volume of the solution. Since there are 56.1 g of KOH in one mole, we have

$$250\ \cancel{\text{g KOH}} \left[\frac{1.00\text{ mole KOH}}{56.1\ \cancel{\text{g KOH}}}\right] = 4.46\text{ moles KOH}$$

Dividing by the total volume gives the molarity:

$$\frac{4.46\text{ moles KOH}}{3.00\text{ liters of solution}} = 1.49\text{ moles KOH/liter of solution} = 1.49\ M\text{ KOH}$$

The problem can also be solved in the following one-step process:

$$\left[\frac{250 \not{g}\ \cancel{KOH}}{3.00 \text{ liters}}\right]\left[\frac{1.00 \text{ mole KOH}}{56.1 \not{g}\ \cancel{KOH}}\right] = 1.49\ M \text{ KOH}$$

Example 14–2: Molarity *What is the molarity of a solution that contains 10.0 g of KOH in a total volume of 200 ml?*

The problem gives us the factor

$$\frac{10.0 \text{ g KOH}}{200 \text{ ml solution}}$$

What we need to know is the number of moles per liter—that is, the molarity. We can obtain the answer by multiplying by two conversion factors:

$$\left[\frac{10.0 \not{g}\ \cancel{KOH}}{200\ \cancel{ml}}\right]\left[\frac{1.00 \text{ mole KOH}}{56.1 \not{g}\ \cancel{KOH}}\right]\left[\frac{1000\ \cancel{ml}}{1.00 \text{ liter}}\right] = 0.891\ M \text{ KOH}$$

The examples illustrate how molarity, *M*, can be calculated if we know three things about the solution: the mass of solute in g, the molecular mass of the solute in g/mole, and the volume, V, of the solution in liters. We can summarize the relationships used in the calculations as follows:

$$\textbf{number of moles of solute} = \frac{\textbf{mass of solute in g}}{\textbf{molecular mass of solute in g/mole}} \qquad \mathbf{n} = \frac{\mathbf{g}}{\mathbf{g/mole}} \qquad (1)$$

and

$$\textbf{molarity} = \frac{\textbf{number of moles of solute}}{\textbf{volume of solute in liters}} \qquad \mathbf{M} = \frac{\mathbf{n}}{\mathbf{V}} \qquad (2)$$

Another important type of problem involves calculating the mass of solute in a given volume of a solution of known molarity. Here the sequence of calculations is to solve first for the number of moles of solute and then for the mass in grams. The equations needed are:

$$\mathbf{n} = VM \quad \textbf{or} \quad \textbf{number of moles of solute} = \textbf{volume in liters} \times \textbf{molarity} \qquad (3)$$

This is obtained from Equation (2) by multiplying both sides by *V*. The second equation is

$$\textbf{mass of solute in g} = \textbf{number of moles of solute} \times \textbf{molecular mass of solute in g/mole} \qquad (4)$$

This expression is obtained from Equation (1) by multiplying both sides by the molecular mass of solute in g mole.

Example 14–3: Preparing molar solutions *What mass of $CuSO_4$ is needed to prepare 3.00 liters of a 0.500 M solution?*

Essentially this problem asks for the same thing as the question, *What mass of $CuSO_4$ is contained in 3.00 liters of a 0.500 M solution?*

The two-step solution to the problem is

Step 1. $VM = n$ (Equation 3)

$$(3.00\ \cancel{\text{liters}})\left(\frac{0.500\ \text{mole CuSO}_4}{1.00\ \cancel{\text{liter}}}\right) = 1.50\ \text{moles CuSO}_4 = n$$

Step 2. $n(\text{g/mole}) = \text{g}$ (Equation 4)

$$(1.50\ \cancel{\text{moles CuSO}_4})\left(\frac{160\ \text{g CuSO}_4}{1.00\ \cancel{\text{mole CuSO}_4}}\right) = 240\ \text{g CuSO}_4$$

The calculations can be combined into one step; note the sequence of conversion factors and the cancellation of units.

$$(3.00\ \cancel{\text{liters}})\left(\frac{0.500\ \cancel{\text{mole CuSO}_4}}{1.00\ \cancel{\text{liter}}}\right)\left(\frac{160\ \text{g CuSO}_4}{1.00\ \cancel{\text{mole CuSO}_4}}\right) = 240\ \text{g CuSO}_4$$

To prepare 3.00 liters of the 0.500 *M* solution, 240 g of $CuSO_4$ is dissolved in sufficient water (the exact amount is not calculated) to make 3.00 liters of solution. Put another way, this value means that one can recover 240 g of solid $CuSO_4$ by evaporating the water from 3.00 liters of the 0.500 *M* solution.

From time to time, we have seen that two conversion factors can be derived from any one statement of equality. For example, 1 hr = 60 min yields the factors 1 hr/60 min and 60 min/1 hr, each of which equals unity. Likewise, a statement of the molarity of a solution yields two conversion factors. For example, when we say that a solution is 0.500 *M* in $CuSO_4$, we say that 1 liter of solution contains (≙) 0.500 mole $CuSO_4$. The conversion factors are 0.500 mole $CuSO_4$/1.00 liter of solution and 1.00 liter of solution/0.500 mole $CuSO_4$. Example 14–3 required the first of these factors, and Example 14–4 will use the second.

Example 14–4: Preparing molar solutions *What volume of 0.500 M $CuSO_4$ solution can be prepared from 90.5 g of $CuSO_4$?*

The sequence of calculations is

1. convert grams of $CuSO_4$ to moles of $CuSO_4$.
2. convert moles of $CuSO_4$ to liters of solution.

The one-step calculation is

$$(90.5\ \cancel{\text{g CuSO}_4})\left(\frac{1.00\ \cancel{\text{mole CuSO}_4}}{160\ \cancel{\text{g CuSO}_4}}\right)\left(\frac{1.00\ \text{liter of solution}}{0.500\ \cancel{\text{mole CuSO}_4}}\right) = 1.13\ \text{liters of solution}$$

To prepare the solution, the 90.5 g of $CuSO_4$ is dissolved in sufficient water (the exact amount is not calculated) to make 1.13 liters of solution.

Example 14–5: Calculating moles of solute in molar solutions *How many moles of NaOH are contained in 6.00 liters of a 3.00 M solution?*

The solution is:

VM = number of moles of solute
(6.00 liters)(3.00 moles/liter) = number of moles = 18.0 moles

---

Example 14–6: Preparing molar solutions *What volume of 0.300 M solution can be prepared from 50.5 g of potassium nitrate?*

The sequence of the conversions is

(g of KNO_3) to (moles of KNO_3) to (liters of 0.300 M solution)

$$50.5\ \cancel{\text{g } KNO_3}\left[\frac{1.00\ \cancel{\text{mole } KNO_3}}{101\ \cancel{\text{g } KNO_3}}\right]\left[\frac{1.00\ \text{liter solution}}{0.300\ \cancel{\text{moles } KNO_3}}\right] = 1.67\ \text{liters}$$

When a solution is diluted—that is, when more solvent is added—the number of moles of solute does not change. Hence we can write:

(number of moles of solute)$_1$ = (number of moles of solute)$_2$

Since the number of moles of solute = VM, we get

$$\mathbf{V_1M_1 = V_2M_2}$$

where the subscript 1 refers to the solution before dilution and the subscript 2 to the solution after dilution.

Example 14–7: Diluting molar solutions *What volume of 0.100 M solution can be prepared from 60.0 ml of 6.00 M HCl? How much water must be added to make the dilution?*

This is typical of the dilution problems one encounters in the laboratory when preparing less concentrated solutions from stock reagents by adding the proper amount of additional solvent. The important thing to remember in solving a dilution problem is that the total number of moles of solute does not change. First, solve for the number of moles of HCl

$$60.0\ \cancel{\text{ml}}\left[\frac{1.00\ \cancel{\text{liter}}}{1000\ \cancel{\text{ml}}}\right]\left[\frac{6.00\ \text{moles HCl}}{1.00\ \cancel{\text{liter of solution}}}\right] = 0.360\ \text{mole HCl}$$

Since the total number of moles of HCl remains constant, we find

$$0.360\ \cancel{\text{mole HCl}}\left[\frac{1.00\ \text{liter of solution}}{0.100\ \cancel{\text{mole HCl}}}\right] = 3.60\ \text{liters of } 0.100\ M\ \text{HCl}$$

To prepare the 0.100 *M* solution, the 60.0 ml of 6.00 *M* HCl must be diluted to 3.60 liters. To do this, it is necessary to add 3540 ml of water (3600 ml − 60 ml) to make the total volume 3.60 liters.

The problem can also be solved by using the $V_1M_1 = V_2M_2$ relationship.

$$(0.0600 \text{ liter})(6.00\ M) = V_2(0.100\ M)$$

$$V_2 = 3.60 \text{ liters.}$$

Again, subtracting the 60 ml starting volume, we find that 3540 ml of water must be added to make the dilution.

Example 14–8: Calculating molarity of diluted solutions *Calculate the molarity of an acetic acid solution prepared by diluting 50.0 ml of a 0.800 M CH_3CO_2H solution to a total volume of 350 ml.*

This problem is solved by first finding the number of moles of solute present and then dividing that value by the final volume of the solution.

The number of moles of solute is

$$50.0\ \cancel{\text{ml}}\left[\frac{1.00\ \cancel{\text{liter}}}{1000\ \cancel{\text{ml}}}\right]\left[\frac{0.800 \text{ mole } CH_3CO_2H}{1.00\ \cancel{\text{liter}}}\right] = 0.0400 \text{ mole}\quad CH_3CO_2H$$

Dividing by the total final volume gives

$$\frac{0.0400 \text{ mole } CH_3CO_2H}{0.350 \text{ liter}} = 0.114\ M\ CH_3CO_2H$$

Alternatively, the problem can be solved by using the formula $V_1M_1 = V_2M_2$. Remember that the number of moles of solute always remains the same in a dilution problem; only the volume of solution and its concentration change.

Substitution into the equation gives

$$(0.050 \text{ liter})(0.800\ M) = (0.050 + 0.300 \text{ liter})(M_2)$$

Solving for M_2 gives

$$M_2 = \frac{(0.050 \text{ liter})(0.800\ M)}{(0.350 \text{ liter})}$$

$$M_2 = 0.114\ M$$

Example 14–9: Stoichiometry example involving solutions *What volume of 0.100 M sodium chloride solution must be added to 340 ml of a 0.200 M silver nitrate solution to precipitate—separate as a solid—all the chloride ions as silver chloride?*

The first step in the solution of any problem involving a chemical reaction is to write and balance the chemical equation.

$$NaCl(aq) + AgNO_3(aq) \rightarrow AgCl(s) + NaNO_3(aq)$$

From the balanced equation, we see that 1 mole of sodium chloride reacts with 1 mole of silver nitrate. We first calculate the number of moles of silver nitrate present.

$$\text{Number of moles of } AgNO_3 = V_{AgNO_3} \cdot M_{AgNO_3}$$
$$= (0.340 \text{ liter})(0.200\ M)$$
$$= 0.0680 \text{ mole}$$

Now we must calculate the volume of 0.100 M NaCl that contains the same number of moles.

$$0.0680 \text{ mole of NaCl} = V_{NaCl} \cdot M_{NaCl}$$
$$= (V_{NaCl})(0.100\ M)$$

Solving for the volume of NaCl gives

$$V_{NaCl} = \frac{0.0680 \cancel{\text{mole}} \text{ NaCl}}{0.100 \cancel{\text{mole}}/\text{liter}}$$

$$= 0.680 \text{ liter or } 680 \text{ ml}$$

Thus we find that 680 ml of 0.100 M NaCl will completely react with 340 ml of a 0.200 M $AgNO_3$ solution.

PROGRAMMED SEQUENCE:

Turn now to Set 8 of the Programmed Sequences, p. 326, and work this series of problems on molarity. The best way to feel comfortable with the quantitative relationships we have been discussing is to work with them until they become second nature. Be sure that you can handle these calculations before proceeding with the chapter.

DIFFERENT UNITS OF CONCENTRATION

A number of methods besides molarity are commonly used to express concentrations of solutions. These other methods, however, are much less important to chemists than molarity. For that reason, we have included the following brief discussions of parts per million, percent by mass, percent by volume, and molality solely for the purposes of reference and completeness. The problems at the end of the chapter entail calculations of molarity only.

Parts per million

A commonly used method for expressing concentrations of dilute solutions is parts per million, or ppm. This unit can be defined as the number of milligrams of solute present in one kilogram of solution. For aqueous solutions, where one kilogram of solution has a volume of approximately one liter, we can use the relationship:

$$\text{parts per million} = \text{ppm} = \text{mg of solute/liter of solution}$$

Example 14–10: Concentration in ppm *A 500 ml sample of water contains 1.80 mg of magnesium ion, Mg^{2+}. What is the Mg^{2+} concentration in ppm?*

The solution is

$$\left[\frac{1.80\text{ mg }Mg^{2+}}{500\ \cancel{\text{ml}}}\right]\left[\frac{1000\ \cancel{\text{ml}}}{1.00\text{ liter}}\right] = 3.60\text{ mg/liter} = 3.60\text{ ppm}$$

Percent by mass

Concentration of solutions can also be expressed in terms of percentages. Mass percent is defined as

$$\text{mass percent} = \frac{\text{mass of solute}}{\text{mass of solvent} + \text{mass of solute}} \times 100 = \frac{\text{mass of solute}}{\text{mass of solution}} \times 100$$

Example 14–11: Percent by mass *What is the percent by mass of a solution prepared by dissolving 15.0 g of sodium chloride in 150 g of water?*

To find percent by mass, divide the mass of solute by the total mass of the solution and multiply by 100.

$$\frac{15.0\text{ g NaCl}}{(15.0 + 150)\text{ g solution}} \times 100 = 9.09\%\text{ by mass}$$

Percent by volume

When the solute and solvent used to prepare a solution are both liquids, it is often convenient to express the concentrations in terms of volume percent. Volume percent is defined as:

$$\text{volume percent} = \frac{\text{volume of solute}}{\text{total volume of solution}} \times 100$$

The solute is usually taken to be the liquid present in the smaller amount. For example, a wine that contains 20 percent alcohol by volume has 20 volume units of alcohol in 100 volume units of the wine solution.

Molality

Molality is a method of expressing concentrations of solutions in terms of the number of moles of solute per kilogram of solute.

$$\text{molality} = m = \frac{\text{number of moles of solute}}{\text{kilogram of solvent}}$$

To find molality of a solution, you first determine the number of moles of solute and then divide that value by the mass of solvent (in kg) used to make the solution. This method of expressing concentrations can be used for working problems related to colligative properties of solutions—the topic of the next section.

Example 14–12: Molality *What is the molality of a solution prepared by dissolving 30.0 g of urea, $CO(NH_2)_2$, in 600 ml of water?*

First we find the number of moles of solute used. The molecular mass of urea is 60.0 g/mole. Hence

$$30.0\ \cancel{g\ urea}\left[\frac{1.00\ \text{mole urea}}{60.0\ \cancel{g\ urea}}\right] = 0.500\ \text{mole urea}$$

Next we must divide by the mass of solute in kg. Assuming the density of water to be 1.00 g/ml, 600 ml equals 600 g or 0.600 kg.

$$\frac{0.500\ \text{mole urea}}{0.600\ \text{kg solvent}} = 0.833\ \text{mole/kg} = 0.833\ m$$

THE COLLIGATIVE PROPERTIES OF SOLUTIONS

The vapor pressure of a solvent in a solution is always different from the vapor pressure of the pure solvent. If the solute is more volatile—has a higher vapor pressure—than the solvent (ethyl alcohol in water, for example), then the combined vapor pressures of the solute and solvent will be greater than the vapor pressure of the solvent alone. When a solid, nonvolatile substance is dissolved in a liquid, however, the vapor pressure is always lowered. This lowering of the vapor pressure can be demonstrated with the simple apparatus pictured in Figure 14–4. If pure water is placed on one side and a sugar solution on the other, the manometer will record a measurable pressure difference. Why the presence of nonvolatile molecules lowers the vapor pressure is not completely understood.

In 1882, François Marie Raoult (1830–1901) showed that at constant temperature, the lowering of the vapor pressure is proportional to the concentration of solutè present in the solution. Two moles of sugar added to 1000 g of water produce twice the vapor-pressure lowering as one mole of sugar. The surprising thing,

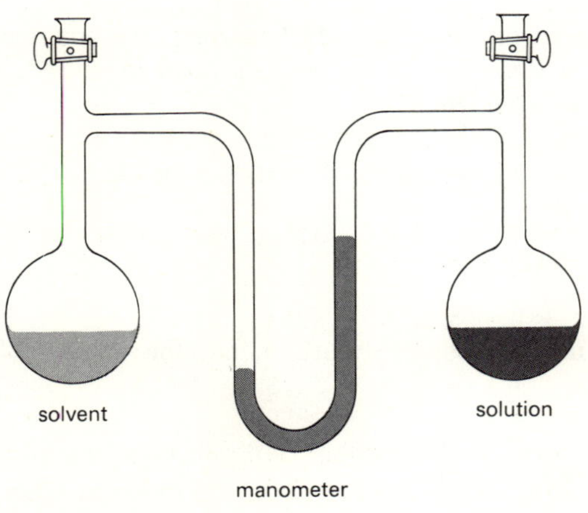

Figure 14–4 Comparison of the vapor pressure of a solution with that of the pure solvent. After the air is removed, the difference in height of the mercury columns gives the difference in vapor pressures of the solvent and the solution.

however, is that one mole of any solute, provided it does not have an appreciable vapor pressure of its own and does not ionize in solution, produces the same vapor-pressure lowering. One mole of sucrose, $C_{12}H_{22}O_{11}$ (molecular mass 342), dissolved in 1000 g of water produces the same vapor-pressure lowering as one mole of urea, $CO(NH_2)_2$ (molecular mass 60). Vapor-pressure lowering and the related freezing-point and boiling-point changes are called **colligative properties,** *properties that depend only on the number of particles present and not on their chemical nature.*

We know that the vapor pressure of a solution of a nonvolatile solute (a solution of salt in water, for example) is lower than the vapor pressure of the pure solvent. Because boiling point is defined as the temperature at which vapor pressure equals atmospheric pressure, it follows that the boiling point of solutions of nonvolatile solutes will always be greater than the boiling point of the solvent. A lowering of the vapor pressure is accompanied by a rise in boiling point. There is a boiling-point elevation constant for every solvent: for water, one mole of solute added to 1000 g water increases the boiling point 0.518°C.

The freezing point of an unsaturated solution of a nonvolatile solute is lower than the freezing point of the solvent. When the temperature of an unsaturated solution is lowered to the freezing point of the solvent, the solution does not solidify. If the temperature is lowered further, however, the solvent freezes. The solution itself does not freeze—it is the solvent that crystallizes when the solution reaches the freezing point. We cannot give an exact theoretical explanation for the phenomenon, but we can picture the solute molecules getting in the way of solvent molecules and making it more difficult for them to take up their fixed positions in the crystal. Freezing-point lowering is a colligative property. For nonionic compounds dissolved in 1000 g of water, the freezing-point lowering is 1.86°C for every mole of solute dissolved. The freezing-point lowering constant for benzene is 5.12°C/mole of solute/1000 g benzene; for cyclohexanol, $C_6H_{11}OH$ (molecular mass 100), it is 37.7°C for every mole of solute added to 1000 g.

The most important application of the freezing-point lowering in chemistry is determining molecular mass. If a substance is soluble in a solute with a known freezing-point lowering constant (in benzene or water, for example), then the molecular mass of the solute can be calculated by finding how far the freezing-point is lowered in a given amount of solvent.

Example 14–13: Calculating molecular mass from freezing-point lowering *The addition of 100 g of an unknown solute to 500 g of water lowers the freezing point to −2.79°C. What is the mass of one mole of the unknown compound?*

First, find how many grams of the unknown solute are contained in 1000 g of water.

$$1000\ \cancel{\text{g H}_2\text{O}}\left[\frac{100\text{ g solute}}{500\ \cancel{\text{g H}_2\text{O}}}\right] = 200\text{ g solute/1000 g H}_2\text{O}$$

One mole of any solute will lower the freezing point −1.86°C; hence the number of moles/1000 g H_2O must be

$$-2.79\ \cancel{^\circ\text{C}}\left[\frac{1\text{ mole/1000 g H}_2\text{O}}{-1.86\ \cancel{^\circ\text{C}}}\right] = 1.50\text{ moles/1000 g H}_2\text{O}$$

Thus 200 g of the unknown equals 1.50 moles. Solving for the mass of 1 mole gives

$$1.00\ \cancel{\text{mole}}\left[\frac{200\ \text{g}}{1.50\ \cancel{\text{moles}}}\right] = 133\ \text{g}$$

1 mole = 133 g

IDEAL SOLUTIONS

The freezing-point lowering produced by the solution of one mole of a substance in 1000 g of water is not exactly the same for every compound. Methyl alcohol lowers the freezing point 1.86°C; ethyl alcohol, 1.83°C; glycerin, 1.92°C. These variations are attributed to intermolecular attractive forces that vary from one compound to another. As the concentration of a solution increases, observed behavior deviates more and more from expected behavior. The properties of dilute solutions, however, do approach ideal behavior. An **ideal solution** is defined as *one in which there are no intermolecular attractions among the solute molecules to cause deviations from the solution laws.*

NEW TERMS

Alloy: a combination of two or more elements that has metallic properties.

Amalgam: alloys containing mercury; they may be either solid or liquid.

Colligative property: a property that depends only on the number of particles present and not on their size or mass.

Concentration of a solution: the specification of the amount of solute in a given amount of solvent or solution.

Dilute solution: a solution containing a relatively small amount of solute per given amount of solvent.

Henry's law: the solubility of a gas in a liquid is directly proportional to the pressure of the gas at a given temperature.

Ideal solution: a solution without intermolecular attractions among the solute molecules to cause deviations from the solution laws.

Immiscible: the term given to a system of two or more components that are insoluble in one another.

Ionic solution: a solution of an electrolyte; a solution that will conduct an electric current.

Miscible: the term given to two or more substances that are soluble in one another in all proportions.

Molality: a method for expressing concentrations of solutions; molality = m = number of moles of solute/kilogram of solvent.

Molarity: a method for expressing the concentration of solutions; a 1 molar solution contains 1 mole of solute per liter of solution; molarity = M = number of moles of solute/volume of solution in liters.

Percent by mass: a method for expressing concentrations of solutions; mass percent = mass of solute/mass of solution times 100.

Percent by volume: a method for expressing concentrations of solutions; volume percent = volume of solute/total volume of solution times 100.

Saturated solution: a solution in equilibrium with excess solute.

Solute: the component of a solution present in amounts less than that of the solvent, the dissolved substance.

Solution: a homogeneous mixture whose composition can vary between certain fixed limits.

Solvent: the medium in which the solute is dissolved; the component of a solution present in the greatest amount.

PROBLEM SET A

1. How does the process by which table salt (NaCl) dissolves in water differ from the process by which table sugar ($C_{12}H_{22}O_{11}$) dissolves in water?
2. All of the following compounds are soluble in water. In the process of dissolving, which ones should resemble NaCl, and which should resemble $C_{12}H_{22}O_{11}$?
   a. glycerol, $CH_2OHCHOHCH_2OH$
   b. potassium nitrate, KNO_3
   c. magnesium sulfate, $MgSO_4$
   d. ethyl alcohol, CH_3CH_2OH
3. By referring to Figure 14–3, estimate the solubility of KCl (in grams of KCl per 100 g of H_2O) at 20°C and at 80°C. If a saturated solution of KCl at 80°C is cooled to 20°C, what percent (approximately) of the solute should precipitate?
4. Calculate both the number of moles and the mass of solute required to prepare each of the following four solutions:

Solution	Volume of solution	Moles of solute	Mass of solute
a. 3.00 *M* KCl	1.00 liter	_____	_____
b. 3.00 *M* NaCl	1.00 liter	_____	_____
c. 1.00 *M* NaCl	3.00 liters	_____	_____
d. 0.500 *M* NaCl	500 ml	_____	_____

5. Calculate both the number of moles and the mass of solute required to prepare 2.00 liters of 1.50 *M* Na_2CO_3 if the solid solute available is
   a. Na_2CO_3 b. $Na_2CO_3 \cdot 7H_2O$ c. $Na_2CO_3 \cdot 10H_2O$
6. Calculate both the number of moles and the mass of solute in the following four solution samples:

Solution	Volume of solution	Moles of solute	Mass of solute
a. 2.50 *M* KCl	235 ml	_____	_____
b. 0.250 *M* NaCl	235 ml	_____	_____
c. 0.332 *M* $CaCl_2$	730 ml	_____	_____
d. 0.222 *M* $ZnCl_2$	2.48 ml	_____	_____

7. Calculate the molarity of each of the following solutions given the information supplied in the table:

Solute	Mass of solute	Volume of solution	Molarity of solution
a. sucrose, $C_{12}H_{22}O_{11}$	10.0 g	250 ml	_____
b. ethanol, CH_3CH_2OH	10.0 g	250 ml	_____
c. NaOH	30.0 g	2.50 liters	_____
d. HCl	30.0 g	2.50 liters	_____

8. Calculate the volume of each solution that can be made from the specified amount of solute (and sufficient water):

Desired solution	Mass of solute weighed	Volume of solution
a. 3.00 *M* KCl	3.50 g	_____
b. 3.00 *M* NaCl	3.50 g	_____
c. 1.00 *M* NaCl	3.50 g	_____
d. 0.500 *M* NaCl	35.0 g	_____

9. If 0.500 liter of 3.60 *M* sulfuric acid is diluted to a volume of 4.50 liters,
   a. what is the molarity of the dilute solution?
   b. how many moles of solute were present in the original solution?
   c. how many moles of solute are present in the dilute solution?

10. If 250 ml of 2.50 *M* NaOH is to be made by diluting a 6.00 *M* NaOH solution, what volume of the more concentrated solution is required?

11. Four dilution problems similar to Problems 9 and 10 are presented here in tabular form; fill in the blanks.

Initial concentrated solution		Final diluted solution		Approximate volume of water added
Molarity	Volume	Molarity	Volume	
a. 3.00 *M* KOH	25.0 ml	1.00 *M*	_____	_____
b. 6.00 *M* KOH	_____	3.50 *M*	5.00 l	_____
c. 0.255 *M* KOH	25.0 ml	_____	100 ml	_____
d. 0.255 *M* KOH	25.0 ml	0.100 *M*	_____	_____

12. What is the molarity of the resulting solution when the following pairs of solutions are mixed? (Assume that volumes are additive; that is, there is no shrinkage or expansion on mixing.)
    a. 50.0 ml of 1.00 *M* NaOH and 100 ml water.
    b. 50.0 ml of 1.00 *M* NaOH and 50.0 ml of 1.00 *M* NaOH.
    c. 50.0 ml of 1.00 *M* NaOH and 50.0 ml of 2.00 *M* NaOH.

13. Both NaOH and $Ba(OH)_2$ break down completely into their ions when dissolved in water. What is the molarity of hydroxide ion in a solution labeled
    a. 1.00 *M* NaOH
    b. 1.00 *M* $Ba(OH)_2$

14. In a solution formed by mixing 50.0 ml of 1.00 *M* NaOH and 50.0 ml of 1.00 *M* $Ba(OH)_2$, what is the concentration of
a. sodium ions b. barium ions c. hydroxide ions

15. Zinc reacts with hydrochloric acid to give hydrogen gas and zinc chloride. Write the balanced equation for the reaction and then calculate the volume of 0.600 *M* HCl needed to react completely with 10.0 g of zinc metal.

16. What volume of hydrogen (STP) is liberated when 50.0 ml of 6.00 *M* hydrochloric acid reacts with an excess of zinc metal?

17. What is the freezing point of an aqueous solution that contains 513 g of sucrose, $C_{12}H_{22}O_{11}$, per 1000 g of water?

18. When 60.0 g of an unknown solute dissolves in 1000 g of water, the freezing point of the water drops 0.62°C. What is the molecular mass of the solute?

PROBLEM SET B

1. By reference to Figure 14–3, estimate the solubility of $Pb(NO_3)_2$ (in grams of $Pb(NO_3)_2$ per 100 g of H_2O) at 20°C and 80°C. If a saturated solution of $Pb(NO_3)_2$ at 80°C is cooled to 20°C, what percent (approximately) of the solute should precipitate?

2. Calculate both the number of moles and the mass of solute required to prepare each of the following solutions:

Solution	Volume of solution	Moles of solute	Mass of solute
a. 1.50 *M* $KMnO_4$	750 ml	______	______
b. 0.175 *M* $Pb(NO_3)_2$	75.0 ml	______	______
c. 1.50×10^{-2} *M* HCl	1.50 liter	______	______
d. 1.75×10^{-4} *M* HF	3.50 liters	______	______

3. Calculate both the number of moles and the mass of solute required to prepare 2.00 liters of 1.50 *M* $CuSO_4$ if the solid solute available is
a. $CuSO_4$ b. $CuSO_4 \cdot 5H_2O$

4. Complete the following table describing four different solutions:

Solute	Molarity of solution	Volume of solution	Moles of solute	Mass of solute
a. NH_3	______	25.0 ml	______	2.55 g
b. $CaCl_2$	______	175 ml	1.00×10^{-3}	______
c. NH_4Cl	3.00 *M*	______	0.667	______
d. $(NH_4)_2SO_4$	3.00 *M*	______	______	0.500 g

5. Perform the dilution calculations needed to supply the data missing below.

Initial concentrated solution		Final diluted solution		Approximate volume of
Molarity	Volume	Molarity	Volume	water added
a. 5.85 *M* H_2SO_4	400 ml	3.00 *M*	______	______
b. 5.85 *M* H_2SO_4	______	3.00 *M*	400 ml	______
c. 16.0 *M* HNO_3	______	0.500 *M*	2.50 liters	______
d. 16.0 *M* HNO_3	10.0 ml	______	1.00 liter	______

6. What is the molarity of hydroxide ion when the following pairs of solutions are mixed? Assume that volumes are additive.
   a. 50.0 ml of 2.00 *M* KOH and 50.0 ml of 4.00 *M* KOH
   b. 50.0 ml of 2.00 *M* $Ba(OH)_2$ and 50.0 ml of water
   c. 50.0 ml of 2.00 *M* KOH and 50.0 ml of 2.00 *M* $Ba(OH)_2$
   d. 25.0 ml of 2.00 *M* KOH and 50.0 ml of 2.00 *M* $Ba(OH)_2$.
7. How many grams of copper metal will react with 500 ml of 2.00 *M* nitric acid? The balanced equation is
   $3Cu(s) + 8HNO_3(aq) \rightarrow 3Cu(NO_3)_2(aq) + 2NO(g) + 4H_2O$
8. What volume of nitrogen oxide (STP) is produced in the reaction described in Problem 7?
9. What is the freezing point of an aqueous solution that contains 60.0 g of sucrose, $C_{12}H_{22}O_{11}$, per 250 g of water?
10. When 16.0 g of a solute dissolves in 500 g of water, the freezing point drops 0.42°C. What is the molecular mass of the solute?

15

solutions of electrolytes, acids, and bases

THIS WILL SAVE YOU TIME:

Refer to this list as you read through the chapter. In answering the questions, you will understand the important concepts you are most likely to be asked about later on.

After reading this chapter, you should be able to:

1. *Distinguish between strong and weak electrolytes, and give examples.*
2. *State the postulates of the Arrhenius theory.*
3. *Explain how the Arrhenius theory accounts for the conduction of an electric current in solutions of electrolytes.*
4. *Explain how the Arrhenius theory accounts for the instantaneous formation of a precipitate when solutions of potassium bromide and silver nitrate are mixed.*
5. *Explain how the results of freezing-point-lowering experiments support the theory of ionization.*
6. *Describe how the Arrhenius theory explains the conductivity of weak electrolytes.*
7. *Discuss the ionization of pure water.*
8. *Tell how the Debye-Hückel theory accounts for the anomalous freezing-point depressions observed for solutions of strong electrolytes.*
9. *Distinguish among a molecular equation, a total ionic equation, and a net ionic equation.*
10. *Write net ionic equations.*
11. *List some of the physical and chemical properties of acids and bases.*
12. *Explain what is meant by the term neutral solution.*
13. *Calculate the hydroxide ion concentration of a solution given the hydronium ion concentration, and vice versa.*
14. *Explain the Brønsted-Lowry definition of acids and bases.*
15. *Explain why a solution of sodium acetate in water is slightly basic.*
16. *Give an example of a salt that hydrolyzes to give an acidic solution.*
17. *Tell what determines the strength of an acid or a base.*
18. *Calculate the pH and pOH of a solution given the hydronium ion or hydroxide ion concentration.*
19. *Tell what feature all acid-base reactions have in common.*
20. *Explain why neutralization reactions go to virtual completion.*
21. *Explain what is meant by polyprotic acids.*
22. *Explain the technique for quantitatively analyzing a solution containing an unknown amount of hydrochloric acid, and be able to work titration problems.*

When an electric current passes through a copper wire, the wire itself does not change. Electrons—the actual electric current—flow into the wire at one point and out at another. The total number of electrons in the copper wire stays the same before and after the flow, and there is no chemical change. This is not the case, however, with an electric current passing through an electrolytic solution. Free electrons as such do not flow through a solution: the current is carried by ions, and chemical reactions occur at each electrode.

In this chapter, we shall study some of the properties of electrolytes in solution and discuss the theories that explain the observed behavior of these solutions. The important topics of acids, bases, and neutralization are also included.

ELECTROLYTIC SOLUTIONS

Solutions are poor conductors compared to metallic conductors. Silver metal, a good conductor of electricity, has a conductance of about 6×10^5 conductance units (ohm^{-1}/cm) at room temperature; solutions of sodium chloride and hydrochloric acid have conductances in the 10^{-2} range (depending on concentration); the value for pure water is 6×10^{-8}; and for an insulator such as glass, it is only 10^{-14}.

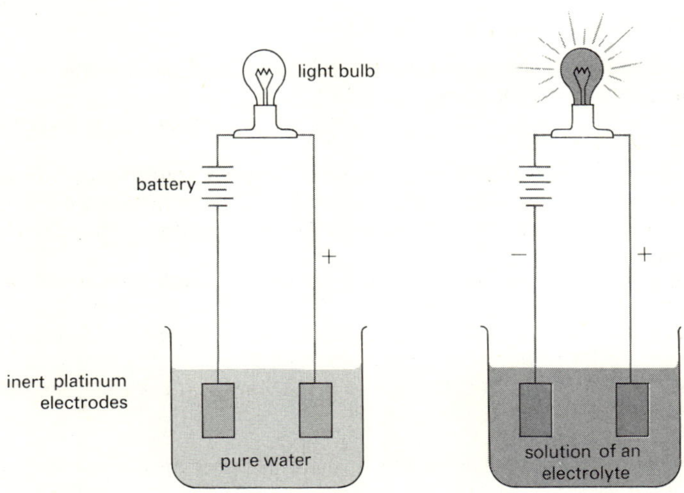

Figure 15–1 Electrical conductivity of electrolytes.

The electrical conductivity of solutions can be tested with the simple apparatus pictured in Figure 15–1. When inert metal electrodes are placed in pure distilled water, too little electric current passes through the circuit for the light bulb to glow. When an **electrolyte,** *a substance whose solution conducts an electric current,* is placed into the beaker, the bulb lights. The brightness of the bulb indicates the conducting power of the solution. If good electrical conductivity is obtained, we say that the substance is a **strong electrolyte;** if it conducts the current poorly, the substance is a **weak electrolyte.**

THE ARRHENIUS THEORY

In 1884, a doctoral student at the University of Uppsala in Sweden named Svante Arrhenius wrote a dissertation explaining the electrical conductivity of electrolytic solutions with a theory of ionization. The idea was met with little enthusiasm from Arrhenius's academic superiors, and his degree was granted reluctantly and "with reservations." In time, though, the scientific community came to recognize the brilliance of the ionization theory, and in 1903, the scientist who nearly failed graduate school was awarded the Nobel prize.

Arrhenius's theory can be summarized in the following postulates:

1. Solutions of electrolytes contain ions.
2. Electrolytes break down, or dissociate, into ions when they are placed in solvents.
3. Ions are responsible for the conduction of an electric current through electrolytic solutions.
4. The conductivity of an electrolytic solution depends on the degree to which the electrolyte dissociates in solution.

THE PROPERTIES OF SOLUTIONS EXPLAINED BY THE ARRHENIUS THEORY

Colors of solutions

The Arrhenius theory accounts for the colors of ionic solutions. All solutions of copper(II) compounds are blue because the hydrated copper(II) ions, $Cu^{2+}(H_2O)_x$, are blue; permanganate ions, $MnO_4^-(aq)$, are purple; dichromate ions, $Cr_2O_7^{2-}(aq)$, are orange; and hydrated nickel(II) ions, $Ni^{2+}(aq)$, are green.

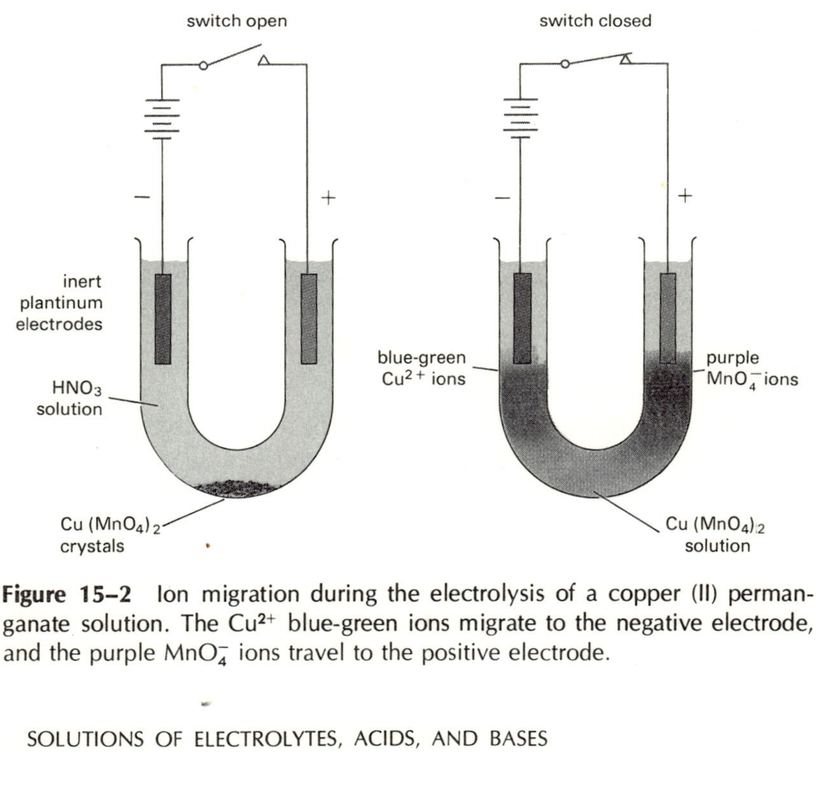

Figure 15–2 Ion migration during the electrolysis of a copper (II) permanganate solution. The Cu^{2+} blue-green ions migrate to the negative electrode, and the purple MnO_4^- ions travel to the positive electrode.

Migration of ions

The experiment in Figure 15–2 illustrates the migration of ions toward electrical charges. A small crystal of copper(II) permanganate, $Cu(MnO_4)_2$, is placed at the bottom of the U tube, where it slowly dissolves to form a deep-purple solution. When the switch is closed and an electric current is permitted to flow through the circuit, it is found that, after an extended period of time, a blue-green solution collects around the cathode (negative electrode), and a purple solution around the anode (positive electrode). The blue-green color is that of the +2 hydrated copper(II) ions, which are attracted to the cathode; and the purple color comes from the permanganate ions, which are attracted to the anode.

Instantaneous reactions

When a solution of sodium chloride is added to a solution of silver nitrate, insoluble silver chloride is formed in an instantaneous reaction:

$$AgNO_3 + NaCl \rightarrow \underline{AgCl} + NaNO_3 \qquad (\underline{AgCl} = AgCl(s) = AgCl \downarrow)$$

The equation as written indicates that both silver nitrate and sodium chloride "molecules" have to break apart before the fragments recombine; that is,

$$AgNO_3 \rightarrow Ag^+ + NO_3^-$$
$$NaCl \rightarrow Na^+ + Cl^-$$

One would not expect this breaking to occur instantaneously. The Arrhenius theory, however, tells us that this process has already occurred, that it happened when the reactant solutions were prepared. Silver nitrate solution is actually a solution of silver and nitrate ions, and sodium chloride solution is a solution of sodium and chloride ions— there are no "molecules" present. When these ionic solutions are mixed, the following reaction occurs instantaneously:

$$Ag^+ + Cl^- \rightarrow AgCl \downarrow$$

Silver ions and chloride ions unite on contact to form the insoluble precipitate of silver chloride; sodium ions and nitrate ions do not react. Thus the theory of ionization adequately explains the rapid rates observed for solution reactions.

Colligative properties

One-tenth of a mole of sodium chloride added to 1000 g of water lowers the freezing point −0.347°C, almost twice the lowering produced by one-tenth of a mole of a nonelectrolyte such as sugar, −0.186°C. If the Arrhenius theory is correct, then this difference is predictable. Freezing-point lowering, like all colligative properties, is a function of the number of particles in the solution. According to the Arrhenius theory, an electrolyte added to a solvent dissociates into its component ions, thus increasing the number of solvated particles. By contrast, a nonelectrolyte remains in its original form, retaining the same number of particles. When one mole of a nonelectrolyte like sugar dissolves, the solution gains 6.02×10^{23} particles. But when a mole of a two-component electrolyte like sodium chloride dissolves, the number of solvated particles is $2(6.02 \times 10^{23})$. The increase in the number of particles—by a factor of 2—roughly equals the increase in freezing-point lowering. Thus the Arrhenius theory explains the colligative property of freezing-point lowering.

Electrical conductivity

The Arrhenius theory can explain electrical conduction in solution. When sodium chloride crystals are placed in water, they separate into solvated sodium and chloride ions. These ions are able to move freely throughout the solution. When the electrodes of the conductivity apparatus (Figure 15–1) are connected to a battery, electrons flow into the cathode and a reaction takes place, giving hydrogen gas:

$$2H_2O(l) + 2\,{}_{-1}^{0}e \rightarrow H_2(g) + 2OH^-$$

Chloride ions migrate toward the anode. On contact, they give up their extra electrons and become neutral atoms:

$$Cl^- \rightarrow Cl + {}_{-1}^{0}e$$

These Cl atoms immediately react with each other, and chlorine gas is produced at the anode:

$$Cl + Cl \rightarrow Cl_2(g)$$

During the overall process, no electrons move through the solution; two reactions occur, one giving off electrons and one taking them on. The net result is a flow of electrons—an electric current—in the external circuit. The current results from the migration of the ions in the solution and from the chemical reactions at each electrode.

Pure water does not readily conduct an electric current. In even the purest water, however, there is a small measurable conductance, which shows that some ions must be present. A minute number of water molecules react with other water molecules to form **hydronium** and **hydroxide** ions:

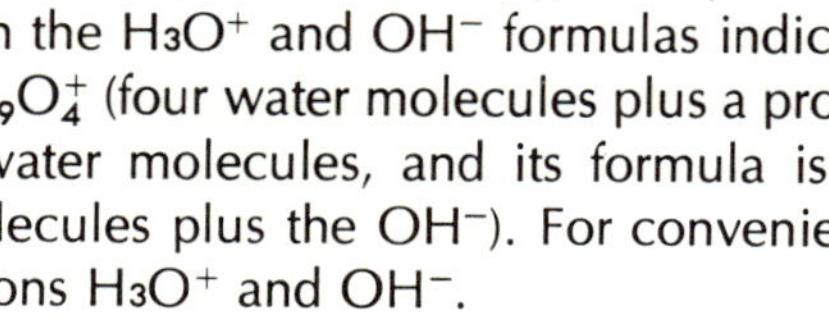

The hydronium ion can be thought of as a water molecule that has acquired a hydrogen ion (often referred to as a proton). Actually, the hydronium and hydroxide ions are more complex than the H_3O^+ and OH^- formulas indicate. The hydronium ion is probably closer to $H_9O_4^+$ (four water molecules plus a proton). The hydroxide ion also contains bound water molecules, and its formula is undoubtedly closer to $H_7O_4^-$ (three water molecules plus the OH^-). For convenience, however, we shall use the standard notations H_3O^+ and OH^-.

Weak electrolytes

Acetic acid, $H(CH_3CO_2)$ or CH_3CO_2H, is an example of a compound called a weak electrolyte (vinegar is a dilute solution of acetic acid). Its solutions do not conduct an electric current as effectively as a solution of sodium chloride (4.7×10^{-4} conductance units for acetic acid versus 1.1×10^{-2} for sodium chloride in solutions of equivalent concentrations). Arrhenius explained this difference by saying that acetic acid molecules are not completely dissociated in solution. Thus a solution of acetic acid contains undissociated $H(CH_3CO_2)$ molecules as well as H_3O^+ and $CH_3CO_2^-$ ions. The expected lowering of the freezing point produced by 1 mole of acetic acid in 1000 g of water should be greater than $-1.86°C$, but less than $2(-1.86°C)$, the lowering expected if each molecule dissociated completely into two ions. The freezing-point lowering found by experiment actually lies between these two values, thus supporting the idea that these molecules are not completely ionized in solution.

Further evidence for the concept of partial ionization of weak electrolytes comes from conductivity measurements. For example, concentrated acetic acid contained between parallel electrodes in an apparatus similar to that shown in Figure 15–1 shows a very low conductance because most of the acid molecules are not dissociated: the bulb glows dimly. Thus conductivity measurements can be used to determine the apparent degree of dissociation.

Example 15–1: Weak electrolytes *Explain why a double arrow is used in writing the following equation:*

$$H_2O(l) + H(CH_3CO_2)(l) \leftrightharpoons H_3O^+ + CH_3CO_2^-$$

Acetic acid is a weak electrolyte and does not dissociate completely when put into solution. An equilibrium is established, with the four species shown in the equation always present; the extent of dissociation of acetic acid depends on a number of factors, one being the concentration of the solution.

Strong electrolytes

Strong electrolytes are compounds that are completely ionized. Strong electrolytes such as sodium chloride, however, show only about 90 percent dissociation when a freezing-point-lowering method is used to calculate the degree of ionization. We know that sodium chloride exists completely in the ionic form even before it is dissolved. The less than 100 percent ionization shown in experiments requires a more careful look at the Arrhenius theory. Peter Debye (1884–1966, Nobel prize

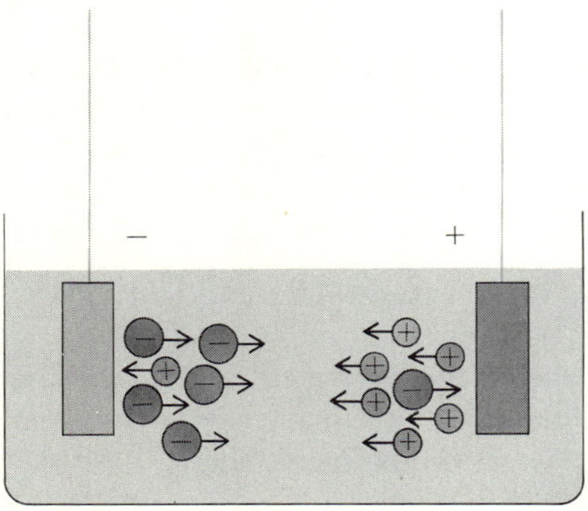

Figure 15–3 Diagram illustrating one of the principles of the Debye-Hückel theory for strong electrolytes. The oppositely charged ions around a given ion retard its movement in solution, thus reducing the conductivity.

in chemistry 1936) and Erich Hückel (1896–) solved the problem in 1923. They said that strong electrolytes are completely ionized, but in solution the mobility of the ions decreases. As a result, the electrolyte appears to be less than 100 percent ionized. According to their theory, an ion attracts and is surrounded by oppositely charged ions; thus the movement of a positive ion to the negative electrode in an electrolysis experiment is hindered by a surrounding cloud of oppositely charged ions. The ion, with its neighbors, tends to be pulled in the opposite direction, and its mobility or movement is slowed (Figure 15–3).

Because of their complexity, the phenomena associated with concentrated solutions are not completely understood. Even the term concentration is difficult to define precisely since the composition of clusters of solvated ions in solutions is unknown. Much theoretical and experimental work remains to be done in this area of chemistry.

WRITING IONIC EQUATIONS

In our discussion of the instantaneous reaction of silver ions with chloride ions on page 223, we found that the reaction could best be described by means of an equation that shows dissociation of the molecules, an ionic equation. Let us consider another example, a solution of silver nitrate added to a solution of potassium bromide to form solid silver bromide. The equation for the reaction can be written in three different forms.

Molecular equation: $AgNO_3(aq) + KBr(aq) \rightarrow AgBr(s) + KNO_3(aq)$

Since silver nitrate, potassium bromide, and potassium nitrate are all ionic compounds that ionize completely when dissolved in water, we can write a total ionic equation.

Total ionic equation: $Ag^+ + NO_3^- + K^+ + Br^- \rightarrow AgBr(s) + K^+ + NO_3^-$

The AgBr is almost completely insoluble in water and does not break down into ions to any appreciable extent. Both sides of this reaction contain K^+ and NO_3^- ions, neither species participates in the reaction. These nonparticipating ions are commonly called **spectator ions.** Removing them from the total ionic equation yields a **net ionic equation.**

Net ionic equation: $Ag^+ + Br^- \rightarrow AgBr(s)$

Some total ionic equations do not have any spectator ions. For example, consider the reaction of sodium metal with water.

Molecular equation: $2Na(s) + 2H_2O \rightarrow 2NaOH(aq) + H_2(g)$
Ionic equation: $2Na(s) + 2H_2O \rightarrow 2Na^+ + 2OH^- + H_2(g)$

Note that for simplicity the formula H_2O refers to liquid water and that ions in aqueous solution are written without the (*aq*) notation.

Ionic equations must be balanced both chemically and electrically. In the two examples above, you will note that the sum total of the electrical charges on all the reactant ions equals the sum total of all the charges on the product ions.

To summarize, the following rules must be followed when writing and balancing ionic equations.

1. Identify all the species present in the reactants and in the products.
   a. All strong electrolytes are written in the ionic form.
   b. Weak electrolytes and nonelectrolytes are written in molecular form.
   c. Gases, insoluble substances, and precipitates are written in molecular form.
2. Write the equation with all the reactants and products and balance it.
3. Delete the spectator ions.
4. Check to make sure the electrical charges balance.

Example 15–2: Net ionic equations *Rewrite the following unbalanced equation as a net ionic equation.*

$BaBr_2(aq) + Na_2SO_4(aq) \rightarrow BaSO_4(s) + NaBr(aq)$

$BaBr_2$, Na_2SO_4, and NaBr are all strong electrolytes. Hence the total ionic equation is

$Ba^{2+} + 2Br^- + 2Na^+ + SO_4^{2-} \rightarrow BaSO_4(s) + Na^+ + Br^-$

Balancing yields

$Ba^{2+} + 2Br^- + 2Na^+ + SO_4^{2-} \rightarrow BaSO_4(s) + 2Na^+ + 2Br^-$

The spectator ions are Na^+ and Br^-. Deleting them gives the net ionic equation.

$Ba^{2+} + SO_4^{2-} \rightarrow BaSO_4(s)$

Finally, check the electrical charges. For the reactants the total of the charges is $(+2)+(-2)=0$; there is no charge on the $BaSO_4$ product. Hence the charges are balanced, $0=0$.

PROGRAMMED SEQUENCE:

Turn now to p. 330 and work Set 9 of the Programmed Sequences. These problems deal with writing and balancing net ionic equations. Working through them will give you the practice needed to be fluent with this kind of chemical calculation.

ACIDS AND BASES

The classical definitions of acids and bases are given in terms of the behavior of the substances in water because all the early work on the chemistry of electrolytes centered on aqueous solutions. According to the Arrhenius theory, an **acid** is a *substance that, when placed in pure water, will increase the concentration of hydronium ions,* H_3O^+, $H^+(aq) = H_3O^+(aq) = H_3O^+$; a **base** is a *substance that will increase the hydroxide ion,* OH^-, *concentration of water.*

Acids have many properties in common. They have a sour taste and will react with bases to form water and a salt. A **salt** is defined as *a compound formed when a metal ion or the ammonium ion,* NH_4^+, *replaces the hydrogen of an acid.*

$HCl(aq)$ +	$NaOH(aq)$ →	$NaCl(aq)$ +	H_2O
acid	base	salt	water

$H_2SO_4(aq)$ +	$2NH_4OH(aq)$ →	$2(NH_4)_2SO_4(aq)$ +	H_2O
acid	base	salt	water

Acids react with active metals to give hydrogen gas and a salt, and they give a definite color to certain organic compounds. Some common acids are hydrochloric, $HCl(aq)$; sulfuric, $H_2SO_4(aq)$; nitric, $HNO_3(aq)$; and acetic, $H(CH_3CO_2)(aq)$.

Bases taste bitter and feel slippery; they react with acids to form water and a salt; and they give characteristic colors to certain organic compounds. Some common bases are sodium hydroxide, NaOH; potassium hydroxide, KOH; ammonium hydroxide, $NH_4OH(aq)$ or $NH_3(aq)$; and calcium hydroxide, $Ca(OH)_2$.

Organic compounds that can be used to test for the presence of acids or bases are called **indicators.** The indicator **litmus** is red in the presence of acids and blue in the presence of bases; **phenolphthalein** is colorless in acids and red in dilute bases.

Water is neutral by definition, for in pure water the concentration of hydronium ions equals the concentration of hydroxide ions. At 25°C, these concentrations are 1×10^{-7} mole/liter. The product of the hydronium ion concentration and the hydroxide ion concentration of pure water, (1×10^{-7} mole/liter)(1×10^{-7} mole/liter) is a constant equal to 1×10^{-14} mole²/liter². This relationship is usually written

$$[H_3O^+]\,[OH^-] = 1 \times 10^{-14}\ \text{mole}^2/\text{liter}^2$$

where the brackets denote the concentrations in molarity, moles per liter of solution. The mole²/liter² units are usually omitted.

Example 15–3: Given OH^- ion concentration, find H_3O^+ concentration *A basic solution has an OH^- ion concentration of 0.001 mole/liter. What is the concentration of hydronium ions?*

A 0.001 mole/liter solution is $1 \times 10^{-3}\ M$. Because the product of the ion concentrations is 1×10^{-14}, the OH^- ion concentration can be substituted into the equation, which can then be solved for the H_3O^+ concentration.

$$[1 \times 10^{-3}]\,[H_3O^+] = 1 \times 10^{-14}$$
$$[H_3O^+] = 1 \times 10^{-11}\ M$$

Example 15–4: Finding $[H_3O^+]$ given $[OH^-]$ *An aqueous solution has an OH^- concentration of 1×10^{-9} mole/liter. Is this solution acidic or basic? What is the hydronium ion concentration?*

First, let us solve for the hydronium ion concentration.

$$[H_3O^+]\,[OH^-] = 1 \times 10^{-14}$$
$$[H_3O^+]\,[1 \times 10^{-9}] = 1 \times 10^{-14}$$
$$[H_3O^+] = 1 \times 10^{-5}$$

Because the hydronium ion concentration is larger than the hydroxide ion concentration (1×10^{-5} is a much larger number than 1×10^{-9}), the solution is acidic. Basic solutions have OH^- concentrations greater than 1×10^{-7} mole/liter, and acidic solutions have H_3O^+ concentrations greater than 1×10^{-7} mole/liter.

THE BRØNSTED-LOWRY DEFINITION OF ACIDS AND BASES

When it was discovered that solvents other than water can form electrolytic solutions, more general definitions for acids and bases were needed. Thomas Martin Lowry (1874–1936) and J. N. Brønsted (1879–1947) with Neils Bjerrum (1879–1958) independently proposed a general definition for acid and base. *An* **acid** *is defined as a substance that can give up a proton, and a* **base** *as a substance that can accept a proton.* The reaction of hydrogen chloride with water shows how this definition works in practice.

$$\underset{\text{acid}_1}{HCl(g)} + \underset{\text{base}_1}{H_2O(l)} \leftrightharpoons \underset{\text{acid}_2}{H_3O^+} + \underset{\text{base}_2}{Cl^-}$$

HCl is an acid, for it gives up a proton; water is a base, for it accepts a proton. The residue left from HCl after it has lost its proton is now a potential proton acceptor; that is, the Cl^- ion is a base. The H_3O^+ ion is an acid, because it is now a potential proton donor. Acid$_2$, which results from base$_1$ by the acceptance of a proton, is said to be the **acid conjugate** to base$_1$; similarly, base$_2$ is the conjugate of acid$_1$.

TABLE 15–1 ACID-BASE REACTIONS ACCORDING TO THE BRØNSTED-LOWRY DEFINITIONS

Acid$_1$		Base$_1$		Acid$_2$		Base$_2$
HCl	+	H_2O	$\rightleftharpoons$	H_3O^+	+	Cl^-
H_2O	+	NH_3	$\rightleftharpoons$	NH_4^+	+	OH^-
H_2O	+	H_2O	$\rightleftharpoons$	H_3O^+	+	OH^-
$H(CH_3CO_2)$	+	H_2O	$\rightleftharpoons$	H_3O^+	+	$CH_3CO_2^-$
NH_3	+	NH_3	$\rightleftharpoons$	NH_4^+	+	NH_2^-
HSO_4^-	+	H_2O	$\rightleftharpoons$	H_3O^+	+	SO_4^{2-}
H_2O	+	$CH_3CO_2^-$	$\rightleftharpoons$	$H(CH_3CO_2)$	+	OH^-
H_2SO_4	+	$H(CH_3CO_2)$	$\rightleftharpoons$	$H_2(CH_3CO_2)^+$	+	HSO_4^-
H_2SO_4	+	H_2SO_4	$\rightleftharpoons$	$H_3SO_4^+$	+	HSO_4^-

A number of acid-base reactions are shown in Table 15–1. Note that most substances called acids or bases according to the classical definitions are also acids and bases under the Brønsted-Lowry definitions. The new definitions are much broader, however, and can be applied to nonaqueous systems.

HYDROLYSIS

When the salt sodium acetate, $Na(CH_3CO_2)$, dissolves in water, the solution is slightly basic; when the salt ammonium chloride, NH_4Cl, dissolves in water, the solution is slightly acidic. Such *an interaction between a salt and water* is called **hydrolysis.** Consider the reaction that occurs when sodium acetate, $Na^+(CH_3CO_2)^-$, is dissolved in water:

$$Na(CH_3CO_2)(aq) + H_2O \leftrightharpoons NaOH(aq) + H(CH_3CO_2)(aq)$$

or in net ionic form

$$CH_3CO_2^- + H_2O \xrightarrow[\text{slightly}]{\text{occurs}} H(CH_3CO_2) + OH^-$$

acetate ion acetic acid

According to the Brønsted-Lowry theory, the water is an acid, for it gives up a proton, and the acetate ion is a base, for it accepts a proton. The acetic acid formed is the conjugate acid, and the OH^- ion is the conjugate base. The resulting solution is basic because of the slight excess of OH^- ions.

Example 15–5: Acids and bases *In the following reaction, which substances are acids and which are bases according to the Brønsted-Lowry definitions?*

$$NH_4^+ + H_2O \rightleftharpoons H_3O^+ + NH_3(aq)$$

NH_4^+ is an acid, for it gives up a proton; and H_2O is a base, for it accepts a proton. H_3O^+ is the acid conjugate to the base H_2O, and $NH_3(aq)$ is the base conjugate to the acid NH_4^+.

Solutions of ammonium chloride in water are slightly acidic because of the above hydrolysis reaction, which produces excess H_3O^+ ions.

THE STRENGTHS OF ACIDS AND BASES

The strength of an acid refers to its ability to give up a proton, and the strength of a base to its ability to accept a proton. Table 15–2 lists the relative strengths of a number of acids and their conjugate bases. Note that the conjugate base of a strong acid is weak, and, conversely, the conjugate base of a weak acid is strong. The very strong bases NH_2^- and O^{2-} cannot exist in aqueous solutions.

TABLE 15–2 RELATIVE STRENGTHS OF ACIDS AND THEIR CONJUGATE BASES

Acid (↑ Strong Acids; Weak Acids at bottom)	Conjugate base (Weak Bases at top; ↓ Strong Bases)
$HClO_4$	ClO_4^-
H_2SO_4	HSO_4^-
HCl	Cl^-
H_3O^+	H_2O
HSO_4^-	SO_4^{2-}
$H(CH_3CO_2)$	$CH_3CO_2^-$
$CO_2(aq)$	HCO_3^-
NH_4^+	NH_3
HCO_3^-	CO_3^{2-}
H_2O	OH^-
OH^-	O^{2-}
NH_3	NH_2^-

pH and pOH

The hydronium-ion concentration of a solution can be expressed by means of a convention called the **pH scale,** which is defined as:

$$pH = \log \frac{1}{[H_3O^+]} = -\log [H_3O^+]$$

Neutral solutions have a pH of 7, acid solutions a pH less than 7, and basic solutions a pH greater than 7.

Example 15–6: Finding pH given molarity *What is the pH of a 0.001 M solution of hydrochloric acid? (Assume complete dissociation.)*

The H_3O^+ concentration is 0.001 *M*, or 1×10^{-3} *M*. That is,

$$[H_3O^+] = 10^{-3}$$

The logarithm of 10^{-3} is -3. Therefore,

$$\begin{aligned} pH &= -\log [H_3O^+] \\ &= -(-3) = 3 \end{aligned}$$

The logarithm of a number is the power to which the base number 10 must be raised to give the number. The logarithm of 1000 is 3; log 100 = 2; log 10 = 1; log 1 = 0; log 0.1 = −1; log 0.01 = −2; and so on.

Note that pH can also be defined as the negative exponent (base 10) of the hydronium-ion concentration.

Example 15–7: Finding pH given molarity *What is the pH of a 0.000001 M solution of nitric acid?*

The hydronium ion concentration is

$$[H_3O^+] = 10^{-6}\ M$$

Hence

$$pH = 6$$

Example 15–8: Finding pH given molarity *What is the pH of a solution whose hydronium ion concentration is 0.005 M?*

Using exponential notation

$$[H_3O^+] = 5 \times 10^{-3}$$

This number is between 10^{-3} and 10^{-2}:

$$\begin{aligned} &10 \times 10^{-3} \quad pH = 2 \\ &5 \times 10^{-3} \\ &1 \times 10^{-3} \quad pH = 3 \end{aligned}$$

The pH is therefore between 2 and 3. The exact value must be taken from logarithm tables. Referring to Table 15–3, which lists the logarithms of the integers, we get

$$\begin{aligned} pH &= -\log(5 \times 10^{-3}) = -(\log 5 + \log 10^{-3}) \\ &= -[0.70 + (-3)] \\ &= -(-2.3) \\ pH &= 2.3 \end{aligned}$$

TABLE 15–3 LOGARITHMS OF THE INTEGERS

Number	Log
1.0	0.00
2.0	0.30
3.0	0.48
4.0	0.60
5.0	0.70
6.0	0.78
7.0	0.84
8.0	0.90
9.0	0.95

A scale similar to the pH scale can be used to express the hydroxide ion concentration of solutions. **pOH** is defined as

$$pOH = \log \frac{1}{[OH^-]} = -\log [OH^-]$$

Because

$$\begin{aligned} &[H_3O^+]\ [OH^-] = 1 \times 10^{-14} \\ &\log ([H_3O^+]\ [OH^-]) = \log(1 \times 10^{-14}) \\ &-\log[H_3O^+] - \log[OH^-] = -\log 10^{-14} \\ &pH + pOH = 14 \end{aligned}$$

Example 15–9: The pH and pOH scales *What are the pH and pOH of a 0.0001 M sodium hydroxide solution?*

The pOH is equal to the negative exponent of the hydroxide ion concentration. Hence

$$\begin{aligned} &[OH^-] = 0.0001 = 10^{-4} \\ &pOH = 4 \end{aligned}$$

The total of pH and pOH in a solution is 14. Thus pH can be determined from a known pOH by subtracting the pOH from 14. In this case,

$$\begin{aligned} &pH = 14 - pOH = 14 - 4 \\ &pH = 10 \end{aligned}$$

Table 15–4 gives the pH and pOH values for a number of common materials. At the neutral point both pH and a pOH equal 7.

The pH values of body fluids must be maintained within certain narrow limits. If the pH of a person's blood goes below 7.0 or above 7.8, he dies.

The pH values of food we eat vary greatly; most, however, have values between 2 and 8. The more acidic foods include cranberry sauce (2.4), grapefruit juice

TABLE 15–4 pH AND pOH

$[H_3O^+]$	$[OH^-]$	pH	pOH	Strength	Examples
10^{0}	10^{-14}	0	14	**Strong acid**	1 *M* HCl
10^{-1}	10^{-13}	1	13		
					gastric juice
10^{-2}	10^{-12}	2	12		
					lemon juice
10^{-3}	10^{-11}	3	11		
					Coca-Cola, vinegar
10^{-4}	10^{-10}	4	10		
					orange juice
10^{-5}	10^{-9}	5	9		black coffee
					urine
10^{-6}	10^{-8}	6	8		coffee with milk
					milk
10^{-7}	10^{-7}	7	7	**Neutral**	saliva
					blood
10^{-8}	10^{-6}	8	6		
					baking soda
10^{-9}	10^{-5}	9	5		milk of magnesia
10^{-10}	10^{-4}	10	4		
					limewater
10^{-11}	10^{-3}	11	3		household ammonia
10^{-12}	10^{-2}	12	2		
10^{-13}	10^{-1}	13	1		caustic soda, 0.1 *M*
10^{-14}	10^{0}	14	0	**Strong base**	1 *M* NaOH

(2.9–3.2), lemon juice (2.3), and Coca-Cola (2.4–2.6). The most basic foods are egg omelette (7.7), boiled shrimp (7.2–7.8), and conchs (7.5–8.4). The initial acidity of foods we eat, however, is no criterion of their effect on the acid-base balance in the body. Digestion of food in the body produces by-product residues that may differ from the original acid-base value of the food. The final reaction products of most fruits and vegetables are alkaline; those of meat, poultry, eggs, cheese, fish, and cereals are acidic. A food showing an acid pH may end up as an alkaline residue in the body, and vice versa.

ACID-BASE REACTIONS

Strong acids plus strong bases

Strong acids such as $HCl(aq)$, $HNO_3(aq)$, and $H_2SO_4(aq)$ react with strong bases such as $KOH(aq)$ and $NaOH(aq)$ to give salts plus water. For example,

$$\underset{\text{acid}}{HCl(aq)} + \underset{\text{base}}{NaOH(aq)} \rightarrow \underset{\text{salt}}{NaCl(aq)} + \underset{\text{water}}{H_2O}$$

Because the reaction actually involves ions in aqueous solution, it can be written

$$H_3O^+ + \cancel{Cl^-} + \cancel{Na^+} + OH^- \rightarrow 2H_2O + \cancel{Na^+} + \cancel{Cl^-}$$

or simply

$$H_3O^+ + OH^- \rightarrow 2\ H_2O$$

This reaction proceeds to virtual completion because the water molecules formed as a product do not ionize to an appreciable extent.

This type of reaction is commonly called a **neutralization reaction.** If just enough base is added to react completely with all the acid present, the resultant solution of the salt is neutral.

Weak acids with strong bases

A typical weak acid is $H(CH_3CO_2)$, acetic acid, which does not completely dissociate in aqueous solution. When it reacts with a strong base such as NaOH, the net change is the disappearance of the weak acid and the formation of water:

$$H(CH_3CO_2)(aq) + Na^+ + OH^- \rightarrow Na^+ + CH_3CO_2^- + H_2O$$

or in net ionic form

$$H(CH_3CO_2)(aq) + OH^- \rightarrow CH_3CO_2^- + H_2O$$

The sodium ions are the spectator ions. If equal molar amounts of acid and base react, the water can be evaporated from the resulting solution to leave solid sodium acetate.

Some weak acids contain more than one acidic hydrogen; carbonic acid, $H_2CO_3(aq)$ or $CO_2(aq)$, is an important example. It reacts with sodium hydroxide as follows:

$$H_2CO_3(aq) + OH^- \rightarrow HCO_3^- + H_2O$$

$$HCO_3^- + OH^- \rightarrow CO_3^{2-} + H_2O$$

If one mole of carbonic acid reacts with one mole of sodium hydroxide, then hydrogen carbonate ion (bicarbonate ion) forms, as the first equation indicates. Increasing the concentration of OH^-, however, by adding another mole of NaOH, results in the formation of the carbonate ion. The first reaction goes almost to completion before the second one starts. The term **polyprotic acid** is sometimes used to describe an *acid that can furnish more than one proton per molecule.*

Example 15-10: Polyprotic acids *What salts would you expect to recover from the evaporation of the solutions left when aqueous solutions containing the following amounts of the listed compounds are reacted?*

a. 1 mole of H_3PO_4 plus 1 mole of NaOH

b. 1 mole of H_3PO_4 plus 2 moles of NaOH

c. 1 mole of H_3PO_4 plus 3 moles of NaOH

Phosphoric acid is a polyprotic acid with three replaceable hydrogens. The net overall reactions are

a. $H_3PO_4(aq) + OH^- \rightarrow H_2PO_4^- + H_2O$

b. $H_3PO_4(aq) + 2OH^- \rightarrow HPO_4^{2-} + 2H_2O$

c. $H_3PO_4(aq) + 3OH^- \rightarrow PO_4^{3-} + 3H_2O$

Evaporation of the solutions would leave the salts: a. monosodium dihydrogen phosphate; b. disodium monohydrogen phosphate; and c. trisodium phosphate.

Strong acids with weak bases

The most important weak base in the laboratory is aqua ammonia, $NH_3(aq)$, also called ammonium hydroxide, $NH_4OH(aq)$. The reaction of this weak base with a strong acid produces a salt and water. With $HCl(aq)$, the reaction is

$$NH_3(aq) + H_3O^+ \rightarrow NH_4^+ + H_2O$$

Chloride is the spectator ion. Because of hydrolysis, the resulting solution of ammonium chloride is slightly acidic.

Strong acids with solid compounds

Many water-insoluble compounds of metals contain such basic anions as hydroxide, phosphate, carbonate, and sulfide; hence they will react with strong acids. Probably the most important reaction of this type is limestone or marble, $CaCO_3$, with acids such as sulfuric or hydrochloric. With excess $H_2SO_4(aq)$, you get

$$CaCO_3(s) + 2H_3O^+ + SO_4^{2-} \rightarrow Ca^{2+} + SO_4^{2-} + 3H_2O + CO_2(g)$$

or in net ionic form

$$CaCO_3(s) + 2H_3O^+ \rightarrow Ca^{2+} + 3H_2O + CO_2(g)$$

The deterioration of marble statues and other art treasures when they are exposed to acid-containing smog is primarily due to this reaction.

Example 15-11: Acid reactions *Why is no carbonic acid, $H_2CO_3(aq)$ formed in the above reactions?*

The reaction of the carbonate ion with acids actually occurs in two steps

$$H_3O^+ + CO_3^{2-} \leftrightarrows HCO_3^- + H_2O$$

$$HCO_3^- + H_3O^+ \leftrightarrows H_2CO_3(aq) \leftrightarrows CO_2(g) + 2H_2O$$

Addition of excess acid to a carbonate results in the liberation of carbon dioxide. Carbonic acid is probably best indicated by the formula $CO_2(aq)$; the species H_2CO_3 exists only in low concentrations.

TITRATION

The name **titration** is given to the process analytical chemists use for **volumetric analyses,** quantitative determinations based on solution techniques. A titration is simply *a determination of the amount of a solution of known concentration, called a* **standard solution,** *needed to completely react with a given amount of a solution of unknown concentration.*

An example of a titration is the reaction of an HCl solution of unknown concentration with a previously prepared 0.2000 *M** standard solution of sodium hydroxide (Figure 15–4). The object of the experiment is to find the concentration of the unknown acid solution. Each solution is placed in a clean, dry **buret,** a long calibrated tube with a stopcock at the bottom that allows accurately measured portions of the solution (to 0.01 ml for a 50-ml buret) to be withdrawn from the bottom. An arbitrary but accurately measured volume of the acid solution is first placed in a flask; we shall assume a measured volume of 48.20 ml. A small amount of the organic indicator phenolphthalein is then added to the acid solution. This indicator is colorless in the presence of acids and red in the presence of bases. During the titration, the standard 0.2000 *M* base solution is slowly and carefully dripped from its buret to the acid solution while the solution is agitated constantly. As the base reacts with the acid, the concentration of hydronium ions decreases. When all the acid has reacted, the neutralization point is reached. The solution contains only sodium and chloride ions and water. The addition of only a minute amount of excess base at this point will turn the solution red. The experimenter then records the volume of standard NaOH solution needed to produce the color change. For our experiment we shall assume this amount to be 23.60 ml.

To solve for the molarity of the HCl solution, we must first examine the balanced chemical equation for the reaction. The equation is

$$HCl(aq) + NaOH(aq) \rightarrow NaCl(aq) + H_2O$$

From this, we see that one mole of hydrochloric acid reacts with one mole of sodium hydroxide. Therefore, if we know how many moles of NaOH reacted, we know how many moles of HCl were in the HCl solution. First, we will solve for the number of moles of NaOH used.

$$23.60 \text{ ml} \left[\frac{1.000 \text{ \cancel{liter}}}{1000 \text{ \cancel{ml}}}\right]\left[\frac{0.2000 \text{ mole NaOH}}{1.000 \text{ \cancel{liter of solution}}}\right] = 0.004720 \text{ mole NaOH used}$$

* In volumetric analysis work, the data are commonly reported to four significant figures.

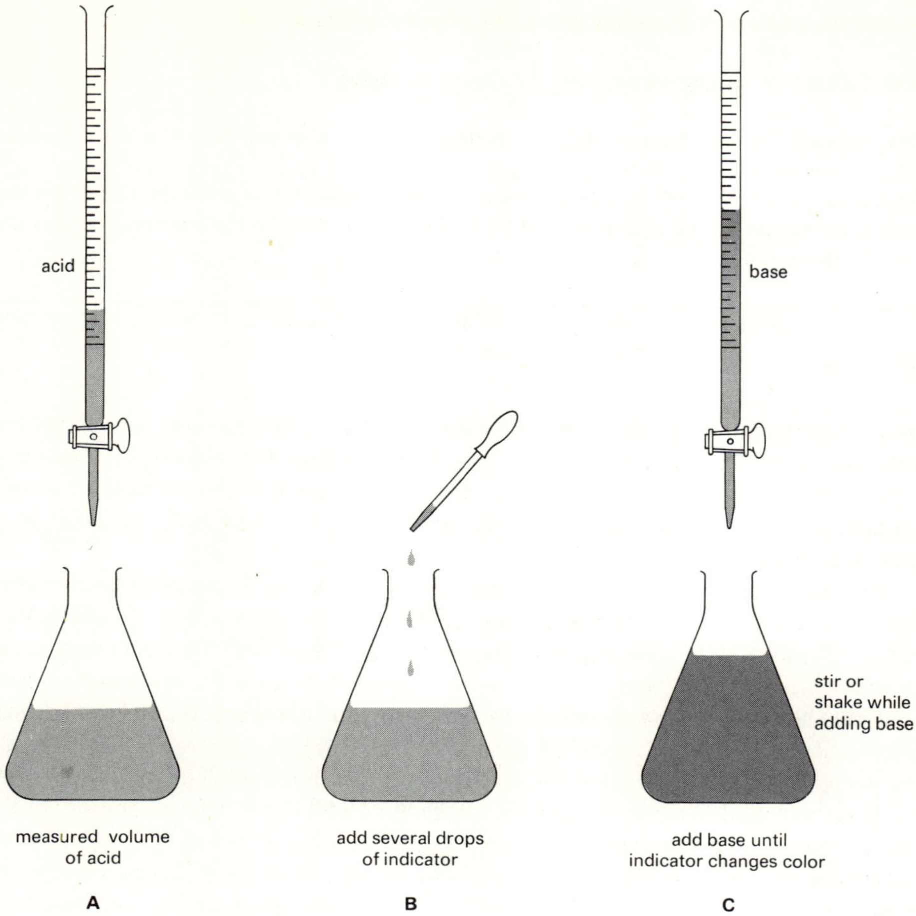

Figure 15–4 Titration of an acid with a base.

Now, to find the molarity of the HCl solution, we divide the number of moles of HCl by the volume of the solution. Since the number of moles of HCl equals the number of moles of NaOH, we have

$$\frac{0.004720 \text{ mole HCl}}{(48.20 \not{ml})\dfrac{1.000 \text{ liter}}{1000 \not{ml}}} = 0.0979 \text{ moles HCl/liter} = 0.0979\ M \text{ HCl}$$

An alternative method of solution uses the equation based on the definition relationships.

$$V_{\text{HCl}}M_{\text{HCl}} = V_{\text{NaOH}}M_{\text{NaOH}}$$
$$(0.04820 \text{ liter})(M_{\text{HCl}}) = (0.02360 \text{ liter})(0.2000\ M)$$
$$M_{\text{HCl}} = 0.0979\ M$$

Example 15–12: Titration *The titration of a 50.0 ml sample of vinegar requires 42.6 ml of 0.974 M NaOH. What is the molarity of the vinegar, and what is the number of grams of acetic acid contained in the sample?*

First, we must examine the balanced equation for the reaction. It is

$$H(CH_3CO_2)(aq) + NaOH(aq) \rightarrow Na(CH_3CO_2)(aq) + H_2O$$

From this, we again see that one mole of acetic acid reacts with one mole of NaOH. Now we find the number of moles of NaOH reacted.

$$42.6\ \cancel{ml}\left[\frac{1.00\ \cancel{liter}}{1000\ \cancel{ml}}\right]\left[\frac{0.974\ \text{mole NaOH}}{1.00\ \cancel{liter}}\right] = 0.0415\ \text{mole NaOH}$$

Therefore, the vinegar sample must contain 0.0415 mole of acetic acid. We find the molarity by dividing the number of moles by the volume.

$$\frac{0.0415\ \text{mole}\ H(CH_3CO_2)}{50.0\ \cancel{ml}\left[\frac{1.00\ \text{liter}}{1000\ \cancel{ml}}\right]} = 0.830\ M\ H(CH_3CO_2)$$

Multiplying the mole factor times the number of moles gives the number of grams of $H(CH_3CO_2)$ in the vinegar sample.

$$0.0415\ \cancel{\text{mole}}\ H(\cancel{CH_3CO_2})\left[\frac{60.0\ \text{g}\ H(CH_3CO_2)}{1.00\ \cancel{\text{mole}}\ H(\cancel{CH_3CO_2})}\right] = 2.49\ \text{g}\ H(CH_3CO_2)$$

If the alternative method based on the definition equation is chosen, the solution is as follows:

Because 1 mole of acetic acid reacts with 1 mole of sodium hydroxide,

$$V_{\text{acid}}\,M_{\text{acid}} = V_{\text{base}}\,M_{\text{base}} = \frac{\text{mass of acid}}{\text{mass of 1 mole of acid}}$$

$$(0.0500\ \text{liter})(M_{\text{acid}}) = (0.0426\ \text{liter})(0.974\ \text{mole/liter}) = \frac{x\ \text{g}}{60.0\ \text{g/mole}}$$

$$M_{\text{acid}} = \frac{(0.0426\ \cancel{\text{liter}})(0.974\ \text{mole/liter})}{(0.0500\ \cancel{\text{liter}})} = 0.830\ M\ \text{acetic acid}$$

$$x = (0.0426\ \cancel{\text{liter}})(0.974\ \cancel{\text{mole}}/\cancel{\text{liter}})(60.0\ \text{g}/\cancel{\text{mole}})$$

$$x = 2.49\ \text{g acetic acid}$$

Example 15–13: Titration *46.8 ml of a 0.100 M NaOH solution is required to titrate a 100 ml sample of a sulfuric acid solution. What is the molarity of the sulfuric acid solution?*

The solution proceeds as follows:

Step 1. Write and balance the chemical equation for the reaction.

$$2\ NaOH(aq) + H_2SO_4(aq) \rightarrow Na_2SO_4(aq) + 2H_2O$$

Step 2. Find out how many moles of the known reactant were needed to perform the titration.

$$46.8\ \cancel{ml}\left[\frac{1.00\ \cancel{liter}}{1000\ \cancel{ml}}\right]\left[\frac{0.100\ \text{mole NaOH}}{1.00\ \cancel{liter}}\right] = 0.00468\ \text{mole NaOH}$$

Step 3. Check the mole ratio indicated by the balanced equation. We find that 2 moles of NaOH are required to neutralize 1 mole of H_2SO_4. The mole-ratio factor is

$$\frac{1.00 \text{ mole } H_2SO_4}{2.00 \text{ moles NaOH}}$$

Step 4. Find the number of moles of the acid. We multiply by the mole ratio.

$$\left[\frac{1.00 \text{ mole } H_2SO_4}{2.00 \cancel{\text{ moles}} \cancel{\text{NaOH}}}\right] 0.00468 \cancel{\text{ mole}} \cancel{\text{NaOH}} = 0.00234 \text{ mole } H_2SO_4$$

Step 5. Solve for the molarity of the acid by dividing the number of moles by the volume.

$$M = \frac{\text{number of moles}}{\text{liters of solution}} = \frac{0.00234 \text{ mole } H_2SO_4}{100 \cancel{\text{ ml}} \left[\frac{1.00 \text{ liter}}{1000 \cancel{\text{ ml}}}\right]} = 0.0234\ M\ H_2SO_4$$

NEW TERMS

Acid: a substance that can give up a proton, a proton donor. In aqueous solutions, acids increase the concentration of hydronium ions.

Arrhenius theory: the theory of ionization.

Base: a proton acceptor. In aqueous solutions, bases increase the concentration of hydroxide ions.

Brønsted-Lowry definitions of acids and bases: acids are proton donors, and bases are proton acceptors.

Buret: a calibrated glass apparatus used to accurately measure the delivered volumes of liquids in titration experiments.

Conjugate acid-base pair: two species so related that one is formed from the other by proton transfer. In the reaction, $HCl + H_2O \rightarrow H_3O^+ + Cl^-$, H_3O^+ is the acid conjugate to H_2O; Cl^- is the base conjugate to HCl.

Debye-Hückel theory: the theory that explains the behavior of strong electrolytes in solution.

Electrolyte: a compound that conducts electricity when in solution or in the molten state.

Hydrolysis: any proton-transfer reaction in which water is a reactant.

Hydronium ion: H_3O^+. It represents a hydrated proton in aqueous solution, $H^+(aq)$.

Indicator: substance that exhibits different colors depending on the acidity or basicity of the solution.

Ion migration: the movement of ions in solutions.

Litmus: an indicator that is red in the presence of acids and blue in the presence of bases.

Net ionic equation: an equation written in ionic form from which all the spectator ions have been eliminated.

Neutralization reaction: the reaction of an acid with a base. In an aqueous solution, the neutralization reaction involves the combination of H_3O^+ ions with OH^- ions to form water; the positive ion of the base and the negative ion of the acid either remain in solution or precipitate as an insoluble solid.

pH: the negative exponent (base 10) of the hydronium-ion concentration. $pH = -\log[H_3O^+]$

Phenolphthalein: an indicator that is colorless in the presence of acids and red in the presence of bases.

pOH: the negative exponent (base 10) of the hydroxide ion concentration. $pOH = -\log[OH^-]$

Polyprotic acid: an acid that can furnish more than one proton per molecule in a neutralization reaction.

Spectator ion: an ion present during a reaction in which it does not participate.

Standard solution: a solution of known concentration.

Strong electrolyte: an electrolyte that is a good conductor of electricity in solution because of the high concentration of ions.

Titration: a procedure for determining the amount of a solution of known concentration needed to completely react with a given amount of a solution of unknown concentration.

Volumetric analysis: quantitative determinations based on the use of solution techniques.

Weak electrolyte: an electrolyte that is a poor conductor of electricity because of the low concentration of ions present when it is in solution.

PROBLEM SET A

1. Which of the following compounds are electrolytes and which are nonelectrolytes?
   a. calcium chloride
   b. sugar
   c. benzene (insoluble)
   d. zinc nitrate
   e. urea ($CO(NH_2)_2$)
2. What is the molarity of sodium ions in a solution of 0.2 *M* Na_2SO_4?
3. What freezing-point lowering might be expected when 2 moles of calcium chloride, $Ca^{2+}Cl_2^-$, are dissolved in 1000 g of water?
4. Complete the following equations and write in net ionic form:
   a. $H_2SO_4(aq) + Zn(s) \rightarrow$
   b. $HNO_3(aq) + KOH(aq) \rightarrow$
   c. $Ca(OH)_2(aq) + HCl(aq) \rightarrow$
5. Explain why a 0.1 *M* solution of a weak electrolyte does not conduct an electrical current as well as a 0.1 *M* solution of a strong electrolyte.
6. Write net ionic equations for the reaction of the following metals with sulfuric acid: Zn, Fe, and Ba.
7. Explain how you could quickly determine whether a sample of an unknown liquid is acidic or basic.
8. The hydronium ion concentration of an aqueous solution is 10^{-8} *M*. What is the concentration of the OH^- ion?
9. A solution has a hydronium ion concentration of 5×10^{-8}. Is this solution acidic or basic?
10. Label the following reactants and products as acids or bases according to the Brønsted-Lowry definitions:
    a. $H_3O^+ + OH^- \rightarrow 2H_2O$
    b. $H_2O + CO_3^{2-} \rightarrow HCO_3^- + OH^-$
    c. $OH^- + NH_4^+ \rightarrow NH_3 + H_2O$

11. Arrange the following in order of increasing acid strength: $CO_2(aq)$, $H_2SO_4(aq)$, H_2O, H_3O^+, and NH_3.
12. The pH of a 0.01 *M* solution of acetic acid is not 2. Explain.
13. The conjugate base associated with the HS^- ion is S^{2-}. Because HS^- is a very weak acid, what can be said about the strength of S^{2-}?
14. What is the pH of a 0.0001 *M* solution of HCl?
15. What is the pH of a solution whose H_3O^+ concentration is 0.003 mole/liter?
16. The hydroxide concentration of a solution is 0.00006 *M*. What are the pH and the pOH of the solution?
17. Which solution would you expect to be more acidic, 1 *M* H_2SO_4 or 2 *M* HCl?
18. What volume of 0.10 *M* KOH solution is needed to titrate 80 ml of 0.040 *M* H_2SO_4 solution? (Note: 1 mole of KOH does not neutralize 1 mole of H_2SO_4.)
19. The pH of a solution is 8. What is the pOH?
20. A 0.600 g sample of pure dry Na_2CO_3 was dissolved in 200 ml of water and neutralized with 32.2 ml of HCl(*aq*). What is the molarity of the acid? The reaction is
$2HCl(aq) + Na_2CO_3(aq) \rightarrow 2NaCl(aq) + CO_2(g) + H_2O$
21. Write the net ionic equation for the reaction in Problem 20.
22. If 38.0 ml of 0.120 *M* HCl is titrated with 46.2 ml of NaOH(*aq*), what is the molarity of the NaOH solution?

PROBLEM SET B

1. Which of the following compounds are electrolytes? $MgCl_2$, CO_2, benzene (C_6H_6), $Pb(CH_3CO_2)_2$, HNO_3.
2. A solution is prepared by dissolving 21.3 g of $Al(NO_3)_3$ in enough water to make 1.00 liter of solution. Predict the species present in the solution and calculate their respective concentrations.
3. A reagent bottle on the laboratory shelf is labeled $NH_4OH(aq)$. When *carefully* tested for odor, it was apparent that $NH_3(g)$ was present. What does this tell you about ammonia as an electrolyte?
4. Write net ionic equations that illustrate the low solubility of three metallic sulfides.
5. Write the formula for both the conjugate acid and the conjugate base of each of the following species: H_2O, NH_3, HSO_4^-.
6. Given two acids HA and HB. Describe what you can do experimentally to determine which of the acids is the stronger.
7. What is the strongest acid that can exist in water solution? The strongest base? (Hint: look at the equations in Table 15–1.)
8. What is the pH of a 0.000001 *M* solution of hydrochloric acid?
9. What is the pH of a 0.000001 *M* solution of sodium hydroxide?
10. pH + pOH = 14. Explain.
11. A hydrochloric acid solution was found to be 3.20×10^{-2} *M*. Calculate the pH.
12. Write net ionic equations for the following reactions:
    a. silver nitrate solution + sodium chloride solution
    b. $BaCl_2(aq) + K_2SO_4(aq)$
    c. carbonic acid + NaOH(*aq*)
    d. phosphoric acid + sodium hydroxide solution

13. Discuss the following statement: The net ionic equation for *all* acid-base reactions is $H_3O^+ + OH^- \rightarrow 2H_2O$.

14. Using Table 15–2 as a reference, would you predict a reaction between H_2SO_4 and $NaNH_2$? Between H_2O and NaCl? Why?

15. Ammonium chloride solutions are slightly acidic. Explain.

16. A student weighed out 0.990 g of an acid (HX), dissolved it in some distilled water, and titrated it with 39.4 ml of 0.195 *M* NaOH solution. Calculate the molecular mass of the unknown acid (g/mole).

17. Given an acid with a molecular mass of 138 g/mole. How much of this acid (HX) in grams should be weighed in order to titrate 40.0 ml of 0.093 *M* NaOH solution?

18. If 43.2 ml of a 0.240 *M* HCl solution is titrated with 38.6 ml of NaOH(*aq*), what is the molarity of the NaOH solution?

16

oxidation-reduction reactions and electrochemistry

THIS WILL SAVE YOU TIME:

Refer to this list as you read through the chapter. In answering the questions, you will understand the concepts you are most likely to be asked about later on.

After reading this chapter, you should be able to:

1. *Explain what is meant by an oxidation-reduction reaction.*
2. *Give examples of oxidation half-reactions and reduction half-reactions.*
3. *Give examples of redox equations and identify the oxidizing and reducing agents.*
4. *Balance oxidation-reduction equations.*
5. *Give an example of a voltaic cell. Draw such a cell, and label the anode and cathode. Give the equations for the oxidation and the reduction reactions, and name the electrode where they take place.*
6. *Tell what is meant by a coulomb, a faraday, an ampere, an ohm, and a volt.*
7. *Describe the construction of a standard hydrogen electrode.*
8. *Work problems involving electrolysis cells.*
9. *Predict battery voltages from oxidation potentials.*
10. *Give examples of displacement reactions.*
11. *Predict whether a redox reaction will occur spontaneously.*
12. *Describe the construction and workings of a lead storage battery.*

One of the most intriguing aspects of chemistry for the beginning student is the spectacular, seemingly magical array of laboratory demonstrations. Magnesium burns with an intense, blinding glare; powdered zinc mixed with sulfur sputters wildly; decomposing ammonium chromate produces an amazingly realistic "volcano"; a small piece of sodium laid on the surface of water sputters crazily; a copper framework immersed in a silver nitrate solution makes a chemical "Christmas tree" of glistening beauty. These displays share something besides showiness; all result from oxidation-reduction reactions. Such reactions also occur outside the laboratory; in fact, they lie behind myriad commonplace and important processes. The rusting of iron, the electric current of storage batteries and dry cells, and the energy metabolism of the human body are all oxidation-reduction reactions.

In this chapter, we will look at this important class of reactions and at closely related topics of electrochemistry.

WHAT ARE OXIDATION-REDUCTION REACTIONS?

Originally, the term oxidation reaction was applied to reactions in which a substance combined with oxygen. For example,

$$2Cu(s) + O_2(g) \rightarrow 2CuO(s)$$

The term reduction reaction was originally applied to reactions in which oxygen was removed from a substance, as in

$$CuO(s) + H_2(g) \rightarrow Cu(s) + H_2O$$

Our present definition of oxidation-reduction reactions, however, has a much broader significance; the reaction does not necessarily require the presence of oxygen.

In the past several chapters, we have seen many reactions in which electrons move from one element to another. When zinc metal reacts with hydrochloric acid, for example, the zinc atoms each lose two electrons, and the hydrogen atoms each gain an electron:

each Zn loses 2 electrons

$$\underset{0}{Zn(s)} + 2\underset{+1\ -1}{HCl}\ (aq) \longrightarrow \underset{+2}{Zn^{2+}} + 2\underset{-1}{Cl^-} + \underset{0}{H_2(g)}$$

← oxidation numbers

each H gains 1 electron

This type of reaction, in which there is a migration of electrons, is called an oxidation-reduction; **oxidation-reduction reactions** can be defined as *reactions in which there is a change in the oxidation number of an element.* In our example, zinc is oxidized. **Oxidation** is *the loss of electrons, or an increase in the oxidation number of an element to a more positive value.*

$Zn(s) \rightarrow Zn^{2+} + 2e^-$ (the notation e^- is used to designate an electron)
$\;\;0 \qquad\quad +2$

oxidation =
loss of electrons =
increase in oxidation number

Reduction is defined as *the gain of electrons, or a decrease in oxidation number to a less positive value.*

$2e^- + 2H^+(H_2O)_x \rightarrow H_2(g) + xH_2O$
$\qquad\quad +1 \qquad\qquad\quad 0$

reduction =
gain of electrons =
decrease in oxidation number

Oxidation and reduction reactions occur simultaneously; every oxidation reaction is accompanied by a reduction reaction. These reactions are commonly called **redox** (*red*uction-*ox*idation) reactions.

THE REDOX TERMINOLOGY

A special vocabulary is used to describe redox equations. As an example, consider the simple oxidation-reduction reaction

$2Na(s) + Cl_2(g) \rightarrow 2NaCl(s)$

As this reaction proceeds, the fate of the Na(*s*) can be described in several ways:

1. It is oxidized.
2. Its oxidation number increases from 0 to +1.
3. It loses one electron per atom. The equation is

   $2Na(s) \rightarrow 2Na^+ + 2e^-$ or $Na(s) \rightarrow Na^+ + e^-$

   This equation, in which sodium gives up an electron, is called a **half-reaction;** it is incomplete in itself, for it does not provide a "home" for the electron. Equations in which electrons are released as reaction products are called **oxidation half-reactions.**

Returning to our original redox equation, we can describe the fate of the $Cl_2(g)$ in the following ways:

1. It is reduced.
2. Its oxidation number decreases from 0 to −1.
3. It gains one electron per atom. The equation is

   $Cl_2(g) + 2e^- \rightarrow 2Cl^-$

   This is a **reduction half-reaction.** Electrons appear as reactants on the left side of the equation.

If Na(*s*) has been oxidized, it must have been oxidized by the $Cl_2(g)$. Therefore, in this reaction, the $Cl_2(g)$ is called the **oxidizing agent.** In the same way, the Na(*s*) is called the **reducing agent.** Species that lose electrons and are oxidized are reducing agents. Species that gain electrons and are reduced are called oxidizing agents.

All of this terminology is consistent and logical, but proper statements like, "The oxidizing agent is itself reduced, and the reducing agent is oxidized" can confuse beginning chemistry students unfamiliar with the jargon. It is important, therefore, to understand the redox terminology from the beginning.

Example 16–1: Redox terminology *For the following reactions (A and B), answer these six questions: What is the oxidation half-reaction? What is the reduction half-reaction? What is oxidized? What is reduced? What is the oxidizing agent? What is the reducing agent?*

Reaction A: the "Christmas tree" reaction

Molecular equation: $Cu(s) + 2AgNO_3(aq) \rightarrow Cu(NO_3)_2(aq) + 2Ag(s)$
Net ionic equation: $Cu(s) + 2Ag^+ \rightarrow Cu^{2+} + 2Ag(s)$
The oxidation half-reaction is $Cu(s) \rightarrow Cu^{2+} + 2e^-$
The reduction half-reaction is $2Ag^+ + 2e^- \rightarrow 2Ag(s)$
Cu(*s*) is the substance oxidized, and Ag^+ is the substance reduced.
Ag^+ is the oxidizing agent, and Cu(*s*) is the reducing agent.

Reaction B: a halogen replacement reaction

Molecular equation: $Cl_2(aq) + 2KBr(aq) \rightarrow 2KCl(aq) + Br_2(aq)$
Net ionic equation: $Cl_2(aq) + 2Br^- \rightarrow 2Cl^- + Br_2(aq)$
When chlorine water, $Cl_2(aq)$, is added to the colorless solution of KBr, the reddish-brown color characteristic of free bromine, Br_2, is produced.
The oxidation half-reaction is $2Br^- \rightarrow Br_2(aq) + 2e^-$
The reduction half-reaction is $Cl_2(aq) + 2e^- \rightarrow 2Cl^-$
Br^- is the substance oxidized, and $Cl_2(aq)$ is the substance reduced.
$Cl_2(aq)$ is the oxidizing agent, and Br^- is the reducing agent.

BALANCING REDOX EQUATIONS

Many simple redox equations can be balanced by trial and error and require no special techniques. A large number, however, are much too complicated and require special techniques. There are two common approaches to balancing redox equations: the oxidation-number method, and the half-reaction, or ion-electron, method.

The oxidation-number method

In Chapter 5, you were invited to try to balance the equation for the reaction of copper metal with dilute nitric acid to produce copper(II) nitrate solution and nitric oxide gas, NO. This reaction will be used as our example.

Step 1. Write a skeletal equation for the reaction. We can write this equation in either molecular or ionic form.

Skeletal molecular equation: $Cu(s) + HNO_3(aq) \rightarrow Cu(NO_3)_2(aq) + NO(g) + H_2O$
Skeletal ionic equation: $Cu(s) + NO_3^- \rightarrow Cu^{2+} + NO(g)$

The reactants of the skeletal ionic equation should be only the oxidizing agent and its reducing agent; the products are the corresponding oxidation and reduction products. Any H_2O, H_3O^+, or OH^- species that participates in the reaction will be added to the equation in one of the later steps. We can use either the molecular or the ionic form for the balancing procedure, but we will restrict our initial discussion to the ionic form.

Step 2. Assign oxidation numbers to the elements that are oxidized or reduced. Write the oxidation numbers under the symbols for the oxidized and the reduced forms of the elements.

$$\underset{0}{Cu}(s) + \underset{+5}{N}O_3^- \rightarrow \underset{+2}{Cu}^{2+} + \underset{+2}{N}O(g)$$

← oxidation numbers

Step 3. Draw tie lines connecting the symbol for the element on the left with the symbol for the same element on the right, and describe the change in terms of electrons gained, ne^-, or electrons lost, $n'e^-$. The *n*s denote the number of electrons involved in the changes.

gain of 3 electrons $ne^- = 3e^-$

$$\underset{0}{Cu}(s) + \underset{+5}{N}O_3^- \longrightarrow \underset{+2}{Cu}^{2+} + \underset{+2}{N}O(g)$$

loss of 2 electrons $n'e^- = 2e^-$

Step 4. Find the least common multiple of the two numbers n and n′, and multiply each by the appropriate factor so that the total number of electrons lost equals the total number of electrons gained. In this case, the least common multiple for n and n′ is 6. The coefficient that goes before the Cu(*s*), therefore, is 3, ($3 \times n' = 6$). The coefficient for the NO_3^- is 2, ($2 \times n = 6$). We now have

$$3Cu(s) + 2NO_3^- \rightarrow 3Cu^{2+} + 2NO(g)$$

Step 5. Balance the number of oxygen atoms on each side of the equation. Since the reaction was carried out in an acid medium, we know that H_3O^+ and H_2O will be needed to balance the final equation. (If the reaction had been in basic medium, OH^- and H_2O would be needed.) To balance the oxygen atoms, we add the required number of water molecules to the side of the equation that needs more oxygens. Since our equation (from Step 4) has 6 oxygens on the left and only 2 on the right, we will need 4 H_2O molecules on the right to balance.

$$3Cu(s) + 2NO_3^- \rightarrow 3Cu^{2+} + 2NO(g) + 4H_2O$$

Step 6. Balance the hydrogens. We will first balance the hydrogens by using H^+ ions, and in Step 7 we will change this to the more acceptable form using hydronium ions, H_3O^+. Our equation from Step 5 has no hydrogen atoms on the left and 8 on the right; hence $8H^+$ are needed to balance.

$$3Cu(s) + 2NO_3^- + 8H^+ \rightarrow 3Cu^{2+} + 2NO(g) + 4H_2O$$

Step 7. Change the H^+ to H_3O^+ notation by adding a H_2O for each H^+ to form H_3O^+. Balance by adding the same number of H_2O molecules to the other side of the equation.

$$3Cu(s) + 2NO_3^- + 8H_3O^+ \rightarrow 3Cu^{2+} + 2NO(g) + 12H_2O$$

Step 8. Double check to make sure everything is balanced. Make sure that there has been no net gain or loss of charge: on the left side $2(-1) + 8(+1) = +6$, on the right side $3(+2) = +6$. The equation is correctly balanced since charge has been conserved.

Example 16–2: Balancing redox equations *The skeletal equation for a redox reaction between permanganate ions and chloride ions in acidic solutions is*

$$MnO_4^- + Cl^- \rightarrow Mn^{2+} + Cl_2(g)$$

Balance the equation.

Step 1. The skeletal equation is given in the problem.

Step 2. Assign oxidation numbers.

$$\underset{+7}{MnO_4^-} - \underset{-1}{Cl^-} \rightarrow \underset{+2}{Mn^{2+}} + \underset{0}{Cl_2(g)}$$

Step 3. Determine the number of electrons lost and gained.

loss of $2e^-$ $n'e^- = 2e^-$

$$\underset{+7}{MnO_4^-} + \underset{-1}{2Cl^-} \longrightarrow \underset{+2}{Mn^{2+}} + \underset{0}{Cl_2(g)}$$

gain of $5e^-$ $ne^- = 5e^-$

Note that we temporarily added the coefficient 2 to the Cl^- to balance the Cl_2 molecule.

Step 4. Balance the number of electrons lost and the number of electrons gained. Since $n = 2$ and $n' = 5$, the least common multiple is 10. Therefore, the coefficient for the Mn is 2, ($2 \times 5 = 10$), and the coefficient for Cl_2 (or 2 Cl^-) is 5, ($5 \times 2 = 10$).

$$2MnO_4^- + 10Cl^- \rightarrow 2Mn^{2+} + 5Cl_2(g)$$

Step 5. Balance the oxygen atoms by adding the proper number of water molecules. There are 8 O's on the left and none on the right; therefore, we add $8H_2O$ to the right.

$$2MnO_4^- + 10Cl^- \rightarrow 2Mn^{2+} + 5Cl_2(g) + 8H_2O$$

Step 6. Balance the hydrogens. The reaction is in acid medium, requiring H^+ ions on the left. The equation needs a total of 16.

$$2MnO_4^- + 10Cl^- + 16H^+ \rightarrow 2Mn^{2+} + 5Cl_2(g) + 8H_2O$$

This equation is now balanced, but the notation H_3O^+ is preferred to H^+.

Step 7. Change H^+ to H_3O^+ ions.
To do this we add 16 H_2O's to both sides of the equation. The 16 on the left are combined with the 16 H^+ to give 16 H_3O^+ ions.

$$2MnO_4^- + 10Cl^- + 16H_3O^+ \rightarrow 2Mn^{2+} + 5Cl_2(g) + 24H_2O$$

Step 8. Check to make sure everything is balanced, including the charges.

The half-reaction method

For our example, we will use the reaction of dichromate ion, $Cr_2O_7^{2-}$, with iron(II) ions in aqueous acid solution to give chromium(III) ions and iron(III) ions.

Step 1. Write the skeletal equation for the reaction. This will usually be given in the statement of the problem in either molecular or ionic form.

Skeletal molecular equation:
$K_2Cr_2O_7(aq) + FeCl_2(aq) + HCl(aq) \rightarrow CrCl_3(aq) + KCl(aq) + FeCl_3(aq) + H_2O$

Skeletal ionic equation:
$Cr_2O_7^{2-} + Fe^{2+} \rightarrow Cr^{3+} + Fe^{3+}$

We will use the skeletal ionic equation for our discussion.

Step 2. Split the reaction into parts, each containing one of the reactants. (One of these reactions will represent an oxidation and the other reduction, but it is not necessary to identify them as such at this time.)

2a. $Cr_2O_7^{2-} \rightarrow Cr^{3+}$
2b. $Fe^{2+} \rightarrow Fe^{3+}$

Step 3. Add coefficients to balance atoms *other than O or H*.

3a. $Cr_2O_7^{2-} \rightarrow 2Cr^{3+}$
3b. no change

Step 4. Balance the oxygens. Do this by adding H_2O (*not* O_2) to the side deficient in oxygen.

4a. $Cr_2O_7^{2-} \rightarrow 2Cr^{3+} + 7H_2O$
4b. no change

Step 5. Balance the hydrogens by adding H^+ (*not* H_2) to the side deficient in hydrogen.

5a. $Cr_2O_7^{2-} + 14H^+ \rightarrow 2Cr^{3+} + 7H_2O$
5b. no change

Step 6. Balance charge by adding electrons to whichever side needs negative charge.

6a. $Cr_2O_7^{2-} + 14H^+ + 6e^- \rightarrow 2Cr^{3+} + 7H_2O$
6b. $Fe^{2+} \rightarrow Fe^{3+} + e^-$

Each half-reaction is now balanced. Note that without recourse to oxidation numbers we have established that 6a is the reduction—electrons were gained—and 6b is the oxidation—electrons were lost. On the other hand, there is nothing wrong with using your knowledge of oxidation numbers in Step 2 to help determine the two sets of reactants undergoing change.

Step 7. It remains to combine the two half-reactions in such a way that the electrons disappear in the overall equation. Or, in more fundamental terms, the electrons gained in the reduction (6a) must equal the electrons lost in the oxidation (6b). To do this, multiply each equation with a number that makes the number of electrons in each half-reaction the same.

7a. no change (multiply by 1)
7b. $6\,[Fe^{2+} \rightarrow Fe^{3+} + e^-]$

Step 8. Add the two modified equations. The electrons will cancel.

$$\begin{array}{rl}
Cr_2O_7^{2-} + 14H^+ + 6e^- & \rightarrow 2Cr^{3+} + 7H_2O \\
6Fe^{2+} & \rightarrow 6Fe^{3+} + 6e^- \\
\hline
Cr_2O_7^{2-} + 6Fe^{2+} + 14H^+ + \cancel{6e^-} & \rightarrow 2Cr^{3+} + 6Fe^{3+} + 7H_2O + \cancel{6e^-}
\end{array}$$

Step 9. If desired, change the H^+ to the more acceptable H_3O^+ and adjust the number of water molecules.

$$Cr_2O_7^{2-} + 6Fe^{2+} + 14H_3O^+ \rightarrow 2Cr^{3+} + 6Fe^{3+} + 21H_2O$$

Check to see that both atoms and charge are conserved. Here the net charge on the left, $-2 + 6(+2) + 14(+1) = +24$, equals the net charge on the right, $2(+3) + 6(+3) = +24$.

Example 16–3: The half-reaction method *Balance the following redox equation:*

$$Cl_2(aq) + KOH(aq) \rightarrow KClO_3(aq) + KCl(aq) + H_2O$$

Step 1. Convert the given skeletal equation to ionic form, omitting ions common to both sides.

$$Cl_2(aq) + OH^- \rightarrow ClO_3^- + Cl^- + H_2O$$

Step 2. In this example, it will be useful to assign oxidation numbers to all the elements to determine what is oxidized and what is reduced.

$$\underset{0}{Cl_2}(aq) + \underset{-2}{O}\,\underset{+1}{H^-} \rightarrow \underset{+5}{Cl}\,\underset{-2}{O_3^-} + \underset{-1}{Cl^-} + \underset{+1}{H_2}\,\underset{-2}{O}$$

Note the unusual feature of this reaction. Chlorine from $Cl_2(aq)$ is both oxidized and reduced. Consequently, the two skeletal half-reactions are

2a. $Cl_2(aq) \rightarrow ClO_3^-$
2b. $Cl_2(aq) \rightarrow Cl^-$

Step 3. Balance atoms other than O or H.

3a. $Cl_2(aq) \rightarrow 2ClO_3^-$
3b. $Cl_2(aq) \rightarrow 2Cl^-$

Step 4. Balance the oxygens.

4a. $Cl_2(aq) + 6H_2O \rightarrow 2ClO_3^-$
4b. no change

Step 5. Balance the hydrogens.

5a. $Cl_2(aq) + 6H_2O \rightarrow 2ClO_3^- + 12H^+$
5b. no change

Note another new feature of this reaction: it takes place in a strong basic solution, KOH(*aq*). Therefore, the extra species allowed are H_2O and OH^-, but not H^+ (or H_3O^+). However, it is simpler to proceed first as if it were an acid medium reaction and then "correct" the balanced equation to basic medium. This will be shown in Step 6.

Step 6. Balance charge.

$$Cl_2(aq) + 6H_2O \rightarrow 2ClO_3^- + 12H^+ + 10e^-$$

This equation is now balanced, but it must be converted to basic medium without destroying balance. Add $12OH^-$ to both sides of the equation.

$$Cl_2(aq) + 6H_2O + 12OH^- \rightarrow 2ClO_3^- + 12H^+ + 12OH^- + 10e^-$$

The H^+ and OH^- combine to form water.

$$Cl_2(aq) + 6H_2O + 12OH^- \rightarrow 2ClO_3^- + 12H_2O + 10e^-$$

Now cancel the extra water molecules.

6*a*. $Cl_2(aq) + 12OH^- \rightarrow 2ClO_3^- + 6H_2O + 10e^-$
6*b*. $Cl_2(aq) + 2e^- \rightarrow 2Cl^-$

Step 7. Make the number of electrons lost equal the number of electrons gained. Here this is accomplished by multiplying equation 6b by the number 5.

7*a*. no change (multiply by 1)
7*b*. $5\ [Cl_2(aq) + 2e^- \rightarrow 2Cl^-]$

Step 8. Add the two equations, canceling electrons.

$$\begin{array}{l} Cl_2(aq) + 12OH^- \rightarrow 2ClO_3^- + 6H_2O + \cancel{10e^-} \\ 5Cl_2(aq) + \cancel{10e^-} \rightarrow 10Cl^- \\ \hline 6Cl_2(aq) + 12OH^- \rightarrow 10Cl^- + 2ClO_3^- + 6H_2O \end{array}$$

This equation is correctly balanced, but correct form dictates that all the coefficients be expressed as the lowest possible numbers. All the coefficients in our equation are divisible by 2; hence the correct equation is

$$3Cl_2(aq) + 6OH^- \rightarrow 5Cl^- + ClO_3^- + 3H_2O$$

VOLTAIC CELLS

If a piece of zinc metal is placed in a dilute solution of zinc sulfate, the zinc has a tendency to dissolve (to be oxidized):

$$Zn(s) \leftrightharpoons Zn^{2+} + 2e^-$$

The electrons produced remain on the metal and give it a negative charge. The process soon reaches an equilibrium, however (Figure 16–1**A**), where zinc dissolves at the same rate that the ions gain electrons to re-form zinc metal, the reverse reaction.

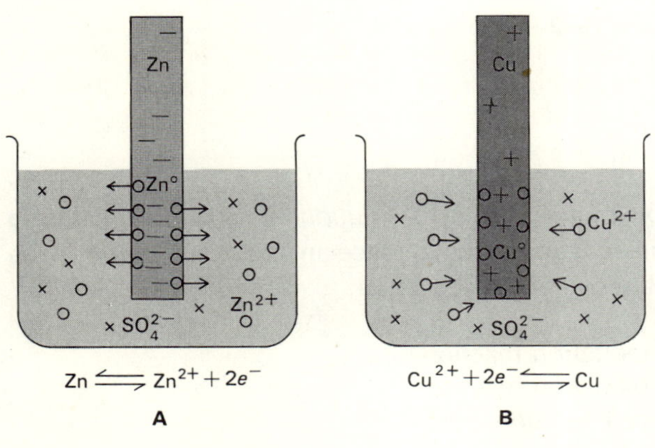

Figure 16–1 Diagram illustrating the tendency for zinc metal to go into solution, **A**, and for copper to plate out, **B**, when they are placed into solutions of their ions.

A piece of copper placed in a solution of copper(II) sulfate reaches an equilibrium favoring the reverse situation, reduction:

$$Cu^{2+} + 2e^- \leftrightharpoons Cu(s)$$

There is a tendency for the copper(II) ions to plate out on the copper metal and thus decrease the electron/proton ratio of the metal, giving it a positive charge (Figure 16–1**B**).

The two systems, zinc plus zinc ions and copper plus copper ions, can be combined to form a **voltaic cell,** Figure 16–2. The solutions must be in contact with each other, separated by a porous divider to prevent rapid diffusion. If the two pieces of metal are connected by a wire, the excess electrons from the zinc can travel through the wire to the electron-deficient copper. This flow of electrons is an electric current. The reaction will continue until either the copper ions have disappeared from the copper sulfate solution or until the zinc metal is gone. The zinc electrode where the oxidation reaction occurs is called the **anode.** The reduction reaction occurs at the copper **cathode.**

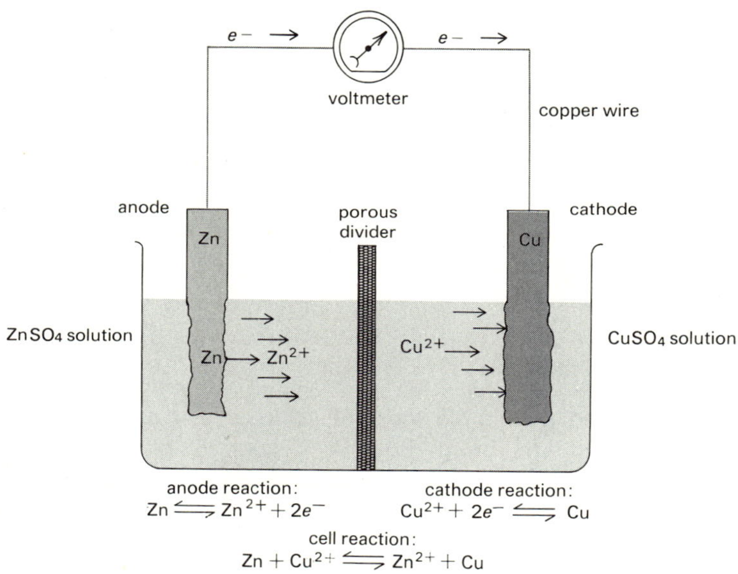

Figure 16–2 A voltaic cell.

The *individual combination of a metal in a solution of its ions* is called a **half-cell.** The equations for the reactions taking place in each half-cell can be added to give the equation for the overall cell reaction:

$$\begin{array}{rl} \cancel{2e^-} + Cu^{2+} \leftrightharpoons Cu(s) & \text{reduction reaction} \\ Zn(s) \leftrightharpoons Zn^{2+} + \cancel{2e^-} & \text{oxidation reaction} \\ \hline Zn(s) + Cu^{2+} \leftrightharpoons Zn^{2+} + Cu(s) & \text{cell reaction} \end{array}$$

ELECTRICAL UNITS

Before continuing the discussion of electrochemical systems, we must define a number of electrical terms and units.

Quantity of electricity

Consider the following half-cell reaction:

$$Ag^+ + e^- \rightarrow Ag(s)$$

One electron reacts with one silver ion to form one silver atom. Because this equation is balanced, we can say that 1 mole of silver ions reacts with 1 mole of electrons to give 1 mole of silver metal. This quantity of electricity, 1 mole, is called a faraday. A **faraday** is defined as *the amount of electricity needed to deposit 107.870 g of silver (1 mole) in an electrolysis experiment.* A smaller unit of electricity, called the **coulomb,** is commonly used; *96,487 coulombs equal one faraday.* Remember that faradays and coulombs refer to an amount of electricity—a precise number of electrons.

Example 16–4: Electrolysis calculations *How many faradays are needed to deposit 10.0 g of silver from $AgNO_3$ solution in an electrolysis experiment?*

This can be treated like a typical stoichiometry problem:

Balanced equation:	Ag^+	$+\ e^-$	$\leftrightharpoons Ag(s)$
Molecular data:	1 mole	1 mole	1 mole
	108 g	1 faraday	108 g

Hence

$$10.0\ \cancel{g} \times \frac{1\ \cancel{\text{mole } Ag^+}}{108\ \cancel{g}} \times \frac{1\ \text{faraday}}{1\ \cancel{\text{mole } Ag^+}} = 0.0926\ \text{faraday}$$

Example 16–5: Electrolysis calculations *Two inert electrodes are placed in a copper(II) sulfate solution, and an electric current is passed through the solution. How many faradays are required to deposit 10.0 g of copper?*

Using the same procedure as above, we have

Balanced equation:	Cu^{2+}	$+\ 2e^-$	$\leftrightharpoons Cu(s)$
Molecular data:	1 mole	2 moles	1 mole
	63.5 g	2 faradays	63.5 g

Hence

$$10.0\ \cancel{g} \times \frac{1\ \cancel{\text{mole } Cu^{2+}}}{63.5\ \cancel{g}} \times \frac{2\ \text{faradays}}{1\ \cancel{\text{mole } Cu^{2+}}} = 0.315\ \text{faraday}$$

Ampere

The **ampere** is a unit used to tell how fast electrons move through an electrical conductor. It is defined as *the flow rate—current—that will move one coulomb of electricity past a given point in one second.*

Ohm

All substances offer some resistance to the flow of electrons. Materials such as copper resist electron flow very little. Other materials, such as glass, are insulators; that is, they offer great resistance to moving electrons. The unit of electrical resistance is called the **ohm;** it is defined as *the resistance offered to a flow of electrons by a uniform section of mercury 106.3 cm long and containing exactly 14.4521 g of mercury at 0°C.* This definition does not have to be memorized; our only use of the definitions for ampere and ohm is to obtain the definition of the volt.

Volt

It takes a push, a force, to get electrons to flow through a wire. This force, called a potential or electromotive force (abbreviated emf), is **voltage;** it can be measured by a simple instrument called a voltmeter. A **volt** is defined as *the electromotive force (push) needed to produce a flow of 1 coulomb of electricity per second through a material with a resistance of 1 ohm.* Mathematically,

$$\text{volts} = \text{amperes} \times \text{ohms}$$

Different voltaic cells have different voltages. These voltages are very important; they tell us whether an oxidation-reduction reaction will go in the forward or reverse direction, and they tell us whether certain electroplating processes can occur.

OXIDATION POTENTIALS

When two half-cells are connected with a voltmeter in the circuit (Figure 16–2), it is an easy matter to decide which electrode is the anode and which is the cathode. The voltmeter also indicates the direction the electrons are flowing and gives the voltage of the complete cell.

It is not possible to measure the voltage of an individual half-cell directly—remember, it takes two to make a battery: the electrons liberated from one half-cell reaction must be taken up by another half-cell reaction. It is possible, however, to measure the voltages of many different voltaic cells in which different half-cells are connected to the same second half-cell. If we assign a voltage of zero to this second, standard half-cell, we can compile a table of relative voltages for the individual half-cells connected to it.

Standard hydrogen electrode

The reference half-cell assigned a zero voltage is the **hydrogen electrode** pictured in Figure 16–3. Hydrogen gas at 1 atm pressure is bubbled over a platinum metal electrode immersed in a solution whose effective hydronium-ion concentration is 1 mole. The reaction at this electrode is

$$2H_2O + H_2(g) \leftrightharpoons 2H_3O^+ + 2e^-$$

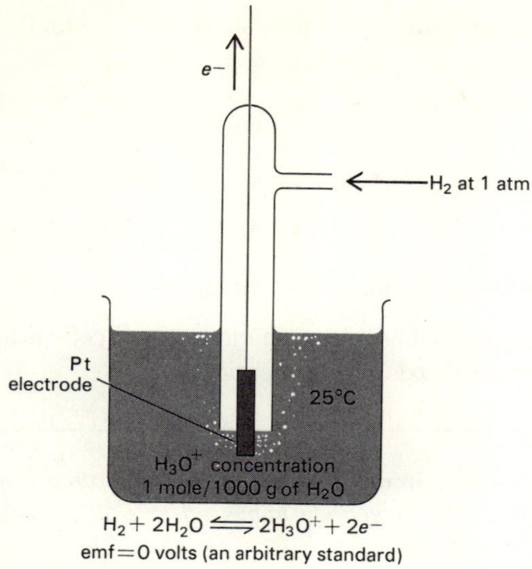

Figure 16–3 The standard hydrogen electrode.

Voltages of half-cells

The emf of any half-cell is the voltage measured when the half-cell is coupled with the standard hydrogen electrode. These emf values, called **oxidation potentials,** are given in Table 16–1 for elements in contact with a 1 *M* solution of their ions. If the value of the emf is positive, an oxidation reaction takes place at the given electrode when it is connected to a hydrogen electrode. Similarly, a negative emf indicates a reduction reaction when the electrode is connected to a hydrogen electrode.

TABLE 16–1 HALF-CELL REACTIONS AND THEIR OXIDATION POTENTIALS
Measured at 25°C for elements in contact with solutions of their ions at a concentration of 1 mole/liter.

Half-reaction	Emf, volts	Half-reaction	Emf, volts
$Li \rightleftharpoons Li^+ + e^-$	3.05	$Cu \rightleftharpoons Cu^{2+} + 2e^-$	−0.34
$K \rightleftharpoons K^+ + e^-$	2.92	$2I^- \rightleftharpoons I_2 + 2e^-$	−0.54
$Ca \rightleftharpoons Ca^{2+} + 2e^-$	2.87	$Fe^{2+} \rightleftharpoons Fe^{3+} + e^-$	−0.77
$Na \rightleftharpoons Na^+ + e^-$	2.71	$2Hg \rightleftharpoons Hg_2^{2+} + 2e^-$	−0.79
$Al \rightleftharpoons Al^{3+} + 3e^-$	1.67	$Ag \rightleftharpoons Ag^+ + e^-$	−0.80
$Zn \rightleftharpoons Zn^{2+} + 2e^-$	0.76	$Hg \rightleftharpoons Hg^{2+} + 2e^-$	−0.85
$Fe \rightleftharpoons Fe^{2+} + 2e^-$	0.44	$2Br^- \rightleftharpoons Br_2 + 2e^-$	−1.06
$Ni \rightleftharpoons Ni^{2+} + 2e^-$	0.25	$6H_2O \rightleftharpoons O_2 + 4H_3O^+ + 4e^-$	−1.23
$Sn \rightleftharpoons Sn^{2+} + 2e^-$	0.14	$2Cl^- \rightleftharpoons Cl_2 + 2e^-$	−1.36
$Pb \rightleftharpoons Pb^{2+} + 2e^-$	0.13	$Au \rightleftharpoons Au^+ + e^-$	−1.68
$2H_2O + H_2 \rightleftharpoons 2H_3O^+ + 2e^-$	0.00	$2F^- \rightleftharpoons F_2 + 2e^-$	−2.87

The reverse of any oxidation half-reaction is a reduction half-reaction, and vice versa. For example,

Oxidation: $Li \rightleftharpoons Li^+ + e^-$ 3.05 volts
Reduction: $Li^+ + e^- \rightleftharpoons Li$ −3.05 volts

Note that the voltage changes sign when a given half-reaction is reversed.

Using oxidation potentials to predict battery voltages

We can determine the emf of any combination of half-cells listed with data from the table of half-reactions and oxidation potentials.

Example 16–6: Cell voltages from oxidation potentials *What is the expected emf of a voltaic cell made from a silver electrode immersed in a 1 M solution of* Ag^+ *ions and a standard hydrogen electrode?*

The two half-reactions and the voltages, from Table 16–1, are

$$Ag(s) \rightleftharpoons Ag^+ + e^- \qquad -0.80 \text{ volt}$$
$$2H_2O + H_2(g) \rightleftharpoons 2H_3O^+ + 2e^- \qquad 0.00 \text{ volt}$$

When these two half-cells are connected to form a voltaic cell, a reduction reaction must occur at one electrode and an oxidation reaction at the other; that is, one of these reactions must go in the reverse direction. The rule is that the oxidation reaction takes place at the electrode with the greater positive oxidation potential (this electrode is the anode); a reduction reaction takes place at the electrode with the lesser positive oxidation potential (the cathode). In this problem, the hydrogen electrode has the greater positive oxidation potential, so it is the anode. Therefore, the two reactions occurring in the battery are:

$$e^- + Ag^+ \rightleftharpoons Ag(s) \qquad +0.80 \text{ volt}$$

and

$$2H_2O + H_2(g) \rightleftharpoons 2H_3O^+ + 2e^- \qquad 0.00 \text{ volt}$$

Adding the two half-cell reactions and their voltages gives the voltage of the battery.

$\cancel{2e^-} + 2Ag^+ \rightleftharpoons 2Ag(s)$	+0.80 volt	reduction
$2H_2O + H_2(g) \rightleftharpoons 2H_3O^+ + \cancel{2e^-}$	0.00 volt	oxidation
$2Ag^+ + H_2(g) \rightleftharpoons 2Ag + 2H_3O^+$	+0.80 volt	

The cell voltage is +0.80 volt. Note that the number of electrons lost by hydrogen must balance the number gained by silver ions. The balanced reduction equation was multiplied by 2 to obtain the net equation for the cell reaction; the emf of the half-cell, however, does not depend on the number of moles considered and is not multiplied by 2.

Example 16–7: Cell voltages from oxidation potentials *What is the voltage of a battery consisting of silver and zinc electrodes in contact with 1 M solutions of their ions?*

The half-reactions and emf's from Table 16–1 are

$$Ag(s) \leftrightharpoons Ag^{+} + e^{-} \qquad -0.80 \text{ volt}$$
$$Zn(s) \leftrightharpoons Zn^{2+} + 2e^{-} \qquad +0.76 \text{ volt}$$

Because zinc has the more positive emf, oxidation takes place at this electrode, and reduction at the silver electrode. Addition of the two half-reactions gives the cell reaction and cell voltage:

$$\begin{array}{rll} Zn(s) \leftrightharpoons Zn^{2+} + \cancel{2e^{-}} & +0.76 \text{ volt} & \text{oxidation} \\ \cancel{2e^{-}} + 2Ag^{+} \leftrightharpoons 2Ag(s) & +0.80 \text{ volt} & \text{reduction} \\ \hline Zn(s) + 2Ag^{+} \leftrightharpoons Zn^{2+} + 2Ag(s) & +1.56 \text{ volts} & \end{array}$$

DISPLACEMENT REACTIONS: WHICH DIRECTION WILL THEY PROCEED?

Will the following reaction occur spontaneously?

$$2I^{-} + Br_2(l) \leftrightharpoons I_2(s) + 2Br^{-}$$

To answer this question, consider the two half-reactions involved:

Oxidation reaction	$2I^{-} \leftrightharpoons I_2(s) + 2e^{-}$	−0.54 volt
Reduction reaction	$Br_2(l) + 2e^{-} \leftrightharpoons 2Br^{-}$	+1.06 volts

The emf of the overall reaction is the sum of −0.54 and +1.06, or +0.52 volt. Because this emf is positive, the reaction will occur spontaneously in the forward direction as written: liquid bromine will displace iodide ions from solution. A negative emf would have indicated that the reverse reaction was spontaneous. This spontaneous reaction, in which bromine displaces iodine from a solution of its ions, is a **displacement reaction.**

Any half-cell listed in Table 16–1 can be combined with the reverse of the one beneath it (one that has a lower emf) to give an overall equation for a spontaneous reaction. From the table, you can see that bromine displaces iodine, chlorine displaces bromine, and fluorine displaces chlorine. The order of these reactions shows the periodic relationships.

An important group of displacement reactions is the series in which metals displace hydrogen. Iron will displace hydrogen from an acid solution because the emf of the cell is a positive 0.44 volt (for a 1 *M* solution). For silver, however, the emf is negative: silver will not displace hydrogen from acid solutions. Metals below hydrogen in the emf series that do not displace hydrogen from acid solutions include Cu, Hg, Ag, Pt, and Au. Some of these metals do react with certain oxidizing acids, but only by means of an oxidation-reduction reaction other than displacement of hydrogen.

Example 16–8: Displacement reactions *Show that copper metal will not displace lead ions from a 1 M solution of lead(II) nitrate.*

Consider the following two half-reactions and emf's from Table 16–1:

$Cu(s) \leftrightharpoons Cu^{2+} + 2e^-$ $\quad -0.34$ volt
$Pb(s) \leftrightharpoons Pb^{2+} + 2e^-$ $\quad +0.13$ volt

When these two half-cells are combined to form a battery, the oxidation reaction will occur at the electrode with the more positive oxidation potential—the lead electrode. Reduction must occur at the copper electrode. Hence the reaction will proceed spontaneously when lead metal displaces copper ions from solution.

oxidation $-2e^-$

$$Pb(s) + Cu^{2+} \xrightarrow[\text{in this direction}]{\text{goes spontaneously}} Pb^{2+} + Cu(s) \qquad +0.47 \text{ volt}$$

reduction $+2e^-$

THE LEAD STORAGE BATTERY

A 12-volt automobile storage battery contains six individual 2-volt batteries wired in series. Each of these batteries contains a half-cell of lead immersed in a moderately concentrated sulfuric acid solution and a half-cell of lead(IV) oxide pressed onto a lead baffle plate immersed in the same solution. When an electric current is drawn from the battery the following reactions occur:

Oxidation at the anode:

$$H_2O + Pb(s) + HSO_4^- \rightarrow PbSO_4(s) + H_3O^+ + 2e^-$$

$-2e^-$

Reduction at the cathode:

$$PbO_2(s) + HSO_4^- + 3H_3O^+ + 2e^- \rightarrow PbSO_4(s) + 5H_2O$$

$+2e^-$

Adding these equations gives the net overall reaction:

$$Pb(s) + 2HSO_4^- + 2H_3O^+ + PbO_2(s) \rightarrow 2PbSO_4(s) + 4H_2O$$

The concentration of sulfuric acid decreases as the reaction proceeds. The extent of discharge of the battery can be determined from the density of the electrolyte: dilute sulfuric acid solutions are less dense than more concentrated solutions.

The electrodes in a discharged lead storage battery have been converted to lead sulfate. If an external emf greater than the voltage of the battery is applied to the battery so that the electrons flow in the opposite direction to that for the discharge, an electrolysis reaction will take place. The great usefulness of storage batteries depends on this reaction, for it regenerates the original lead and lead(IV) oxide as well as the sulfuric acid. Thus the charging reaction is the reverse of the discharge reaction.

Theoretically, storage batteries can be charged and discharged an infinite number of times. Several factors, however, contribute to shorter battery life. Some of the water is electrolyzed during recharging, and if this is not replaced, the upper portions of the electrodes will dry up and lead sulfate will drop off. Bumping and shaking also tends to cause the lead sulfate to fall from the electrodes. The addition of impure water containing foreign ions will produce undesirable side reactions that will shorten the life of the battery. It is an accepted fact of nature, however, that a battery guaranteed for 36 months will last exactly 35 months.

NEW TERMS

Ampere: an electrical unit that tells how fast electrons move through an electrical conductor. An ampere equals 1 coulomb per second.

Anode: the cell electrode where the oxidation reaction takes place. In batteries, the anode is the electrode with a negative charge. In electrolysis, it is the electrode given a positive charge.

Battery: a voltaic cell or a series of voltaic cells used as a source of electricity.

Cathode: the cell electrode where the reduction reaction takes place. It is the negative electrode in an electrolytic cell, and the positive electrode in a battery.

Conductor: a material that offers little resistance to the flow of an electric current.

Coulomb: a unit of electricity equal to amperes times seconds; 96,487 coulombs equals 1 faraday.

Displacement reaction: a reaction in which one element displaces another from solutions of its ions.

emf: abbreviation for electromotive force; voltage.

Faraday: one mole of electrons; 1 faraday equals 96,487 coulombs.

Half-cell: the section of an electrochemical cell in which either the oxidation or the reduction half-reaction takes place.

Half-reaction: an equation containing both electrons and chemical species which is used to show what happens in either the oxidation or the reduction part of a redox reaction.

Hydrogen electrode: the standard half-cell that is used as a reference for the tabulation of oxidation potentials. It is assigned a zero voltage.

Insulator: a material that does not readily conduct an electric current.

Ohm: the unit of electrical resistance.

Oxidation: a reaction in which there is a loss of electrons or an increase in the oxidation number of an element to a more positive value.

Oxidation potential: the voltage of a half-cell when it is coupled with a standard hydrogen electrode. It is a measure of the tendency for a specific oxidation half-cell reaction to occur.

Oxidizing agent: the reactant in a redox equation that increases in oxidation number and gains electrons. It is reduced in the course of the reaction.

Reducing agent: the reactant in a redox reaction that decreases in oxidation number and loses electrons. It is oxidized in the course of the reaction.

Reduction: a reaction in which there is a gain of electrons or a decrease in the oxidation number of an element to a less positive value.

Storage battery: a battery that can be recharged by passing an electrical current through it in a direction opposite to that which occurs on discharge.

Volt: the electromotive force needed to produce a flow of 1 coulomb of electricity per second through a material that has a resistance of 1 ohm.

Voltaic cell: a device that produces an electric current by means of an oxidation-reduction reaction.

PROBLEM SET A

1. Add electrons to complete and balance the following half-reactions. Tell which reactions are oxidations and which are reductions.
   a. $Al \rightarrow Al^{3+}$
   b. $Al^{3+} \rightarrow Al$
   c. $O_2 \rightarrow 2O^{2-}$
   d. $Fe^{2+} \rightarrow Fe^{3+}$
2. Balance the following half-reaction equations by first balancing the number of atoms and then adding electrons. Tell which are oxidation and which are reduction reactions.
   a. $N_2 \rightarrow N^{3-}$
   b. $H^- \rightarrow H_2$
   c. $S_8 \rightarrow S^{2-}$
   d. $Cl_2 \rightarrow Cl^-$
3. The following skeletal equations describe some of the common oxidizing agents used in laboratory experiments. Balance these half-reactions, assuming an acid medium.
   a. $MnO_4^- \rightarrow Mn^{2+}$
   b. $NO_3^- \rightarrow NO_2$
   c. $NO_3^- \rightarrow NO$
   d. $Cr_2O_7^{2-} \rightarrow Cr^{3+}$
4. The following skeletal equations describe some of the common reducing agents used in laboratory experiments. Balance these half-reactions, assuming an acid medium.
   a. $H_2S \rightarrow S$
   b. $H_2O_2 \rightarrow O_2$
   c. $H_2C_2O_4 \rightarrow CO_2$
   d. $SO_3^{2-} \rightarrow SO_4^{2-}$
5. Combine the oxidation and reduction half-reactions indicated below to make complete, balanced, ionic-redox equations. (These are known spontaneous reactions; however it must not be assumed that all such redox pairings result in reaction—see the discussion on displacement reactions.)
   a. Problems 3a and 4c
   b. Problems 3b and 4a
   c. Problems 3d and 4b
6. Balance the following skeletal redox equations. Assume an acid medium.
   a. $MnO_4^- + Fe^{2+} \rightarrow Mn^{2+} + Fe^{3+}$
   b. $Fe^{2+} + H_2O_2 \rightarrow Fe^{3+} + H_2O$
   c. $S_2O_3^{2-} + I_2 \rightarrow S_4O_6^{2-} + I^-$
   d. $Cr_2O_7^{2-} + Cl^- \rightarrow Cr^{3+} + Cl_2$
7. Balance the following skeletal redox equations. Assume an alkaline medium.
   a. $Cr^{3+} + Cl_2 \rightarrow CrO_4^{2-} + Cl^-$
   b. $Cr(OH)_4^- + H_2O_2 \rightarrow CrO_4^{2-} + H_2O$
8. Consider the following balanced equation:
   $3Cu + 2NO_3^- + 8H^+ \rightarrow 3Cu^{2+} + 2NO + 4H_2O$
   What molecular equation corresponds to this net ionic equation? Since there can be more than one answer, assume that the spectator ions were nitrate ions.
9. Convert the following to molecular equations. Identify the oxidizing agent and the reducing agent.
   a. $H_2O_2 + 2H^+ + 2Fe^{2+} \rightarrow 2H_2O + 2Fe^{3+}$
   b. $3Cl_2 + 6OH^- \rightarrow ClO_3^- + 5Cl^- + 3H_2O$
   c. $2Fe^{3+} + 2I^- \rightarrow 2Fe^{2+} + I_2$
10. How many faradays of electricity are theoretically needed to deposit 1 mole of silver when an electric current is passed through a silver nitrate solution using inert electrodes?

How many faradays are needed to deposit 1 mole of copper from copper nitrate solutions? One mole of iron from iron(III) nitrate?

11. a. 0.30 faraday = ___ coulombs.
    b. 10,000 coulombs = ___ faraday.

12. How many faradays are needed to deposit 60.0 g of silver metal from 0.468 *M* silver nitrate solution using inert electrodes?

13. Two inert electrodes are put into a copper(II) sulfate solution and 4.50 faradays of electricity are passed through the solution. How many g of copper are deposited?

14. What mass of silver will be deposited from a solution of $AgNO_3$ by a current of 10.0 amperes flowing for 5 hr? (A coulomb equals amperes times seconds.)

15. A copper, copper-ion half-cell is coupled to a hydrogen electrode. Which electrode is the anode, and which is the cathode? Where does the oxidation reaction take place?

16. Which of the following reactions will proceed in the direction written?
    a. $Cl_2(g) + 2Br^- \rightarrow 2Cl^- + Br_2(l)$
    b. $Br_2(l) + 2Cl^- \rightarrow 2Br^- + Cl_2(g)$
    c. $Cu + Sn^{2+} \rightarrow Sn + Cu^{2+}$
    d. $2Ag(s) + 2HCl(aq) \rightarrow 2AgCl(s) + H_2(g)$
    e. $Cu^{2+} + 2Cl^- \rightarrow Cu(s) + Cl_2$

17. What is the predicted emf of voltaic cells made from the following combinations of half-cells, assuming concentrations of 1 *M* for the electrolytes? Write the equations for the overall reactions.
    a. $Zn(s) \leftrightharpoons Zn^{2+} + 2e^-$ and $Ag(s) \leftrightharpoons Ag^+ + e^-$
    b. $Fe(s) \leftrightharpoons Fe^{2+} + 2e^-$ and $Au(s) \leftrightharpoons Au^+ + e^-$
    c. $Pb(s) \leftrightharpoons Pb^{2+} + 2e^-$ and $Ni(s) \leftrightharpoons Ni^{2+} + 2e^-$

18. Bumping and shaking of lead storage batteries can cause lead sulfate to drop off the electrodes. Explain why this loss decreases the life of the battery.

PROBLEM SET B

1. Which of the following reactions are oxidation reactions, and which are reduction reactions? Add the correct number of electrons to balance the equations.
    a. $2I^- \rightarrow I_2(s)$
    b. $2H_3O^+ \rightarrow H_2(g) + 2H_2O$
    c. $Sn(s) \rightarrow Sn^{2+}$
    d. $Br_2(l) + 12OH^- \rightarrow 6H_2O + 2BrO_3^-$
    e. $K^+ \rightarrow K(s)$

2. For the following equations, (1) write the oxidation number under each element (2) identify the element that is oxidized and the one that is reduced, (3) show the number of electrons lost and gained.
    a. $MnO_2(s) + 4HCl(aq) \rightarrow MnCl_2(aq) + Cl_2 + 2H_2O$
    b. $3P(s) + 5HNO_3(aq) + 2H_2O \rightarrow 5NO(g) + 3H_3PO_4(aq)$
    c. $2IO_3^- + 6H_2O \rightarrow I_2(s) + 12OH^-$
    d. $3Fe(s) + 6H_3O^+ \rightarrow 3H_2(g) + 2Fe^{3+} + 6H_2O$
    e. $2H_2S(g) + SO_2(g) \rightarrow 3S(s) + 2H_2O$
    f. $2Na(s) + H_2(g) \rightarrow 2NaH(s)$

3. Balance the following half-reaction equations.
    a. $Sn^{4+} \rightarrow Sn^{2+}$
    b. $Hg_2^{2+} \rightarrow Hg$
    c. $O_3 \rightarrow O_2 + O^{2-}$
    d. $H_2O_2 \rightarrow H_2O$
    e. $HClO \rightarrow Cl^-$
    f. $I_2 \rightarrow IO_3^-$
    g. $NO_2^- \rightarrow NO_3^-$

4. Balance the following skeletal redox equations.
    a. $Zn + NO_3^- \rightarrow Zn^{2+} + NH_4^+$
    b. $MnO_4^- + H_2O_2 \rightarrow Mn^{2+}$
    c. $I_2 + H_2S \rightarrow I^- + S$
    d. $Al + NO_3^- \rightarrow Al(OH)_4^- + NH_3$
    e. $Cl_2 + IO_3^- \rightarrow IO_4^- + Cl^-$

5. Convert the following ionic equations to molecular equations.
   a. $MnO_4^- + NO_2 + 2OH^- \rightarrow MnO_4^{2-} + NO_3^- + H_2O$
   b. $3CN^- + 2MnO_4^- + H_2O \rightarrow 3CNO^- + 2MnO_2 + 2OH^-$

6. A strip of zinc metal placed in a copper sulfate solution changes to a metallic copper color. Explain.

7. a. 0.650 faraday = ___ coulombs.
   b. 5200 coulombs = ___ faradays

8. How many faradays of electricity are needed to deposit 124 g of silver metal from 0.850 *M* $AgNO_3$ solution using inert electrodes?

9. What mass of silver will be deposited from a solution of $AgNO_3$ by a current of 5.0 amperes flowing for 16 hours?

10. A copper-copper ion half-cell is coupled to a silver-silver ion half-cell. Which electrode is the anode, and which the cathode? Where does the oxidation reaction take place? Write the balanced net ionic equation for the reaction.

11. Which of the following reactions will proceed in the direction written?
   a. $Cl_2(g) + 2I^- \rightarrow 2Cl^- + I_2(s)$
   b. $Br_2(l) + 2F^- \rightarrow 2Br^- + F_2(g)$
   c. $2Au(s) + 2H_3O^+ \rightarrow 2Au^+ + H_2(g) + 2H_2O$
   d. $Ni(s) + 3H_3O^+ \rightarrow Ni^{2+} + H_2(g) + 2H_2O$
   e. $Pb(s) + Hg^{2+} \rightarrow Pb^{2+} + Hg(l)$

12. What is the predicted emf of the voltaic cells made from the following combinations of half-cells, assuming concentrations of 1 *M* for the electrolytes? Write the equations for the overall reactions.
   a. $2H_2O + H_2(g) \rightleftharpoons 2H_3O^+ + 2e^-$ and $Ag(s) \rightleftharpoons Ag^+ + e^-$
   b. $Au(s) \rightleftharpoons Au^+ + e^-$ and $2H_2O + H_2(g) \rightleftharpoons 2H_3O^+ + 2e^-$

17

reaction rates and chemical equilibrium

THIS WILL SAVE YOU TIME:

Refer to this list as you read through the chapter. In answering the questions, you will understand the important concepts you are most likely to be asked about later on.

After reading this chapter you should be able to:

1. *State some of the information about a chemical reaction not available from the balanced equation.*
2. *Explain what is meant by reaction rates.*
3. *List some of the techniques used to measure reaction rates.*
4. *Discuss the four factors that affect the rate of a chemical reaction.*
5. *Explain why a small increase in temperature can produce a marked increase in the rate of a chemical reaction.*
6. *Explain the law of mass action and write the rate equations for elementary reactions.*
7. *Give a general explanation for catalysis in terms of energy considerations, activated complexes, and surface compounds.*
8. *Write the equations for the equilibrium constants for elementary reactions.*
9. *Work problems involving equilibrium constants.*
10. *Explain the effects of concentration changes, temperature, and catalysts on chemical equilibrium.*
11. *Discuss the reaction conditions used to prepare ammonia commercially.*
12. *Work problems involving ionic equilibria.*
13. *Work problems involving the common ion effect.*
14. *Work solubility product problems.*

Although a balanced chemical equation gives much quantitative data about the reactants and products of a reaction, it still omits a vast amount of information of interest to both theory and practice. Consider the following reaction as an example:

$$\underset{\text{ethyl alcohol}}{CH_3-CH_2-OH} + \underset{\text{acetic acid}}{HO-\underset{\underset{\displaystyle O}{\|}}{C}-CH_3} \leftrightharpoons \underset{\text{ethyl acetate}}{CH_3-CH_2-O-\underset{\underset{\displaystyle O}{\|}}{C}-CH_3} + H_2O$$

This type of reaction—the reaction of an alcohol with an acid to yield water and one of a class of compounds known as esters—is important to organic chemistry and to the commercial manufacture of products as diverse as soap and synthetic fibers. The equation does tell us what goes into the reaction and what comes out, but it leaves a number of significant questions unanswered.

1. How fast does the reaction proceed? If it goes slowly, how can it be speeded up to give maximum yield of the product in a given time?
2. What is the actual mechanism for the reaction? Does the oxygen atom in the water molecule come from the alcohol or the acid? Do the alcohol and acid react to produce the water and ester directly, or do they first yield an intermediate product, which then dissociates to ester and water?
3. Will the ester react with water to regenerate the initial reactants? If so, how can this reverse reaction be prevented?

In this chapter, we will look at a way of answering these questions—the study of the rates of chemical reactions and of chemical equilibria.

RATES OF CHEMICAL REACTIONS

The rate of a chemical change, the **reaction velocity** or **reaction rate,** is defined as *the amount of reactant consumed or product produced in a given period of time.* In the laboratory, the rate of any given reaction can be measured in a variety of ways. As a matter of fact, there probably are as many ways to measure reaction rates as there are reaction rates to be measured. The rate of decomposition of ammonia,

$$2NH_3 \rightarrow N_2 + 3H_2$$

can be studied by observing the reaction in a constant-volume apparatus and measuring the change in pressure. Other reaction rates can be measured by following color changes in reactants or products, changes in absorption spectra, variations in electrical conductivity, and so on. In a much-used technique, a sample is periodically withdrawn from the reaction system for analysis of the relative amounts of reactant and product. Removal of reactants or products, however, alters the concentration of the remaining species and makes the interpretation of the result more difficult.

Experiments with a great many reactions have shown that four factors influence the rate of a chemical reaction: the nature of the reactants, the concentration of reactants, temperature, and the presence or absence of a catalyst.

THE NATURE OF REACTANTS

The nature of the reactants is a factor we cannot control. Zinc metal reacts slowly with hydrochloric acid, sodium reacts very vigorously, and copper does not react at all. Iron rusts (reacts with oxygen) very slowly; magnesium burns very rapidly in air. These properties depend on the atomic and molecular structures of the substances and on the strengths of the chemical bonds broken and formed during the reaction.

THE LAW OF MASS ACTION: THE EFFECT OF CONCENTRATION ON REACTION RATES

A match burns slowly in air, but in pure oxygen the reaction is extremely fast: an increase in the concentration of one of the reactants—oxygen—greatly increases the rate of the reaction. This observation shows how concentration affects the rate of a chemical reaction. To account for this phenomenon, C. M. Guldberg (1836–1902) and Peter Waage (1833–1900) formulated the **law of mass action:** *The velocity or rate of a chemical reaction at a given instant is proportional to the concentration of reactants, where each concentration is raised to a power equal to the number of moles of that species participating in the reaction.*

For a general elementary reaction of the type

$$aA + bB \rightleftharpoons cC + dD$$

the law of mass action says that the rate of the forward reaction of A with B to give C plus D is equal to

$$\text{rate} = k[A]^a[B]^b$$

where the brackets indicate concentrations in moles per liter and k is a constant called the **specific rate constant.**

Consider the following general reaction scheme:

$$A_2 + B_2 \rightleftharpoons 2AB$$

For the forward reaction to take place, the A_2 molecules have to collide with the B_2 molecules. Increasing the concentration of either A_2 or B_2 molecules will make it possible for more collisions to occur in a given period of time—the rate of the reaction will increase. For this reason, it has been found that

(concentration of A_2 molecules) (concentration of B_2 molecules)

is proportional to the rate of the reaction. In mathematical form,

$$k_f[A_2][B_2] = \text{rate of the forward reaction}$$

The constant k_f is needed to provide the mathematical equality. As this reaction proceeds, the rate of the forward reaction decreases. This decrease must occur, for the concentrations of both A_2 and B_2 decrease as more and more product is formed.

The reverse reaction

$2AB \rightleftharpoons A_2 + B_2$ or $AB + AB \rightleftharpoons A_2 + B_2$

involves the combination of two molecules of AB to produce one molecule each of A_2 and B_2. For this reaction, the rate depends on the square of the concentration of AB molecules:

$k_r[AB]^2 =$ rate of reverse reaction (the constant k_r differs from k_f)

Example 17–1: Mass action *Explain why things burn faster in pure oxygen than in air.*

Reactions can occur only when the two reacting species come into contact with one another: they must make effective collisions before they can react. Because air is only 21 percent oxygen, there are approximately five times as many oxygen molecules in a given volume of oxygen as there are in the same volume of air. Thus the chance of a substance finding oxygen molecules to react with is much greater when it is placed in pure oxygen, and the reaction rate is greatly increased.

TEMPERATURE AND REACTION RATE

A rise in temperature increases the rate of almost all chemical reactions; conversely, a drop in temperature decreases the rate. For many reactions, a 10°C rise in temperature can double or triple the rate.

The effect of temperature on the rate of a homogeneous gas reaction can be visualized as follows: the molecules of the gaseous reactants collide with each other at a very rapid rate, but most of these collisions are elastic—involving no net energy change—and produce no reaction. Some molecules, however, are moving with greater than average speeds. These molecules have sufficient energy to overcome the repulsive forces of the molecules they strike, as well as the cohesive forces of the bonds within the molecules, and a reaction ensues. If the temperature of the gaseous reactants increases 10°C, the average velocity of the molecules increases by a relatively small amount, but the reaction velocity more than doubles. The rate of the reaction does not depend on the average velocity or the average energy of the molecules but on the number with the greater than average amount of energy needed for reaction. In Figure 17–1, the graph shows how molecules are distributed as a function of kinetic energy at two temperatures, T_1 and T_2, 10° apart. At the lower temperature, the number of molecules with sufficient energy for reaction is indicated by the curve below dashed line *a;* at the higher temperature, by the curve below line *b.* Even though the average energy of the molecules has increased by only a small percent, the number with the needed energy for reaction has more than doubled.

Reactions cannot take place until the reactants have acquired a certain amount of energy. Consider the following general scheme for an exothermic reaction:

$A + B \rightleftharpoons AB +$ heat

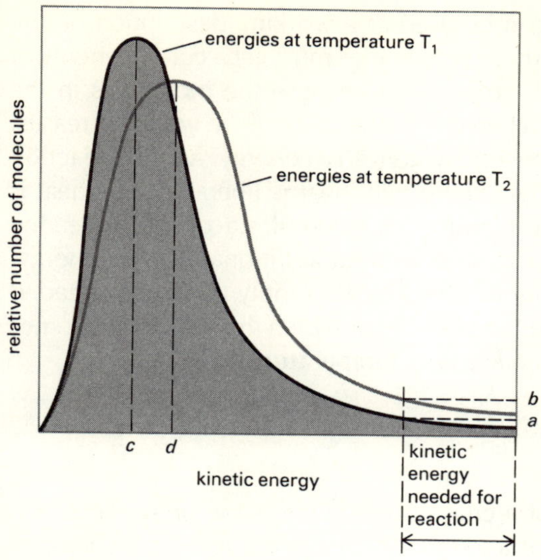

Figure 17–1 The relative energy distribution for gaseous molecules at two different temperatures (not to scale). A small temperature increase produces a large increase in the number of molecules that have a greater-than-average energy.

No reaction can occur until the molecules (atoms or ions) of reactant A collide with molecules (atoms or ions) of reactant B to form an intermediate species called an **activated complex:**

$$A + B \rightleftharpoons [A\cdots B] \rightarrow AB + \text{heat}$$

The activated complex, indicated by $[A\cdots B]$, may have a lifetime of from 10^{-14} to 10^{-15} seconds. During this time, it either breaks apart to re-form the reactants or stabilizes to produce the product AB. *The energy needed to form the activated complex* is called the **activation energy.** The energy relationships are illustrated in Figure 17–2. Reactants A and B have a certain amount of energy E_{A+B}. The molecules with enough energy to react on collision form an activated complex of greater energy $E_a + E_{A+B}$. If this complex stabilizes to yield the reaction product AB, heat is given off. The difference between the energy of the reactants E_{A+B} and the energy of the product E_{AB} is the exothermic heat of the reaction.

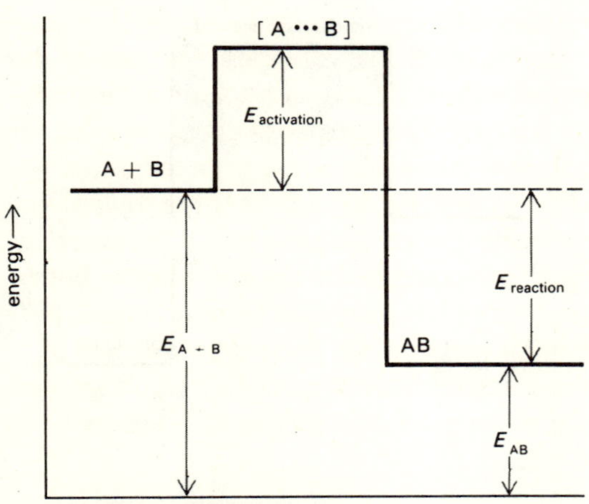

Figure 17–2 Energy relationships in an exothermic reaction.

This rather theoretical discussion can be used as a working description of any chemical reaction. Before wood will burn, the molecules must get a certain energy of activation. No reaction occurs at room temperature because the molecules in the wood lack this extra amount of energy. When the temperature of the wood increases, however, a few molecules acquire the necessary activation energy, and the reaction begins. Because this oxidation reaction is exothermic, heat is liberated. The heat is absorbed by other wood molecules; more and more molecules acquire the needed energy of activation; and the process continues without additional outside energy. The reaction is self-sustaining; the first molecules to react supply the energy needed to activate other molecules to react. The temperature at which the reaction becomes a self-sustaining combustion is called the **kindling temperature.**

CATALYSIS

The fourth factor that affects the rate of a chemical reaction is the presence or absence of a catalyst. A **catalyst** is *a substance that changes the rate of a chemical reaction but does not itself appear in the reaction products.* The presence of a catalyst will alter the rate of a reaction.

The common laboratory method for preparing oxygen can be used as an example. When heated, potassium chlorate, $KClO_3$, decomposes slowly to give potassium chloride and oxygen:

$$2KClO_3(s) \rightarrow 2KCl(s) + 3O_2(g)$$

A small amount of manganese dioxide, MnO_2, added to the potassium chlorate greatly accelerates the rate of the reaction. After all the potassium chlorate has decomposed, the manganese dioxide catalyst can be recovered unchanged. Catalysts do participate in some step of the reaction, however; and many are chemically altered during the course of the reaction.

In general, a catalyst lowers the activation-energy requirements for a reaction (Figure 17–3). A catalyst furnishes an alternative, lower-energy path for the reaction.

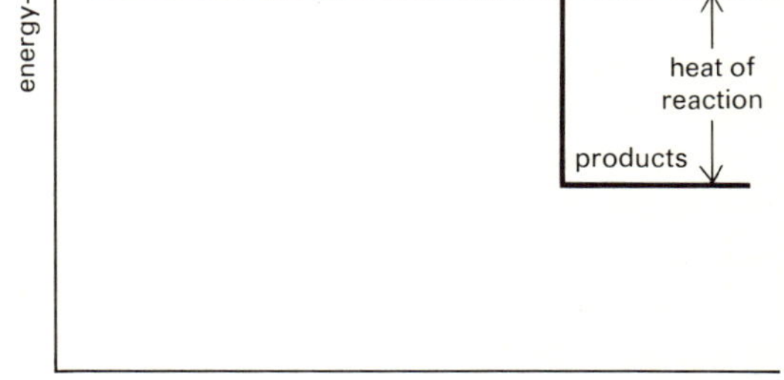

Figure 17–3 Energy relationships for an exothermic reaction with and without a catalyst.

This alternative route undoubtedly involves the formation of an intermediate complex that includes the surface of the catalyst, probably a **surface compound.** If this activated-complex, surface-containing intermediate has a lower energy than does the activated complex formed when the catalytic surface is not present, the reaction will follow the path of least resistance. The catalyst speeds the rate of the reaction by lowering the energy requirements.

CHEMICAL EQUILIBRIA

Many reactions are reversible. If 1 mole of nitrogen and 3 moles of hydrogen are mixed in a vessel at 200°C at a constant pressure of 1 atm, a 15.3 percent yield of ammonia is obtained as a product. If 2 moles of pure ammonia alone are placed in a similar vessel at the same temperature and pressure, all but 15.3 percent of the ammonia decomposes to hydrogen and nitrogen. These data show that the reaction is reversible. The double arrow in the equation indicates reversibility:

$3H_2(g) + N_2(g) \rightleftharpoons 2NH_3(g)$
forward reaction

$2NH_3(g) \rightleftharpoons 3H_2(g) + N_2(g)$
reverse reaction

If these two reactions are allowed to continue for a long period of time, the distribution of the reaction products will not change. The mixtures will contain the same amounts of hydrogen, nitrogen, and ammonia, whether we start with hydrogen and nitrogen or with ammonia. This situation, in which a reaction does not go to completion and the forward reaction competes with the reverse reaction, is called a state of chemical equilibrium. Alexander W. Williamson (1824–1904) studied the reaction of alcohols with acids and concluded that **chemical equilibrium** is *a dynamic situation in which both the forward and reverse reactions are going on at the same time: the rate of the forward reaction is equal to the rate of the reverse reaction.*

EQUILIBRIUM CONSTANTS

For any general elementary reaction

$a\mathrm{A} + b\mathrm{B} \rightleftharpoons c\mathrm{C} + d\mathrm{D}$

the law of mass action permits us to write

rate of forward reaction $= k_f[\mathrm{A}]^a[\mathrm{B}]^b$
rate of reverse reaction $= k_r[\mathrm{C}]^c[\mathrm{D}]^d$

where the exponents, *a, b, c,* and *d,* have the same numerical values as those in the balanced chemical equation.

Because the rate of the forward reaction equals the rate of the reverse reaction at equilibrium, we can write rate of forward reaction = rate of reverse reaction

$$k_f[A]^a[B]^b = k_r[C]^c[D]^d$$

or

$$\frac{k_f}{k_r} = \frac{[C]^c[D]^d}{[A]^a[B]^b} = K$$

This combined constant, k_f/k_r, is called the **equilibrium constant,** K. This constant is determined by experiment, and its value can range from very small to very large numbers.

Example 17–2: Equilibrium constants *Write the equilibrium-constant expression for the reaction of hydrogen with iodine.*

From the balanced equation

$$H_2(g) + I_2(g) \rightleftharpoons 2HI(g)$$

the equilibrium constant expression can be written

$$K_f = \frac{[HI]^2}{[H_2][I_2]}$$

The magnitude of the equilibrium constant tells how far a reaction proceeds in the forward or reverse direction. If the constant is large, for example, 10^5, the reaction proceeds almost to completion in the forward direction. If the constant is small, for example, 10^{-5}, the reverse reaction goes almost to completion, and the progress of the forward reaction is almost negligible.

Example 17–3: Equilibrium constants *The equilibrium constant for the hydrogen plus iodine reaction at 444°C is 48 (the concentrations of the reactants and products are expressed in pressure units; the constant K in this case is unitless). How far will the reaction proceed in the forward direction?*

The value of the equilibrium constant is greater than 1; hence the forward reaction proceeds farther than does the reverse reaction. Because the value of the equilibrium constant is quite small, the reaction does not go to completion but reaches an equilibrium somewhat in favor of the production of hydrogen iodide.

Example 17–4: Equilibrium constants *If the equilibrium constant for the hydrogen plus iodine reaction at 444°C is 48, what is the equilibrium constant for the decomposition of hydrogen iodide at that same temperature?*

The decomposition of hydrogen iodide is the reverse reaction,

$$2HI(g) \rightleftharpoons H_2(g) + I_2(g)$$

The equilibrium constant for this reaction is

$$K_r = \frac{[H_2][I_2]}{[HI]^2}$$

This equilibrium constant is the reciprocal of the equilibrium constant for the hydrogen plus iodine reaction,

$$K_r = \frac{1}{K_f}$$

Hence the numerical value of K_r must be

$$K_r = \frac{1}{48} = 0.020$$

This equilibrium constant is less than 1; hence the reverse reaction, $H_2 + I_2$, proceeds farther. Thus we arrive at the same conclusion as was obtained in Example 17–3. Note that the reciprocal of the equilibrium constant for the forward reaction is always the equilibrium constant for the reverse reaction.

Example 17–5: Equilibrium constant *A reaction vessel in which the following reaction had reached a state of equilibrium,*

$$CO(g) + Cl_2(g) \rightleftharpoons COCl_2(g),$$

was found to contain 0.300 mole of CO, 0.200 mole of Cl_2, and 0.800 mole of $COCl_2$ (phosgene). Calculate the value of the equilibrium constant for this reaction.

Let K be the equilibrium constant; it equals

$$K = \frac{[COCl_2]}{[CO][Cl_2]}$$

The notation $[COCl_2]$ means concentration in moles of $COCl_2$/liter; similarly, [CO] and $[Cl_2]$ refer to their concentrations in moles/liter. Substituting the given values into the equilibrium expression gives

$$K = \frac{0.800}{(0.300)(0.200)} = 13.3$$

It is not necessary to worry about the units for K; it is always assumed that concentrations are expressed in moles/liter unless other units are specified.

THE EFFECT OF CONCENTRATION ON CHEMICAL EQUILIBRIUM

Consider the general reaction

$$A + B \rightleftharpoons C + D$$

When the concentration of a reactant, B for example, increases, the rate of the forward reaction increases. As more and more product forms, the reaction reaches a point where the rate of the forward reaction equals the rate of the reverse—equilibrium has been attained. The increase in concentration of a reactant not only makes the reaction reach equilibrium more quickly, but also alters the position of the equilibrium. Consider the equilibrium constant for the reaction,

$$K_f = \frac{[C][D]}{[A][B]}$$

If the concentration of B increases and the concentration of A remains constant, the numerical value of the denominator in the equilibrium expression increases; that is, [A][B] gets larger. The equilibrium constant K_f, however, must remain constant; therefore, it is necessary for the concentration of C or D, or both, to increase so that the numerical value of the numerator balances the increased value of the denominator.

LE CHÂTELIER'S PRINCIPLE

Le Châtelier's principle, named for Henri Le Châtelier (1850–1936), says that *when a stress is placed on a system in equilibrium, the system will readjust itself in a direction that will tend to relieve the stress and restore equilibrium.* The principle can be applied to the effect of concentration changes on chemical reactions at equilibrium. An increase in concentration of a reactant places a stress on the system. The system then tries to overcome the stress by producing more products: the concentration of the reactants decreases, and equilibrium is again obtained.

Example 17–6: Le Châtelier's principle *The reaction of zinc metal with hydrochloric acid goes to completion. Can you explain this in terms of the effect of concentration on chemical equilibrium?*

The removal of a reaction product shifts the position of equilibrium in the direction of more products. This behavior can be explained by a consideration of the reaction and its equilibrium constant.

$$Zn(s) + 2H_3O^+ + \leftrightharpoons Zn^{2+} + H_2(g) + 2H_2O$$

$$K = \frac{[Zn^{2+}][H_2O]^2[H_2]}{[Zn][H_3O^+]^2}$$

The hydrogen gas escapes from the vicinity of the reaction, and its concentration in the reaction system becomes very low. To counteract this effect, the denominator of the equilibrium expression must also decrease; hence more and more reaction occurs, and the reaction goes to completion.

THE EFFECT OF TEMPERATURE AND CATALYSTS ON CHEMICAL EQUILIBRIUM

The magnitude of the equilibrium constant increases with a rise in temperature if the reaction is endothermic, and decreases if the reaction is exothermic. The equilibrium constant and the position of equilibrium are not affected, however, by the presence of a catalyst. A catalyst alters the rate of both the forward and reverse reactions, but does not change the position of equilibrium.

AMMONIA SYNTHESIS: AN EXAMPLE OF RATE AND EQUILIBRIUM CONSIDERATIONS

What experimental conditions are used to produce the best yields in the commercial production of ammonia? To answer this question, we must consider how concentration, temperature, and catalysts affect the reaction. The balanced equation and the equilibrium constant are

$$3H_2(g) + N_2(g) \rightleftharpoons 2NH_3(g) + 22 \text{ kcal/mole (exothermic)}$$

$$K = \frac{[NH_3]^2}{[H_2]^3[N_2]} = 0.0088 \text{ (pressure units) at } 450°\text{C and } 300 \text{ atm}$$

An increase in the concentration of a reactant will shift the position of equilibrium to the right; this increase can be accomplished by forcing more reactants into the same volume. Compression of the reaction mixture will increase the pressure of the reactants as well as the pressure of the products. Le Châtelier's principle applies readily; if a volume change occurs during a reaction, a pressure increase will shift the position of equilibrium in the direction of the smaller volume. In the ammonia-synthesis reaction, the change is from four to two volumes; thus a compression of the reaction system will shift the equilibrium to the right (product formation), in the direction of the smaller volume. Some manufacturers use pressures as high as 1000 atm, but in this country most use pressures from 200 to 350 atm.

An increase in temperature will favor the reverse reaction, because the forward process is exothermic. In practice, however, it is necessary to operate at a temperature of 500°C or more for the reaction to proceed at a reasonable rate. This increase in temperature lowers the ammonia yield. At 450°C and 300 atm, only 35.5 percent ammonia is present at equilibrium.

An iron catalyst (containing small amounts of potassium and aluminum oxides) increases the rate of reaction, but it does not change the amount of ammonia present at equilibrium conditions. In industrial plants where the reactants are recirculated to react again and again, the overall yield of ammonia approaches 85 to 90 percent.

Example 17-7: Chemical equilibrium *What effect does an increase in pressure of the total reaction system have on the position of equilibrium for the following reaction?*

$$CO_2(g) + H_2(g) \rightleftharpoons CO(g) + H_2O(g)$$

All the reactants and products are gases, and there is no volume change during the reaction: 1 mole of each reactant produces 1 mole of each product. A pressure change does not affect the position of equilibrium in this reaction.

IONIC EQUILIBRIA

The equilibrium constant for the dissociation of water into hydronium and hydroxide ions is

$$2H_2O(l) \rightleftharpoons H_3O^+ + OH^-$$

$$K = \frac{[H_3O^+][OH^-]}{[H_2O]^2} \quad \text{or} \quad K[H_2O]^2 = [H_3O^+][OH^-]$$

The concentration of liquid water is a constant, 55.5 moles/liter, the mass of one liter divided by the mass of one mole, 1000 g/liter/18 g/mole. Therefore, a new equilibrium constant can be written:

$$K_w = K[H_2O]^2 = [H_3O^+][OH^-]$$

or

$$K_w = [H_3O^+][OH^-] = 1 \times 10^{-14}$$

This constant is the same one we discussed in Chapter 15. It is called the **ion product of water.**

For a weak acid or base, the equilibrium constant for dissociation in water is called the **ionization constant.** At room temperature, the ionization constant for the weak base ammonium hydroxide, $NH_3(aq)$, is

$$NH_3(g) + H_2O \rightleftharpoons NH_3(aq) \rightleftharpoons (NH_4OH(aq)) \rightleftharpoons NH_4^+ + OH^-$$

$$K_i = \frac{[NH_4^+][OH^-]}{[NH_3(aq)]} = 1.8 \times 10^{-5}$$

The very small value of the constant shows that ammonium hydroxide dissociates very slightly in aqueous solution.

Example 17-8: Ionic equilibrium *What is the hydroxide ion concentration of a 0.10 M solution of ammonium hydroxide?*

Let x equal the ammonium and hydroxide ion concentrations at equilibrium. The concentration of the undissociated ammonium hydroxide at equilibrium, $[NH_3(aq)]$, equals the initial concentration minus the number of dissociated moles per liter. Because the number dissociated must equal x, the concentration of ammonium hydroxide at equilibrium is $0.10 - x$.

Also, because x is very small compared to 0.10, it can be neglected to give $[NH_3(aq)] = 0.10$. Substituting these values into the equilibrium equation gives

$$\frac{(x)(x)}{0.10} = 1.8 \times 10^{-5}$$

$$x^2 = 0.10(1.8 \times 10^{-5}) = 1.8 \times 10^{-6}$$

$$x = 1.3 \times 10^{-3}$$

Thus the concentration of OH^- ions in 0.10 *M* ammonium hydroxide is only 1.3×10^{-3} mole/liter.

COMMON ION EFFECT

A weak base or a weak acid ionizes in aqueous solution until an equilibrium is established. A change in the concentration of either of the ions present will shift the position of the equilibrium as described by Le Châtelier's principle. For example, the equilibrium, or ionization, constant for acetic acid is

$$H_2O + H(CH_3CO_2) \rightleftharpoons H_3O^+ + CH_3CO_2^-$$

$$K_i = \frac{[H_3O^+][CH_3CO_2^-]}{[H(CH_3CO_2)]} = 1.8 \times 10^{-5}$$

If acetate ions are added to a solution of acetic acid, the equilibrium shifts toward the left; therefore, in an attempt to maintain the equilibrium, the concentration of hydronium ions must decrease. This *displacement of equilibrium by the addition of an ion involved in the equilibrium* is called the **common ion effect.**

Example 17–9: Common ion effect *Calculate the hydronium ion concentration of a 0.10 M solution of acetic acid to which sodium acetate has been added in an amount sufficient to make the acetate ion concentration 0.20 M. Sodium acetate completely dissociates in aqueous solution.*

The given concentration values are placed in the equilibrium constant equation, and the equation is solved for $[H_3O^+]$. Let $[H_3O^+]$ equal x; then the acetate ion concentration is $0.20 + x$ (0.20 *M* from the added sodium acetate and x *M* from the dissociation of the acid). The concentration of the undissociated acid is $0.10 - x$ (x moles of acid dissociate for every x moles of acetate or hydronium ions formed).

$$\frac{[H_3O^+][CH_3CO_2^-]}{[H(CH_3CO_2)]} = \frac{(x)(0.20 + x)}{0.10 - x} = 1.8 \times 10^{-5}$$

To solve for x, you must solve a quadratic equation, but this work can be avoided if you assume that the value of x is very small compared with 0.10 and 0.20. This assumption is valid, because, by definition, a weak acid is only slightly ionized. Making the assumption, we get

$$\frac{(x)(0.20)}{0.10} = 1.8 \times 10^{-5}$$

Solving for x,

$$x = \frac{(0.10)(1.8 \times 10^{-5})}{0.20} = 9.0 \times 10^{-6} \text{ mole/liter} = [H_3O^+]$$

The hydronium-ion concentration in a 0.10 M acetic acid solution to which no sodium acetate has been added is 1.3×10^{-3} mole/liter. The addition of the common ion in the form of sodium acetate lowers the hydronium-ion concentration from 1.3×10^{-3} to 9.0×10^{-6}—over a hundredfold decrease in acidity.

SOLUBILITY PRODUCT

Theoretically, no compound is completely insoluble in water; even the most insoluble ionic compounds are ionized to some extent. *The equilibrium constants for the solubility in water of relatively insoluble salts* are called **solubility-product constants.** Silver chloride is an example important to analytical chemistry. The equilibrium constant is

$$K = \frac{[Ag^+][Cl^-]}{[AgCl]}$$

for the reaction

$$AgCl(s) + \cancel{H_2O} \rightleftharpoons Ag^+ + Cl^- + \cancel{H_2O}$$

TABLE 17–1 SOLUBILITY-PRODUCT CONSTANTS FOR SOME COMPOUNDS AT ROOM TEMPERATURE

Compound	Equilibrium	K_{sp}
Aluminum hydroxide	$Al(OH)_3 \leftrightharpoons Al^{3+} + 3OH^-$	1×10^{-33}
Barium sulfate	$BaSO_4 \leftrightharpoons Ba^{2+} + SO_4^{2-}$	1×10^{-10}
Calcium carbonate	$CaCO_3 \leftrightharpoons Ca^{2+} + CO_3^{2-}$	4.8×10^{-9}
Calcium hydroxide	$Ca(OH)_2 \leftrightharpoons Ca^{2+} + 2OH^-$	1.3×10^{-6}
Calcium sulfate	$CaSO_4 \leftrightharpoons Ca^{2+} + SO_4^{2-}$	2.4×10^{-5}
Copper(II) sulfide	$CuS \leftrightharpoons Cu^{2+} + S^{2-}$	4×10^{-38}
Iron(II) hydroxide	$Fe(OH)_2 \leftrightharpoons Fe^{2+} + 2OH$	1.6×10^{-15}
Iron(II) sulfide	$FeS \leftrightharpoons Fe^{2+} + S^{2-}$	1×10^{-19}
Lead(II) chloride	$PbCl_2 \leftrightharpoons Pb^{2+} + 2Cl^-$	1.7×10^{-5}
Lead(II) sulfide	$PbS \leftrightharpoons Pb^{2+} + S^{2-}$	1×10^{-28}
Magnesium hydroxide	$Mg(OH)_2 \leftrightharpoons Mg^{2+} + 2OH^-$	8.9×10^{-12}
Mercury(I) chloride	$Hg_2Cl_2 \leftrightharpoons Hg_2^{2+} + 2Cl^-$	1.1×10^{-18}
Mercury(II) sulfide	$HgS \leftrightharpoons Hg^{2+} + S^{2-}$	3×10^{-53}
Nickel(II) sulfide	$NiS \leftrightharpoons Ni^{2+} + S^{2-}$	3×10^{-21}
Silver bromide	$AgBr \leftrightharpoons Ag^+ + Br^-$	3.3×10^{-13}
Silver chloride	$AgCl \leftrightharpoons Ag^+ + Cl^-$	1.7×10^{-10}
Silver iodide	$AgI \leftrightharpoons Ag^+ + I^-$	8.5×10^{-17}
Zinc sulfide	$ZnS \leftrightharpoons Zn^{2+} + S^{2-}$	4.5×10^{-24}

The concentration of AgCl(s) is a constant: mass, or moles, per unit volume of a solid is equal to its density, and density is a constant. Hence the equilibrium constant can be written

$$K_{sp} = K[\text{AgCl}] = [\text{Ag}^+][\text{Cl}^-]$$

$$K_{sp} = [\text{Ag}^+][\text{Cl}^-] = 1.7 \times 10^{-10}$$

K_{sp} is the solubility-product constant. The product of the ion concentrations of a relatively insoluble salt in water is a constant at a given temperature.

The magnitude of the solubility-product constant tells how soluble the salt is in pure water. Table 17–1 lists the solubility-product constants for a number of important salts.

Example 17–10: Solubility product *Which salt is more soluble in water, mercury(I) chloride or mercury(II) sulfide?*

Reference to Table 17–1 shows that the solubility product constants are these: mercury(I) chloride, 1.1×10^{-18}; and mercury(II) sulfide, 3×10^{-53}. From the constants, we see that the mercury(I) chloride is much more soluble than the mercury(II) sulfide: 10^{-53} is an extremely small number.

Example 17–11: Solubility product *What is the concentration of calcium ions in a saturated solution of calcium carbonate?*

From Table 17–1, we get the data for the equilibrium

$$\text{CaCO}_3 \rightleftharpoons \text{Ca}^{2+} + \text{CO}_3^{2-}$$

$$K_{sp} = [\text{Ca}^{2+}][\text{CO}_3^{2-}] = 4.8 \times 10^{-9}$$

Because the concentration of the calcium ions must equal the concentration of the carbonate ions, we can let each equal x, to give

$$(x)(x) = 4.8 \times 10^{-9}$$
$$x^2 = 4.8 \times 10^{-9} = 48 \times 10^{-10}$$
$$x = \sqrt{48} \times \sqrt{10^{-10}} = 6.9 \times 10^{-5}$$
$$[\text{Ca}^{2+}] = [\text{CO}_3^{2-}] = 6.9 \times 10^{-5} \text{ mole/liter}$$

Example 17–12: Solubility product *The solubility product of silver chloride is 1.7×10^{-10}. What is the solubility of AgCl(s) in moles/liter and g/liter?*

Let x equal the solubility of AgCl(s) in moles/liter. Then

$$[\text{Ag}^+] = [\text{Cl}^-] = x$$

Substituting in the solubility product expression gives

$$[\text{Ag}^+][\text{Cl}^-] = (x)(x) = 1.7 \times 10^{-10}$$
$$x^2 = 1.7 \times 10^{-10}$$
$$x = \sqrt{1.7} \times \sqrt{10^{-10}} = 1.3 \times 10^{-5} \text{ mole/liter}$$

To convert 1.3×10^{-5} mole/liter to g/liter, multiply by the mass of one mole of AgCl:

$$(1.3 \times 10^{-5} \text{ mole/liter})(143 \text{ g/mole}) = 1.9 \times 10^{-3} \text{ g/liter}$$

Hence the solubility of AgCl(*s*) is 1.3×10^{-5} mole/liter, or 1.9×10^{-3} g/liter.

NEW TERMS

Activated complex: an intermediate species formed in a chemical reaction. It has excess energy and usually quickly decomposes to either reactants or products.

Activation energy: the amount of energy needed to form the activated complex in a chemical reaction.

Catalyst: a substance that changes the rate of a chemical reaction but does not itself appear in the reaction products.

Chemical equilibrium: the state of balance that exists when the rate of a forward reaction equals the rate of the reverse reaction.

Common ion effect: the reduction in the degree of ionization of a weak electrolyte caused by the presence of a common ion from another electrolyte.

Equilibrium constant: a numerical constant characteristic of a given reaction at equilibrium at a given temperature. It is equal to the specific rate constant of the forward reaction divided by the specific rate constant of the reverse reaction.

Ion product of water: the equilibrium constant K_w, which is equal to $[H_3O^+][OH^-]$ for any aqueous solution.

Ionization constant: the equilibrium constant for the ionization reaction of weak electrolytes in water.

Kindling temperature: the temperature at which an oxidation reaction becomes a self-sustaining combustion—the temperature at which a substance bursts into flame.

Law of mass action: the velocity or rate of a chemical reaction at a given instant is proportional to the concentration of reactants, where each concentration is raised to a power equal to the number of molecules of that species participating in the reaction.

Le Châtelier's principle: when a stress is placed on a system in equilibrium, the system will readjust itself in a direction that will tend to relieve the stress and restore equilibrium.

Reaction mechanism: the series of steps in a chemical reaction by which the reactants are converted to products.

Reaction rate: the amount of reactant consumed (or product produced) in a given period of time.

Reaction velocity: reaction rate.

Solubility product: the equilibrium constant, K_{sp}, for the solubility in water of relatively insoluble salts; the equilibrium constant for the equilibrium between a solid salt and its saturated solution.

Specific rate constant: the proportionality constant, *k*, in the rate expression for a reaction.

Surface compound: an intermediate complex that includes the surface of a catalyst.

PROBLEM SET A

1. Devise a method for studying the rates of the following reactions:
   a. $2H_2(g) + 2NO(g) \rightleftharpoons 2H_2O(g) + N_2(g)$
   b. $5C_2O_4^{2-} + 2MnO_4^- + 16H_3O^+ \rightleftharpoons 10CO_2(g) + 2Mn^{2+} + 24H_2O$
2. Oily rags left in a poorly ventilated spot can suddenly burst into flame. Explain this phenomenon of spontaneous combustion.
3. Draw a diagram similar to Figure 17–2 to illustrate the energy relationships for an endothermic reaction.
4. Explain what is meant by the statement, "A chemical equilibrium is a dynamic equilibrium."
5. Show how the expression for the equilibrium constant can be derived from reaction rates. (Given the equation $aA + bB \leftrightarrows cC + dD$.)
6. Write equilibrium-constant expressions for the following reactions:
   a. $PCl_5(g) \rightleftharpoons PCl_3(g) + Cl_2(g)$
   b. $NH_4OH(aq) \rightleftharpoons NH_4^+ + OH^-$
   c. $2HgO(s) \rightleftharpoons 2Hg(l) + O_2(g)$
   d. $3O_2(g) \rightleftharpoons 2O_3(g)$
   e. $4NH_3(g) + 5O_2(g) \rightleftharpoons 4NO(g) + 6H_2O(g)$
7. Predict the effect of an increase in temperature, pressure, concentration of one reactant, and a catalyst on the following reactions at equilibrium:
   a. $2CO(g) + O_2(g) \rightleftharpoons 2CO_2(g) + \text{heat}$
   b. $2SO_2(g) + O_2(g) \rightleftharpoons 2SO_3(g) + \text{heat}$
   c. $N_2(g) + O_2(g) \rightleftharpoons 2NO(g) - \text{heat}$
   d. $CO_2(g) + H_2(g) \rightleftharpoons H_2O(l) + CO(g) + \text{heat}$
8. All the complicated reactions occurring in your body are accomplished at atmospheric pressure and about 37°C. Few if any of these can be carried out in the laboratory without the use of higher temperatures and pressures. Explain.
9. The equilibrium constant for the reaction
   $SO_2(g) + NO_2(g) \rightleftharpoons SO_3(g) + NO(g)$
   at a given temperature is 5. What is the equilibrium constant for the reaction
   $SO_3(g) + NO(g) \rightleftharpoons SO_2(g) + NO_2(g)$
   at the same temperature?
10. What is the hydronium-ion concentration of a solution that contains 1 mole of acetic acid and 1 mole of sodium acetate per liter? What is the hydronium-ion concentration after 4.0 g of sodium hydroxide is dissolved in the liter of solution? (Assume no volume changes when NaOH is added.)
11. The solubility product for silver bromide is 3.3×10^{-10}. What is the solubility of silver bromide in grams per liter?
12. When solutions of barium chloride and zinc sulfate are mixed, is there a precipitate?
13. Calculate the solubility in grams per liter for each of the following salts (room temperature):
   a. nickel(II) sulfide
   b. calcium sulfate
   c. lead(II) chloride

PROBLEM SET B

1. A good Boy Scout does two things to start his campfire more quickly: he cuts the wood into small shavings and blows on the shavings once they have started to burn. Why do these procedures increase the rate of the reaction?
2. Explain why an increase in temperature causes an increase in the rate of a chemical reaction.
3. Draw a diagram similar to Figure 17–3 to illustrate the energy relationships for a catalyzed endothermic reaction.
4. Two reactions have equilibrium constants equal to 10^{-4} and 10^{4}. Which reaction goes to near completion in the forward direction?
5. Write the equilibrium expression for the reaction of an organic acid with an alcohol to form an ester and water.
6. Predict which of the following reactions will go almost to completion in the forward direction. Explain why.
   a. $NaOH(aq) + HCl(aq) \rightleftharpoons NaCl(aq) + H_2O(l)$
   b. $CaCO_3(s) \rightleftharpoons CaO(s) + CO_2(g)$
   c. $2KClO_3(s) \rightleftharpoons 2KCl(s) + 3O_2(g)$
   d. $Ag^+ + Cl^- \rightleftharpoons AgCl(s)$
7. Why is the concentration in moles per liter of a pure solid or liquid always constant?
8. The concentrations for A, B, C, and D at equilibrium are 0.4 *M*, 0.6 *M*, 0.8 *M*, and 0.3 *M*, respectively, for the following reaction:
   $A + B \rightleftharpoons C + D$
   a. What is the numerical value of the equilibrium constant?
   b. What is the concentration of D when the concentrations of A, B, and C are 0.4 *M*, 0.5 *M*, and 0.4 *M*, respectively?
9. Hypochlorous acid has an ionization constant of 3.2×10^{-8}. The constant for nitrous acid is 4×10^{-4}. Which is the stronger acid?
10. What is the hydronium-ion concentration of a 0.10 *M* solution of ammonium hydroxide?
11. List the following salts in order of increasing solubility: silver chloride, silver iodide, and silver bromide. Do they follow a predictable periodic-table relationship?
12. Calculate the solubility-product constant for the following compounds from their known solubilities:
    a. Ag_2S, solubility 1.1×10^{-17} mole/liter
    b. $CaCO_3$, solubility 5.9×10^{-5} mole/liter
    c. CaF_2, solubility 1.6×10^{-3} g/liter

18

introduction to organic chemistry

THIS WILL SAVE YOU TIME:

Refer to this list as you read through the chapter. In answering the questions, you will understand the concepts you are most likely to be asked about later on.

After reading this chapter you should be able to:

1. *Name the various functional groups, and write their general formulas.*
2. *Explain the meaning of these classifications of hydrocarbons: aromatic, aliphatic, saturated, unsaturated, alkanes, alkenes, alkynes, and acetylenes.*
3. *Tell what is meant by structural isomerism, and give an example.*
4. *Name and give formulas for the first ten members of the homologous series* $n\text{-}C_nH_{2n+2}$.
5. *Give the names and formulas for the common alkyl groups.*
6. *Name simple hydrocarbons from their formulas, and give the formulas from their names.*
7. *Give equations for a commercial preparation of ethene and ethyne.*
8. *Illustrate the resonance structures of benzene.*
9. *Give examples of addition and substitution reactions.*
10. *Describe the polymerization reaction for polyethylene.*
11. *Show two ways of drawing the formula for glucose.*
12. *Give equations for the reactions used to manufacture rayon, Dacron, and nylon.*
13. *Write the equation for the general reaction between an organic acid and an alcohol and between an organic acid and an amine.*
14. *Tell what proteins are, where they are found, and why they are important.*
15. *Give the names and formulas for several essential amino acids.*
16. *Discuss the basic structure of a protein molecule.*

Organic chemistry is the chemistry of carbon-containing compounds. The vast majority of organic compounds contain a carbon-carbon bond, a carbon-hydrogen bond, or both. By convention, many carbon-containing compounds, such as carbon dioxide, carbon disulfide, and the carbonates, however, are usually included in discussions of inorganic chemistry. Known organic molecules now number over a million, and new ones are identified every day.

HOW ORGANIC CHEMISTRY IS ORGANIZED

Fortunately, the numerous organic compounds can be divided into a relatively small number of groups. Each group is characterized by the presence of a particular atom or group of atoms known as a **functional group.** The important functional groups and the structural arrangements of their atoms are listed in Table 18–1.

TABLE 18–1 ORGANIC FUNCTIONAL GROUPS

Name	Group	General formula*	Example	
Hydrocarbon	—H	R—H	$CH_3—CH_3$	Ethane
Acid	$-\underset{\underset{O}{\|}}{C}-OH$	$R-\underset{\underset{O}{\|}}{C}-OH$	CH_3CO_2H	Ethan*oic acid* (acetic acid)
Alcohol	—OH	R—OH	CH_3CH_2OH	Ethan*ol* (ethyl alcohol)
Aldehyde	$-\underset{\underset{O}{\|}}{C}-H$	$R-\underset{\underset{O}{\|}}{C}-H$	CH_3CHO	Ethan*al* (acetaldehyde)
Amide	$-\underset{\underset{O}{\|}}{C}-NH_2$	$R-\underset{\underset{O}{\|}}{C}-NH_2$	CH_3CONH_2	Ethan*amide* (acetamide)
Amine	$—NH_2$	$R—NH_2$	CH_3NH_2	*Amino*methane (methylamine)
Ester	$-\underset{\underset{O}{\|}}{C}-O-$	$R-\underset{\underset{O}{\|}}{C}-O-R'$	$CH_3CO_2CH_3$	Methyl ethan*oate* (methyl acetate)
Ether	—O—	R—O—R′	$CH_3CH_2OCH_2CH_3$	Eth*oxy*ethane (ethyl ether)
Halide	—X (Cl, Br, I)	R—X	C_2H_5I	*Iodo*ethane (ethyl iodide)
Ketone	$-\underset{\underset{O}{\|}}{C}-$	$R-\underset{\underset{O}{\|}}{C}-R'$	CH_3COCH_3	Propan*one* (acetone)

* R or R′ refers to a group of atoms that has a carbon atom forming the indicated attachment.

Consider —OH as an example of a functional group. By definition, all compounds with an —OH group attached to a carbon atom are known as **alcohols.** (The carbon atom carrying the —OH should not be bonded to another oxygen atom; also, phenol (C_6H_5—OH) and its derivatives are usually not classified as alcohols.) Some examples of alcohols are

CH_3—OH

methanol
(methyl alcohol)

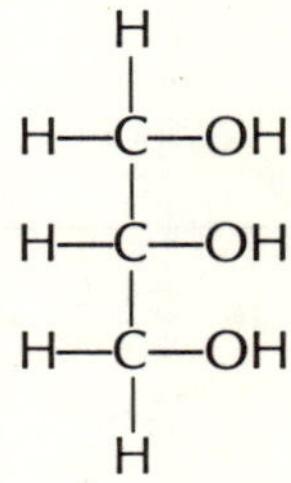

glycerol
(glycerin)

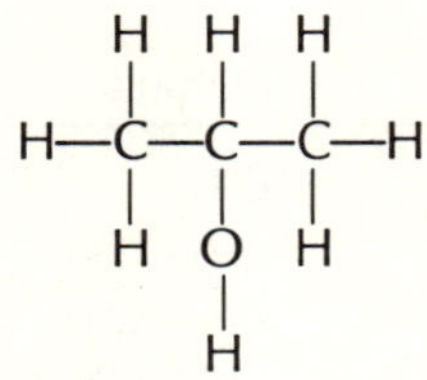

2-propanol
(isopropyl alcohol)

These and other compounds that contain the alcohol functional group have certain common chemical properties; for example, all alcohols react with organic acids to give esters. If we know something about the chemistry of one alcohol, we know something about the chemistry of all alcohols. Thus organic chemistry is the chemistry of the various functional groups.

HYDROCARBONS

Hydrocarbons are *compounds containing only carbon and hydrogen*. They are widely distributed in nature, primarily in natural gas, petroleum, and coal. Natural gas, the volatile fraction of crude petroleum, varies in composition but contains approximately 80 percent methane, CH_4; 7 percent ethane, CH_3CH_3; and 6 percent propane, $CH_3CH_2CH_3$. The balance consists of various C_4H_{10} and C_5H_{12} hydrocarbons. Crude petroleum is a complex mixture of hydrocarbons that includes small amounts of oxygen-, sulfur-, and nitrogen-containing compounds.

Hydrocarbons find use as fuel, and they provide the raw material for such diverse products as synthetic fibers, elastomers, alcohol, drugs, plastics, and thousands of other important items.

Hydrocarbons can be classified into main divisions: the **aromatic hydrocarbons** *have structures related to the benzene molecule;* the **aliphatic hydrocarbons** *have structures not related to benzene*. The aliphatic group can be subdivided into saturated and unsaturated hydrocarbons. **Saturated hydrocarbons** *contain only single carbon–carbon bonds,* while **unsaturated hydrocarbons** *have double or triple carbon–carbon bonds.*

ALKANES

Saturated hydrocarbons are called **alkanes.** A series of formulas can be written for these compounds in which each compound differs from the next by one —CH_2— group. Such a series of related compounds is called a **homologous series.** The first ten members of the alkane series, which have the general formula C_nH_{2n+2}, are listed in Table 18–2. The melting points, boiling points, densities, and many other physical properties of the compounds in an homologous series show a systematic dependence on the molecular mass. All the saturated hydrocarbons have similar chemical properties.

TABLE 18–2 ALKANES: THE HOMOLOGOUS SERIES[a] n-C_nH_{2n+2}

Name	Formula	Melting point, °C	Boiling point, °C	Specific gravity[b] (as liquids)
Methane	CH_4	−182.6	−161.7	0.424
Ethane	CH_3—CH_3	−172	−88.6	0.546
Propane	CH_3—CH_2—CH_3	−187.1	−42.2	0.582
Butane	CH_3—$(CH_2)_2$—CH_3	−135	−0.6	0.579
Pentane	CH_3—$(CH_2)_3$—CH_3	−129.7	36.1	0.626
Hexane	CH_3—$(CH_2)_4$—CH_3	−94	68.7	0.659
Heptane	CH_3—$(CH_2)_5$—CH_3	−90.6	98.4	0.684
Octane	CH_3—$(CH_2)_6$—CH_3	−56.8	125.7	0.703
Nonane	CH_3—$(CH_2)_7$—CH_3	−53.7	150.7	0.718
Decane	CH_3—$(CH_2)_8$—CH_3	−29.7	174.0	0.730

[a] These data are for normal (*n*-), open-chain hydrocarbons—those in which no one carbon is bonded to more than two other carbons.

[b] Specific gravity is defined as the ratio of the mass of a given volume of liquid to the mass of an equal volume of water at the stated temperature (20°C for these data). It is approximately equal numerically to the density in grams/ml.

STRUCTURAL ISOMERISM

How many ways can 5 carbon and 12 hydrogen atoms be combined? This question can be easily answered by constructing ball and stick models. The three possible combinations for C_5H_{12} are:

CH_3 — CH_2 — CH_2 — CH_3

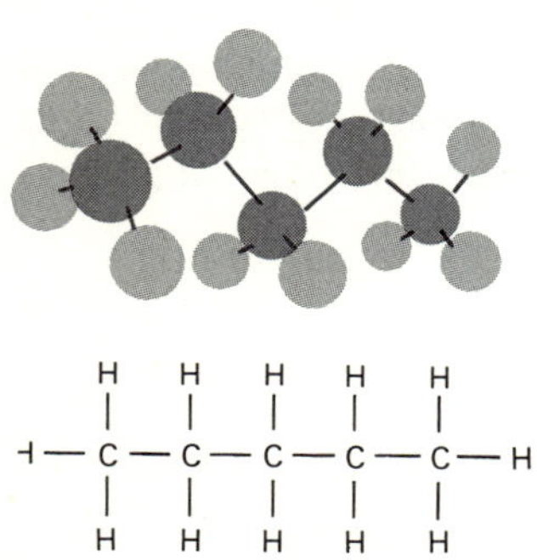

n- pentane

A

CH_3 — $CH(CH_3)$ — CH_2 — CH_3

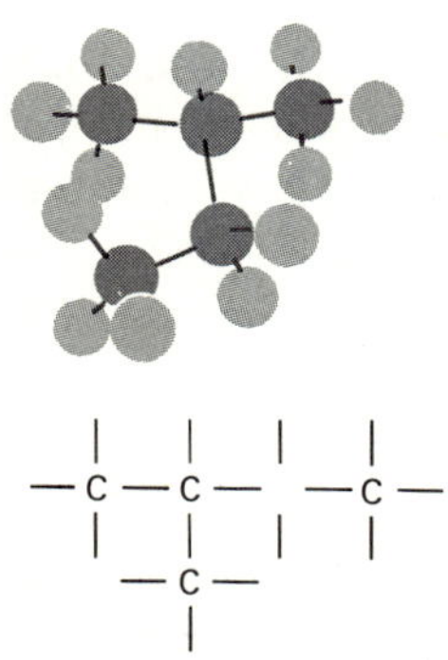

2-methylbutane

B

CH — $C(CH_3)_2$ — CH_3

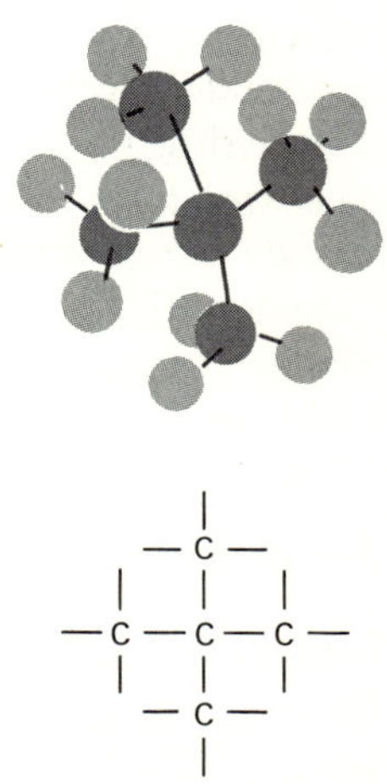

2, 2-dimethylpropane

C

Compound **A** has 5 carbon atoms connected together in an **open-chain** structure; the angles between the bonds are all approximately 109.5°. The second compound, **B,** has 4 atoms in the longest carbon chain; and unlike **A,** one of its carbon atoms has bonds to each of three other carbon atoms and only one bond to a hydrogen atom. The third possibility, **C,** has a chain of only 3 carbon atoms and contains one carbon atoms with bonds to four other carbon atoms and no bonds to a hydrogen. Molecules **B** and **C** are examples of **branched-chain** structures. These three molecules represent different compounds, even though they all have the same molecular formula, a phenomenon known as **structural isomerism.** *Compounds with the same molecular formula but different structural formulas are called* **isomers.**

The number of structural isomers possible for a given molecular formula increases very rapidly as the number of carbon atoms increases. For $C_{10}H_{22}$, a total of 75 different isomers is possible; for $C_{15}H_{32}$, there can be 4347 isomers; and for $C_{40}H_{82}$, 62,491,178,805,831 different structural arrangements are possible. Theoretically, the number of organic compounds is infinite.

THE NOMENCLATURE FOR ALKANES

Organic chemistry would be extremely difficult if there were no method for naming such a large number of compounds. Fortunately such a method does exist. The system in general use throughout the world is known as the IUPAC (International Union of Pure and Applied Chemistry) system. With this system, you only have to memorize a few basic names and rules, and then you can name thousands of compounds or, conversely, write their formulas from their names. The following rules summarize the application of the system to the alkanes.

1. Locate the longest continuous chain of carbon atoms. Count the number of carbon atoms in this chain and use the name of the corresponding open-chain alkane (Table 18–2) as the base name for the compound. Note that these names consist of prefixes showing the number of carbon atoms plus an -ane ending. Do not be fooled by formulas written with bends and corners. The base name for the carbon skeleton is octane; the

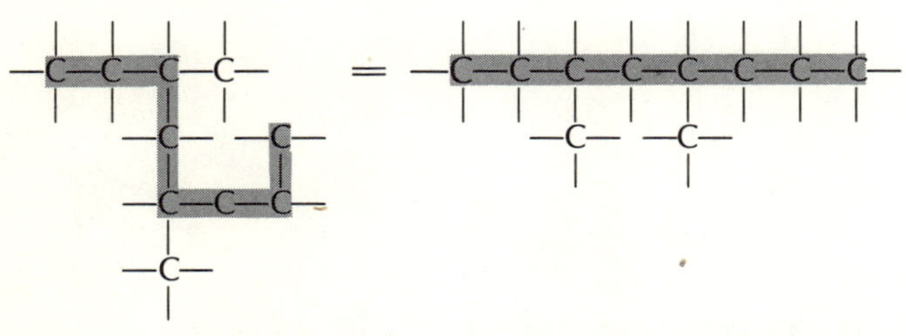

longest chain, whether straight or bent, has eight carbon atoms.

2. Number the longest chain beginning at the end closest to the point of branching:

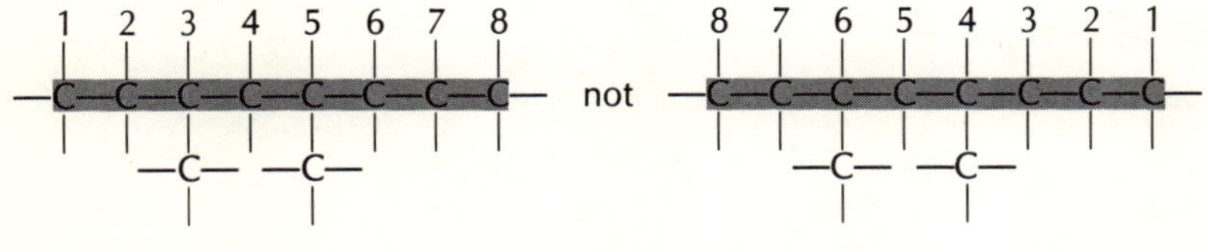

3. Name the attachments to the longest chain and locate them by using the number of the carbon atom to which they are attached. The CH_3— units attached to the octane base are called methyl groups. Hence the name for the compound

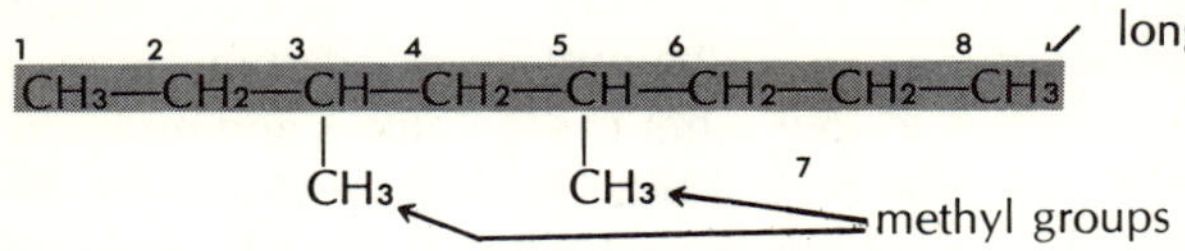

is 3-methyl-5-methyloctane, shortened to 3,5-dimethyloctane.

When the attachment of the carbon-chain base is one derived from an alkane (an alkane minus a hydrogen atom), it is named by changing the -ane ending of the particular alkane to -yl. These groups are called **alkyl groups.** Some of them are

methyl	$CH_3—$	*sec*-butyl	$CH_3—CH—CH_2—CH_3$ (attached at CH)
ethyl	$CH_3—CH_2—$	*iso*-butyl	$CH_3—CH(CH_3)—CH_2—$
n-propyl	$CH_3—CH_2—CH_2—$	*tert*-butyl	$(CH_3)_3C—$
iso-propyl	$CH_3—CH—CH_3$ (attached at CH)		
n-butyl	$CH_3—CH_2—CH_2—CH_2—$		

Note the use of the prefixes: normal (*n*-), *iso*- (or *i*-), secondary (*sec*- or *s*-), and tertiary (*tert*- or *t*-). They are used to designate the isomers.

Example 18–1: Organic nomenclature *Name the compound that has the following formula:*

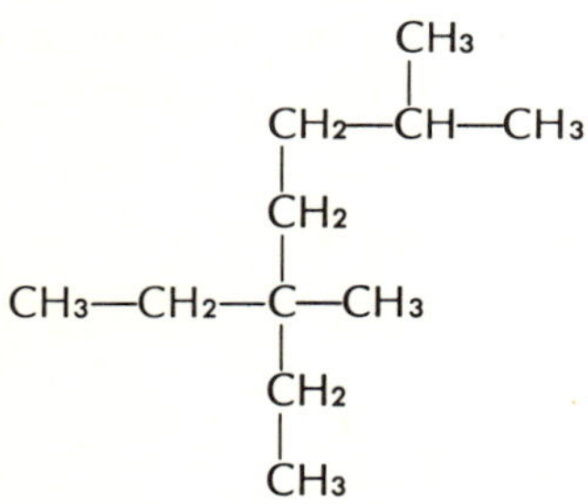

The longest chain contains seven carbon atoms and is numbered as follows:

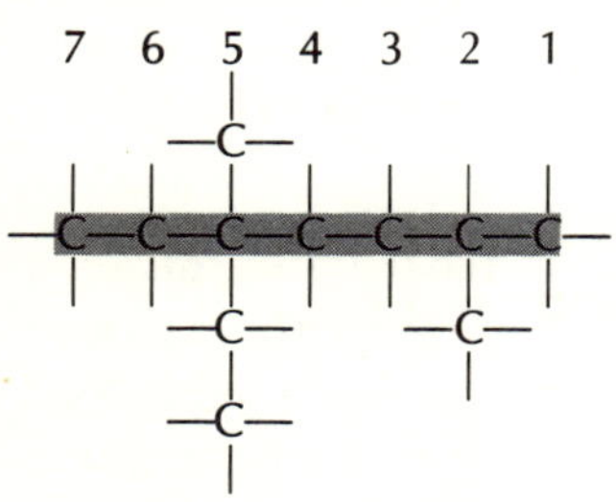

Methyl groups are attached to carbons 2 and 5, and an ethyl group to carbon 5. The name of the compound is 2,5-dimethyl-5-ethylheptane.

ALKENES AND ALKYNES

Unsaturated hydrocarbons with double bonds between carbon atoms are called **alkenes;** *those with triple bonds* are called **alkynes,** or **acetylenes.** The unsaturated hydrocarbons are very similar to the saturated hydrocarbons in physical properties. The low-molecular mass compounds are gases, whereas the heavier members of the homologous series are liquids and solids. They are only slightly soluble in water.

Ethene

Ethene (common name, ethylene), the simplest alkene, is produced by the thermal decomposition of ethane:

$$CH_3CH_3 \rightarrow CH_2{=}CH_2 + H_2$$

Most of the ethene produced commercially is used to manufacture the plastic polyethene (polyethylene).

Ethyne

The first member of the alkyne homologous series is ethyne (common name acetylene), commonly manufactured from inorganic raw materials. Heating calcium carbonate (limestone, oyster shells) gives calcium oxide (lime):

$$CaCO_3 \rightarrow CaO + CO_2$$

Calcium carbide, CaC_2, is then formed by heating a 60-to-40 mixture of calcium oxide and coke (carbon) to 2000°C in an electric furnace:

$$3C + CaO \rightarrow CaC_2 + CO$$

The hot liquid calcium carbide is then cooled to a solid and reacted with water to form ethyne:

$$CaC_2 + H_2O \rightarrow H{-}C{\equiv}C{-}H + Ca(OH)_2$$

Ethyne is one of the most important industrial raw materials. About 10 percent of the ethyne produced goes to oxyacetylene torches used for welding and cutting metals. Most, however, serves as a raw material to make many products such as plastics, fibers, resins, synthetic rubber, and so on.

Nomenclature for unsaturated hydrocarbons

The rules for naming the alkenes and alkynes are similar to those for the alkanes. First, locate the longest carbon atom chain with the double or triple bond. Number this chain beginning with the end closest to the double or triple bond rather than the first attachment. Then name the carbon chain; use the name for the corresponding alkane chain of the same length, but change the -ane ending to -ene for compounds with the double bond and -yne for those with a triple bond. The attachments are then labeled and numbered in the same manner as the alkanes.

Example 18–2: Organic nomenclature *Name the compound*

$CH_2{=}CH{-}CH_2{-}CH(CH_3){-}CH_3$ or 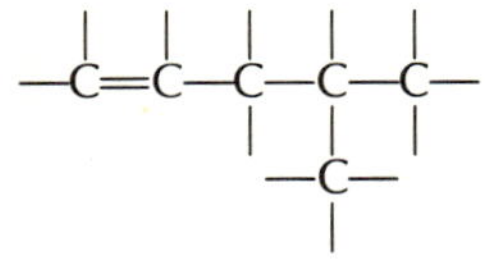

The longest carbon chain contains five carbons. It is numbered from the end that gives the double-bond carbons the lowest numbers:

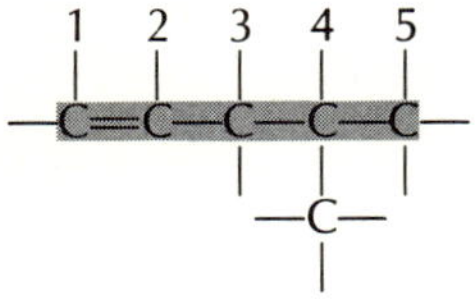

not

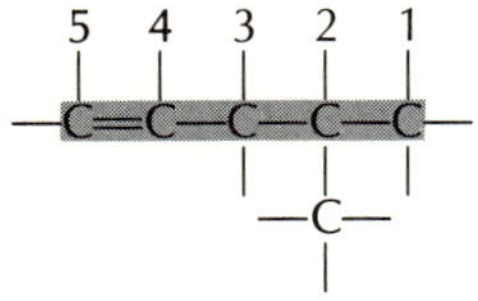

The base name is pentene. The location of the double bond is indicated by placing the number of the lower-numbered carbon atom of the double bond before the base name, 1-pentene. The methyl group is then numbered and named to give the complete name for the compound, 4-methyl-1-pentene.

BENZENE

As we have seen, aromatic compounds have structures related to the compound benzene, C_6H_6. Benzene is a clear, colorless liquid with a pleasant odor. It is flammable, and mixtures from 1.5 to 8.0 percent benzene vapor by volume in air are explosive. Benzene is soluble in most organic liquids but only very slightly soluble in water. It boils at 80.1°C and freezes at 5.56°C. Benzene is produced from coal and petroleum and used for the manufacture of plastics, synthetic rubbers, nylon, dyes, pesticides, explosives, pharmaceuticals, and a host of other important products.

In 1865, Friedrich A. Kekulé (1829–1896) proposed a ring structure for benzene in which the bonds between the carbons were alternately double and single covalent linkages:

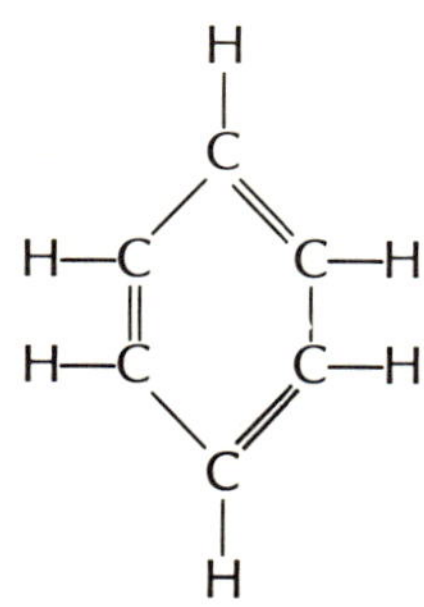

A more modern view is that the structure is a resonance hybrid of the two forms

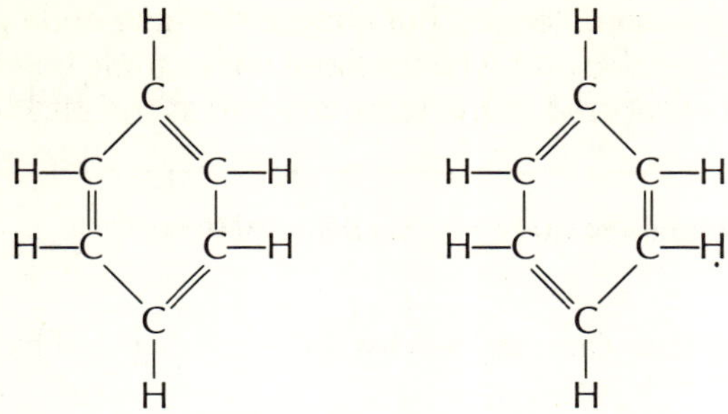

The bonds are neither single nor double, but intermediate between the two types. The two equivalent resonance structures for benzene can be represented by the notations

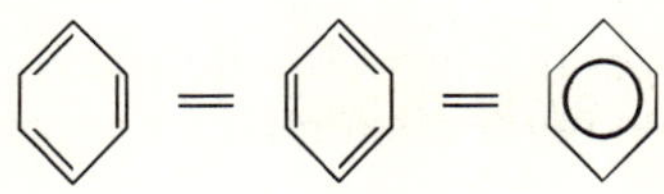

SOME CHEMICAL PROPERTIES OF HYDROCARBONS

Unsaturated hydrocarbons are characterized by their ability to participate in **addition reactions,** *reactions in which atoms or groups of atoms attach to the carbons at the multiple bonds.* In contrast, the saturated hydrocarbons undergo **substitution reactions** in which *hydrogen atoms of the alkanes are replaced by other atoms or groups of atoms.* The reactions of bromine with ethane and ethene illustrate the two types:

$CH_3—CH_3 + Br_2 \rightarrow CH_3—CH_2Br + HBr$
substitution reaction of alkanes

$CH_2═CH_2 + Br_2 \rightarrow CH_2Br—CH_2Br$
addition reaction of alkenes

The rate of addition of halogens to unsaturated hydrocarbons follows a periodic-table relationship: fluorine reacts extremely rapidly, and iodine very slowly, the order being $F_2 > Cl_2 > Br_2 > I_2$. The hydrogen halides, however, add with increasing ease in the order $HI > HBr > HCl$.

Some examples of addition reaction are:

$CH_2═CH—CH_3 + HI \rightarrow CH_3—CHI—CH_3$
propene 2-iodopropane

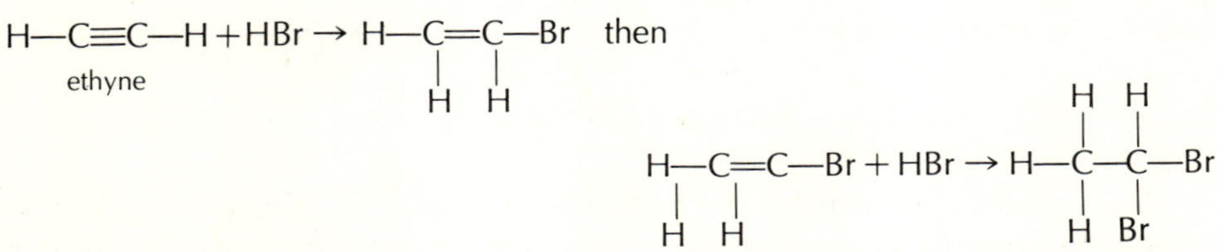

ethyne

1,1-dibromoethane

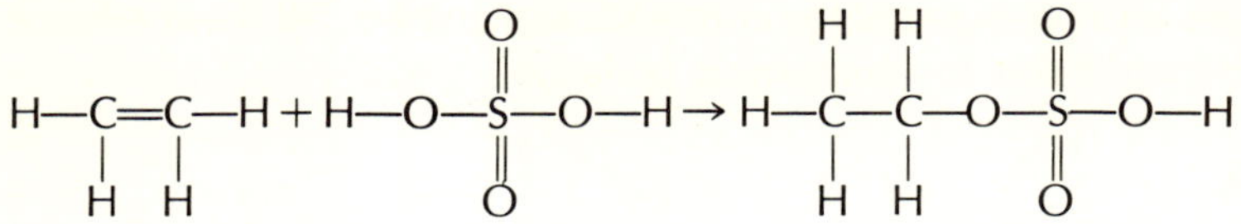

ethene — sulfuric acid — ethyl hydrogen sulfate

The ethyl hydrogen sulfate can then react with steam to give ethanol (ethyl alcoh

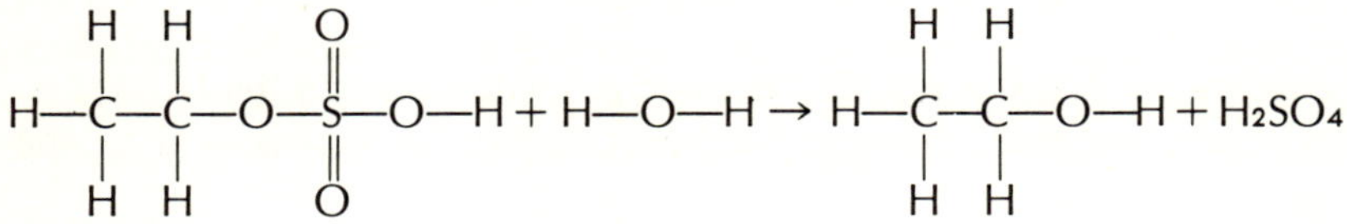

This reaction is used for preparing alcohols industrially.

Halogenation of benzene produces a mixture of products:

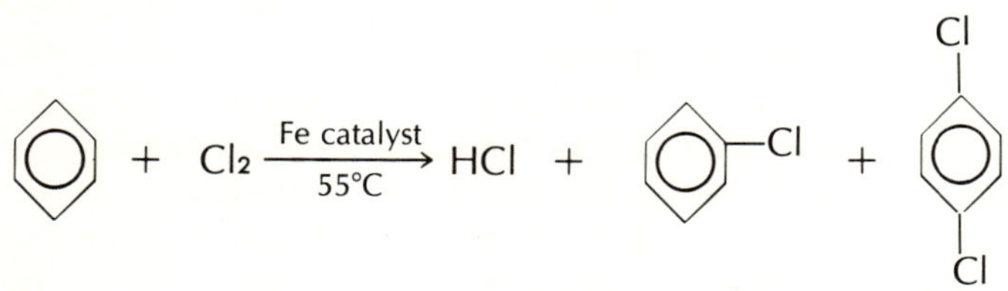

chlorobenzene 80% — 1,4-dichlorobenzene 15%

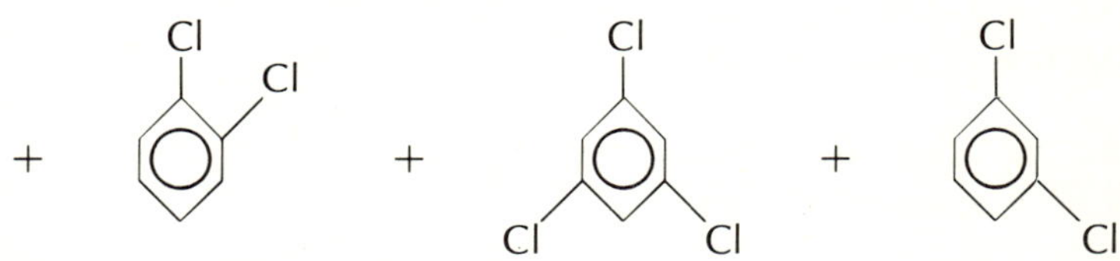

1,2-dichlorobenzene 5% — 1,3,5-trichlorobenzene — 1,3-dichlorobenzene ~0%

The oxidation of hydrocarbons is the important reaction that makes our cars run and supplies us with heat and power. In the presence of excess oxygen, all hydrocarbons react to produce carbon dioxide and water. The skeletal reaction is

$$C_nH_{2n+2} + O_2 \rightarrow CO_2 + H_2O + \text{heat}$$

Using C_8H_{18} as an example,

$$2C_8H_{18} + 25O_2 \rightarrow 16CO_2 + 18H_2O + 1307 \text{ kcal}$$

When gasoline is burned in the automobile engine, the hydrocarbon oxidation reaction does not go to completion. Leaving the exhaust pipe along with carbon dioxide and water are unburned hydrocarbons, carbon monoxide, and oxides of nitrogen. The latter are formed from nitrogen of the air reacting with oxygen under the high temperatures and pressures of the combustion chambers. All these exhaust products are major contributors to atmospheric pollution.

GIANT MOLECULES

Molecules that can contain many thousands of atoms are called **macromolecules,** or **polymers** (Greek: *poly, meros*—many, parts). Giant molecules are extremely important to everyday life. You do not have to look very far to discover that the world is a wonderland of giant molecules. Consider this list of a few of the polymers around you: your body (proteins), your clothing (silk, wool, synthetic fibers, elastomers), your shoes (leather, rubber), the chair you are sitting on (wood, plastics), the floor covering (rugs, plastic tiles), the book you are reading (paper), and even the paint on the walls. The study of giant molecules is one of the most important topics of modern chemistry.

POLYETHENE (POLYETHYLENE), AN IMPORTANT ADDITION POLYMER

Unsaturated hydrocarbons unite to form large molecules by a mechanism called **addition polymerization.** In this type of polymerization, the small molecules, called the **monomers,** combine with one another to form large molecules. No product other than the polymer is formed. The polymerization of ethene (C_2H_4, common name ethylene) will serve as an example. When ethene is heated under pressure in the presence of a peroxide initiator such as benzoyl peroxide,

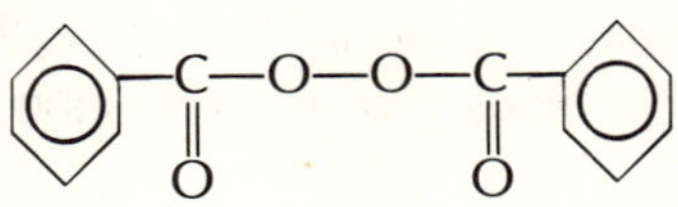

the peroxide first decomposes to form two free radicals; a **free radical** is a *substance with an unpaired electron*. The reaction is

$$\underset{\text{peroxide}}{R{-}O{-}O{-}R} \rightarrow \underset{\text{peroxide free radicals}}{2R{-}O\cdot}$$

Then the ethene monomer molecules react according to the following scheme:

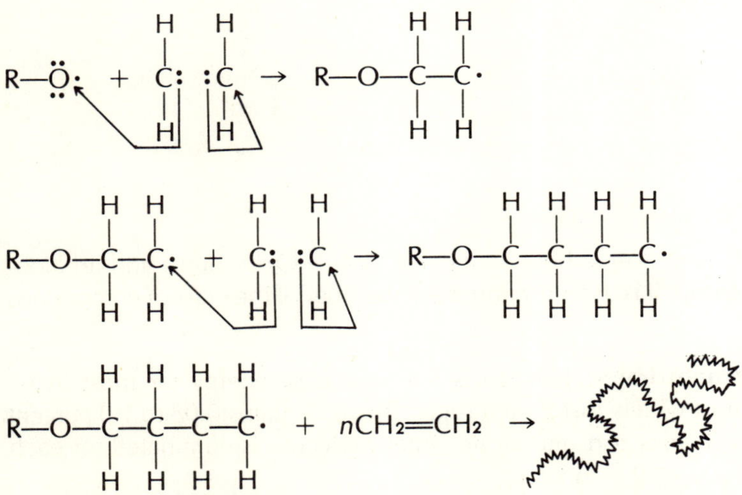

long-chain polymer

The peroxide free radical adds to the ethene monomer molecule to produce another free radical; the arrows show the way the electrons are redistributed. This free radical then reacts with another ethene molecule to produce a larger free radical. This process is repeated very rapidly, and in a short time the polymer chain will have grown to a molecule with a molecular mass of around 50,000. The overall reaction is usually written by the abbreviated equation:

$$n\mathrm{CH_2{=}CH_2} \xrightarrow[\text{1000–2500 atm, peroxide initiator}]{\text{100–300°C}} \mathrm{+\!\!\!\!\!\!\;\left[CH_2{-}CH_2\right]}_n$$

ethene (monomer) → polyethene or polyethylene (polymer)

Polyethene can be a viscous liquid, a wax, or a translucent solid, depending on the size of the polymer molecules, and the length of the growing chain can be controlled by various techniques. The solid polyethylene plastic has approximately 2000 monomer units united to form one polymer molecule.

CELLULOSE

Cellulose is *a polymer molecule found in the cell walls and fibers of most plants.* Chemically, it is a polymer composed of 1800 to 3000 units of glucose.

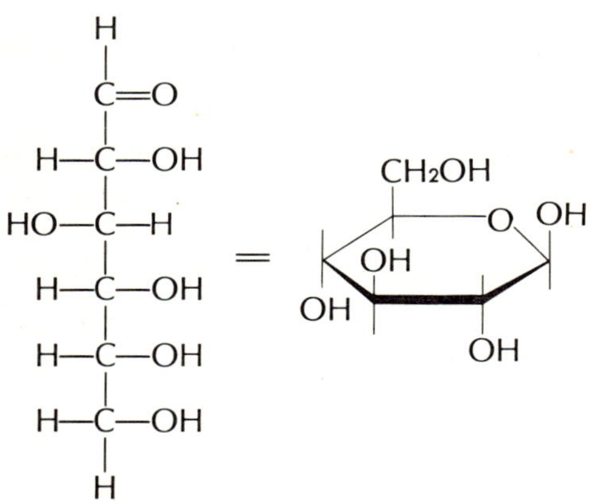

Glucose (also called dextrose) is both an alcohol (five —OH groups) and an aldehyde (one —C(H)=O) group. It is the most important member of the class of compounds called **sugars** or **carbohydrates;** glucose is the source of energy for most living organisms and the most widely distributed sugar. Blood contains 0.08 to 1.0 percent glucose. Glucose molecules can unite (one water molecule is eliminated for each link formed) in the following manner:

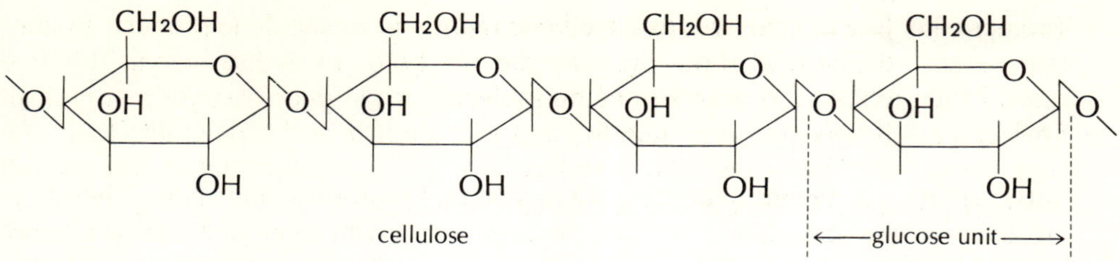

cellulose

Cotton is approximately 95 percent cellulose. Cotton treated with ether to remove the fats and waxes present is called **absorbent cotton. Mercerized cotton** is cotton whose fibers have been treated with sodium hydroxide while under tension, increasing their strength and giving them a silky luster. **Linen** is a less pure form of cellulose obtained from flax.

Wood contains cellulose, cellulose-like compounds, lignin (another polymer material), resins, and mineral matter. Spruce, a softwood, consists of approximately 60 percent cellulose, 10 percent cellulose-like compounds, 28 percent lignin, and 2 percent mineral ash. Hardwoods like maple and oak contain less lignin. Inexpensive papers are prepared from pulped, untreated wood fibers.

Rayon, the first important manmade fiber, is reconstituted cellulose. In the manufacture of rayon, prepared wood pulp is treated with sodium hydroxide and then carbon disulfide:

$$\underset{\text{cellulose}}{\mathrm{R{-}OH}} + \underset{\text{sodium hydroxide}}{\mathrm{NaOH}} + \underset{\text{carbon disulfide}}{\mathrm{CS_2}} \rightarrow \underset{\text{cellulose xanthate}}{\mathrm{R{-}O{-}\underset{\underset{\displaystyle S}{\|}}{C}{-}S^-\ Na^+}} + \mathrm{H_2O}$$

The cellulose xanthate is soluble. When a viscous solution of this xanthate (called **viscose**) is forced through small holes into an acid solution, the cellulose is regenerated in the form of long fibers.

$$\mathrm{R{-}O{-}\underset{\underset{\displaystyle S}{\|}}{C}{-}S^-\ Na^+} + \mathrm{H_3O^+} \rightarrow \underset{\text{rayon (regenerated cellulose)}}{\mathrm{R{-}OH}} + \mathrm{CS_2} + \mathrm{Na^+} + \mathrm{H_2O}$$

Viscose rayon is the most absorbent of the synthetic fibers and one of the most economical. It is used extensively for all types of fabrics. If the viscose is extruded in the form of a film, the product is called **cellophane.**

DACRON, A POLYESTER FIBER

Esters are a group of organic compounds that contain the $(\mathrm{{-}\underset{\underset{\displaystyle O}{\|}}{C}{-}O{-}})$ functional group. They are formed when an organic acid reacts with an alcohol:

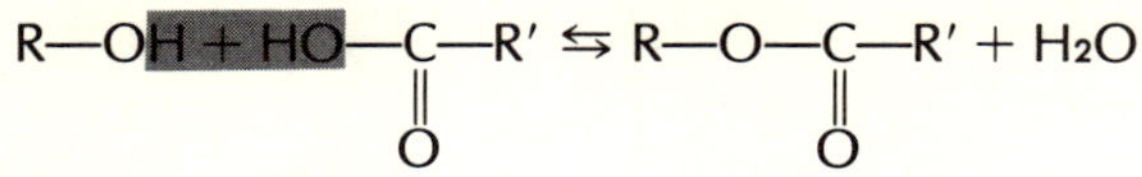

alcohol acid ester

This reaction is a very important one. One of its minor uses is in the aging of whiskey; on standing, the undesirable acids in the brew react with some of the ethyl alcohol to produce esters that give the whiskey its bouquet. Esters are the compounds that lend flavor and aroma to fruits such as apples, oranges, bananas, pears, and apricots.

If you react a compound containing two acid groups (a diacid) with a compound containing two alcohol groups (a dialcohol), it is possible to form long-chain polymer molecules.

This reaction is the basic one used to prepare Dacron.

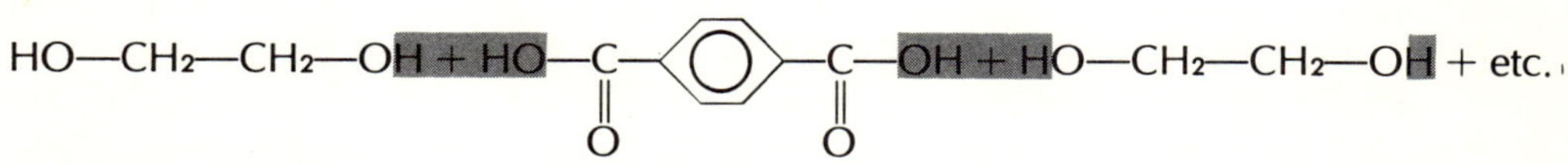

ethylene glycol terephthalic acid

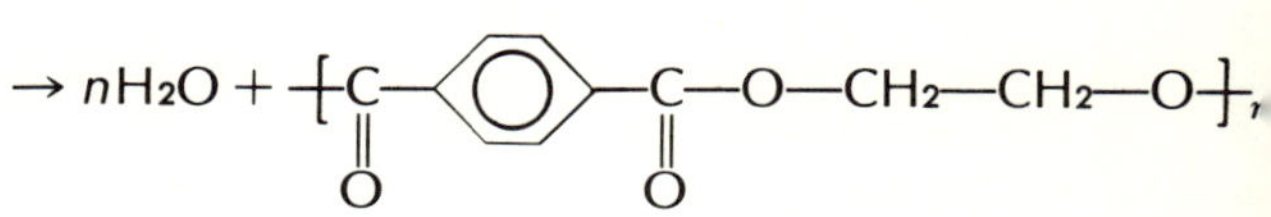

Dacron (a polyester)

The dialcohol used is called ethylene glycol, and the diacid is terephthalic acid. Note that the repeating units in the polymer are held together by ester-group linkages; hence the polymer is called a **polyester.**

Dacron fibers have high resilience. Because they soften on heating, they can be "permanently" pleated, and fabrics woven from them need no ironing.

NYLON

Nylon is a generic term for polyamide fibers, polymers that contain the amide (—C(=O)—N(H)—) functional group. Amides are formed when a compound containing an amine functional group, $—NH_2$, reacts with an organic acid.

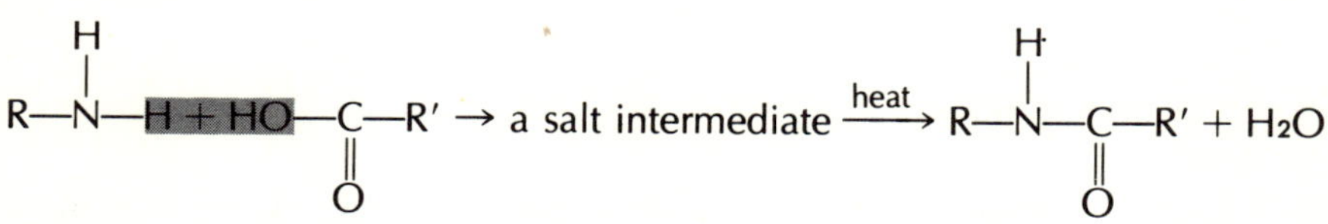

amine acid amide

If a diamine is reacted with a diacid, it is possible to form long-chain polymers:

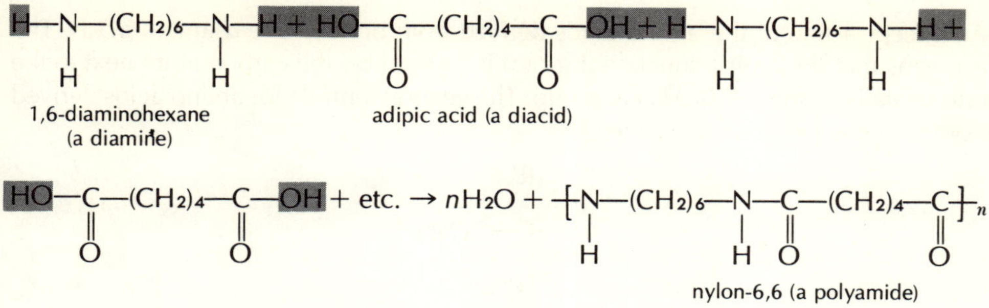

The diacid, adipic acid, and the diamine, 1,6-diaminohexane, contain six carbon atoms each. It is possible to use diamines and diacids with different numbers of carbon atoms to produce different nylons; hence the nylon in our example is designated nylon-6,6.

Nylon was the first completely synthetic polymer fiber. It is a strong durable fiber with good abrasion resistance, it absorbs little moisture, and it has extremely good stretch and recovery properties.

PROTEINS

Proteins are complex organic molecules found in every living organism. They constitute approximately 50 percent of the dry weight of most plant and animal cells. Proteins constitute a major portion of the body in the form of skin, muscles, and other soft tissues, and hair. Enzymes (body catalysts) are proteins, and the disease-producing viruses also contain protein molecules. Proteins find use in industry and the home in the form of wool, glues, silk, and leather.

Proteins form an essential part of our diet. The minimum daily intake for an average man is 44 g while an optimum daily intake varies from 60 g/day for an average woman to 100 g/day for a nursing mother. Generally, malnutrition refers to a diet deficient in protein.

The study of proteins is one of the major topics of modern biology and biochemistry, for all life processes depend on them.

Amino acids

When proteins react with strong acids or in the presence of certain enzymes, they add water to form amino acids:

$$\text{proteins} \xrightarrow[\text{hydrolysis}]{\text{strong acids}} \text{amino acids}$$

This reaction indicates that, chemically, proteins are polymers of amino acids.

Amino acids are organic molecules that contain an amine and an acid functional group. The simplest amino acid is glycine (aminoacetic acid):

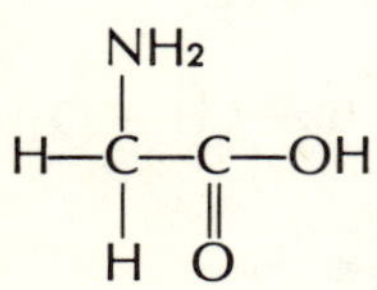

glycine
α-aminoacetic acid

All except a few of the amino acids derived from proteins are α-amino acids. The α means that the amino functional group is located on the carbon atom next to the one containing the —CO_2H acid group. The general formula for amino acids derived from proteins can thus be written

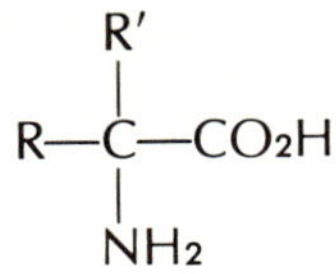

where R and R′ designate hydrogen or a carbon-containing radical.

Amino acids are colorless, crystalline solids at room temperature; most are moderately soluble in water. Table 18–3 lists the eight protein-derived amino acids essential to man's diet. These amino acids cannot be synthesized by the body, but must be contained in the food we eat; they are called **essential amino acids.** The minimum daily nutritional requirement for each of these compounds is indicated on the table, along with the abbreviations for the molecules used by biochemists. The specific essential amino acids needed in the diet varies from one species to another. Many microorganisms can manufacture all the needed amino acids from inorganic raw materials, whereas rats, for example, require the same eight as humans plus one more called histidine.

TABLE 18–3 THE EIGHT ESSENTIAL AMINO ACIDS
The minimum daily nutritional requirements are shown in grams following the abbreviation for the molecule.

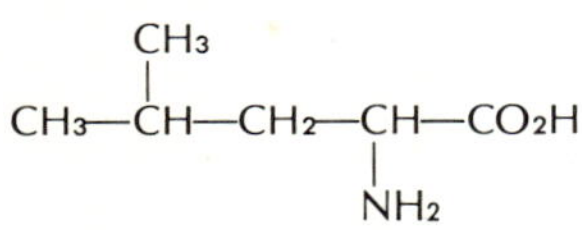 leucine Leu (1.1 g)	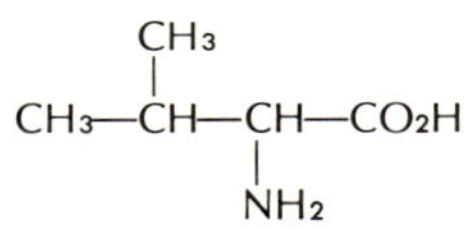 valine Val (0.8 g)	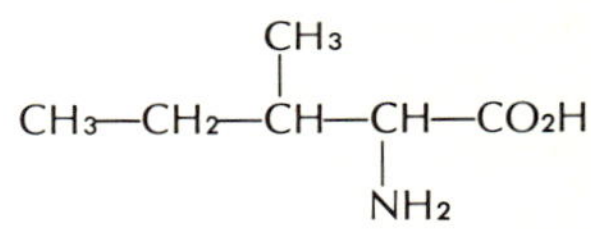 Isoleucine Ile (0.7 g)
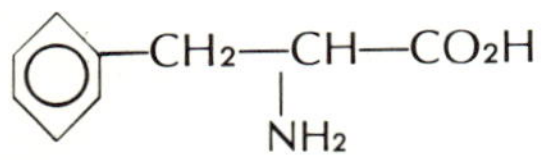 phenylalanine Phe (1.1 g)	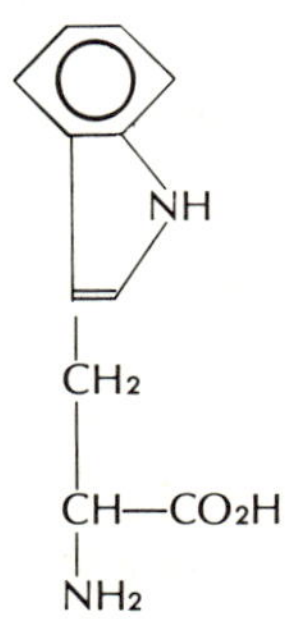 tryptophan Try (0.25 g)	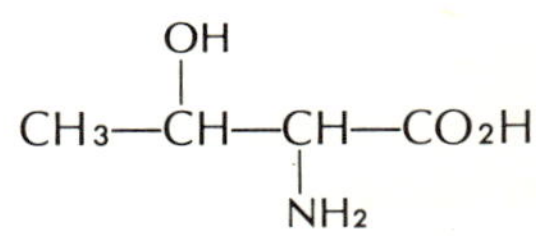 threonine Thr (0.5 g)
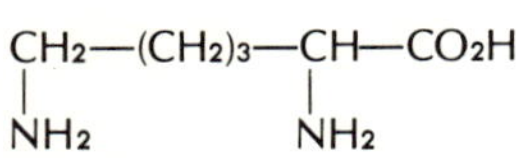 lysine Lys (0.8 g)		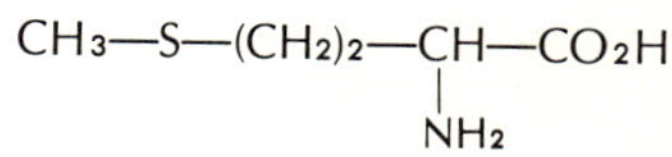 methionine Met (1.1 g)

Polymer Structure of Proteins

Proteins are solid polymers of amino acids. They have indefinite melting points and molecular masses from a few thousand (insulin, 6000 amu) to many millions (glutamate dehydrogenase, 1×10^6 amu). The amino acid monomer units in proteins are linked together by **peptide bonds,** bonds formed between nitrogen and carbon atoms when an amine combines with an acid to form an amide and water.

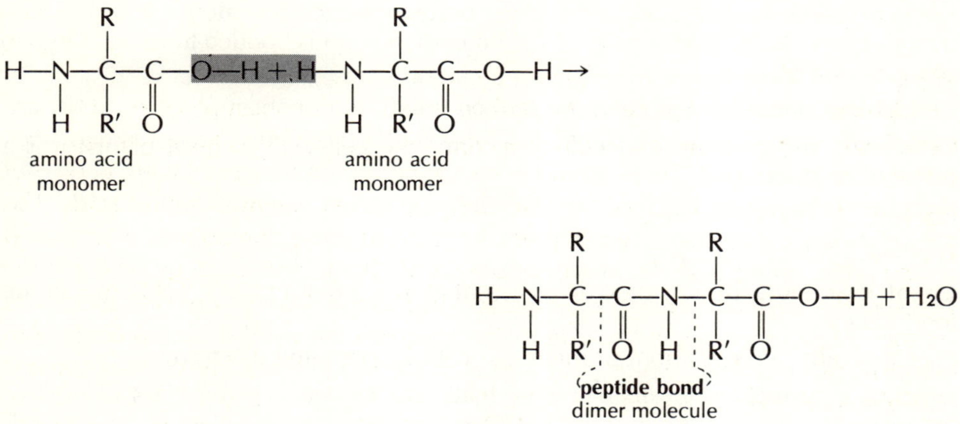

This reaction is the same as that discussed for nylon: the peptide bond is actually an amide functional group. Because the dimer molecule (*di* means two) still has an amine group at one end and an acid group at the other, it can continue to add more monomer units to produce a long-chain polymer. The overall reaction is

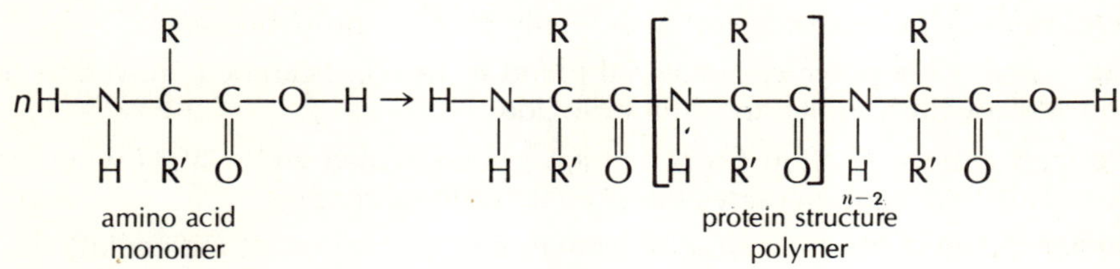

NEW TERMS

Absorbent cotton: cotton that has been treated with ether to remove fats and waxes.

Addition polymerization: the uniting of many small molecules to form giant molecules; no product other than polymer is formed.

Addition reaction: a type of reaction characteristic of unsaturated compounds; atoms or groups are added to the carbons of the multiple bond, thus converting double bonds to single bonds.

Aliphatic compound: compounds that have a noncyclic structure—not aromatic.

Alkanes: hydrocarbons of the homologous series C_nH_{2n+2}; all carbon-carbon bonds are single covalent bonds.

Alkenes: hydrocarbons that contain carbon-carbon double bonds.

Alkyl group: a group of atoms containing one less hydrogen atom than the corresponding alkane: CH_3— is the methyl group, CH_3—CH_2— is the ethyl group.

Alkynes: hydrocarbons containing a carbon-carbon triple bond; they are also called acetylenes.

Amino acid: organic compounds that contain an amine and an acid functional group.

Aromatic compounds: compounds with structures related to that of the benzene molecule.

Branched-chain structure: a noncyclic arrangement in a molecule in which at least one carbon atom is bonded to three or four other carbon atoms; this is in contrast to open-chain structures in which no carbon atom is bonded to more than two other carbon atoms.

Cellophane: regenerated cellulose formed into a film or sheet.

Cellulose: the polymer molecule that forms the cell wall in most plants: it is a polymer of glucose.

Dacron: a polyester fiber.

Enzyme: a body catalyst.

Essential amino acids: amino acids essential to life that must be contained in the food we eat.

Free radical: an atom, ion, or molecule with an unpaired electron.

Functional group: a group of atoms that characterize a given class of organic compounds; the functional group $—CO_2H$ characterizes the class of compounds known as carboxylic acids.

Glucose: the most important sugar,

$$CH_2OH—CHOH—CHOH—CHOH—CHOH—CHO.$$

Homologous series: a series of compounds in which each member differs from the preceding member by one $—CH_2—$ unit.

Hydrocarbons: compounds containing only carbon and hydrogen.

Lignin: a complex polymeric material found in wood. It can be removed by treating the wood with bisulfite and sulfur dioxide.

Mercerized cotton: cotton fibers that have been treated with NaOH while under tension. This treatment increases the strength of the thread.

Monomer: one of the individual units that unite to form giant molecules.

Natural gas: the volatile fraction of crude petroleum; it is primarily methane.

Nylon: a polyamide polymer.

Open-chain structure: a noncyclic arrangement of carbon atoms in which no carbon atom is bonded to more than two other carbon atoms.

Peptide bond: the term given to the linkage that unites amino acids in protein molecules; it is an amide functional group, —NH—CO—.

Peroxides: compounds with the general formula R—O—O—R that can be used as a source of free radicals, R—O·.

Polyamide: polymers containing amide groups; formed from diamines and diacids or from amino acids.

Polyesters: polymers containing ester functional groups; formed from diacids and dialcohols.

Polymers: giant molecules.

Protein: polymers of amino acids; they are complex organic molecules found in every living organism.

Rayon: regenerated cellulose.

Saturated hydrocarbons: hydrocarbons that contain no double or triple bonds.

Structural isomerism: two or more compounds with the same molecular formula but different structural arrangements.

Substitution reactions: reactions characteristic of benzene and the alkanes, in which a hydrogen atom is removed from the molecule and replaced by another kind of atom or group of atoms.

Unsaturated hydrocarbons: hydrocarbons that contain double and/or triple bonds.

Viscose: the cellulose xanthate solution that is forced through small holes into an acid bath to make rayon fibers.

Xanthate: the class of compounds produced when cellulose is reacted with NaOH and CS_2; R—O—CS—S—Na.

PROBLEM SET A

1. Classify the following compounds according to the functional groups they contain: CH_3CH_2CHO, $CH_3CH_2CHOHCH_3$, CH_3COCH_3, HO_2CCH_3, CH_3I.
2. Give the structural formulas for a 6-carbon-atom molecule representative of the following classes of organic compounds: acid, alcohol, amine, ester, amide, aldehyde, and halide.
3. Name the following compounds:
   a. $CH_3CH_2CH_2CH_2Br$
   b. *n*—C_8H_{18}
   c. CH_3CHICH_3
   d. $CH_3—CH_2—C(CH_3)_2—CH_2—CH_3$
   e. $(CH_3)_2—CH—CH_2—C(C_2H_5)_2—CH_3$
   f. Br—CH_2—CH_2—I
4. Write structural formulas for:
   a. 2-methylbutane
   b. 2,2,4-trichloropentane
   c. 1,1,1-trichloropropane
   d. 2-bromo-4,4-dimethyl-6-ethylnonane
   e. *tert*-butyl alkyl group
5. Name the following:
   a. $CH_2{=}CH—CH_2—CH_2—CH_3$
   b. $HC{\equiv}C—CH_2—CH_2—CH_2—Cl$
6. Give equations for the following reactions:
   a. preparation of ethyne from calcium carbonate and coke.
   b. propene + Br_2
   c. preparation of ethyl hydrogen sulfate from ethene.
   d. 1-pentyne + HCl
7. Give the reaction scheme for the polymerization of $CH_2{=}CHCl$ (vinyl chloride, or chloroethene). Hint: use a peroxide initiator and refer to the discussion for the polymerization of ethene (ethylene).
8. How many ethene monomer units are in a polyethene molecule with a molecular mass of 50,000?
9. Explain why cellulose is both an alcohol and an ether.
10. Give an equation illustrating the reaction of an acid with an alcohol to give an ester.
11. Give an equation to illustrate the combination of three amino acids with peptide bonds.
12. How are the molecular structures of nylon and protein molecules similar?

PROBLEM SET B

1. Classify the following compounds according to the functional groups they contain: $C_6H_5—O—CH_3$, C_2H_5Br, $CH_3CH_2CH_2NH_2$, CH_3CO_2H.
2. Write the names and the structural formulas for all the isomers of hexane.
3. Name the following:
   a. $CH_3CH_2CH_2CHClCH_2CH_2CH_3$
   b. $CH_3CH_2CH{=}CHCH_2CH_2CHBrCH_3$
   c. $CH_3CH_2C{\equiv}CCH_3$
4. Give equations to show how ethanol (ethyl alcohol) can be prepared from ethene (ethylene).
5. Name all the products that might be produced from the chlorination of ethane (hint: different compounds can be formed, depending on the number of hydrogens replaced).
6. Write a balanced equation for the complete oxidation of *n*-butane.
7. Show in an equation how two glucose molecules can unite to form a dimer molecule.
8. A cellulose molecule contains 3000 glucose monomer units. What is its molecular mass?
9. What kind of polymer structure might be expected from the reaction of glycerol, $CH_2OH—CHOH—CH_2OH$ (three alcohol groups) with the diacid terephthalic acid?
10. Show by an equation how three glycine molecules can unite by peptide bonds to form a trimer.

appendix A programmed sequences

INSTRUCTIONS FOR USING PROGRAMMED SEQUENCES

Programmed instruction is a self-teaching aid designed to help you over troublesome areas of a subject. All answers are provided, and there are no "trick" questions. Since there is no time limit, you may proceed at your own rate.

You will need a few sheets of paper and a pen or pencil, and sometimes a sliderule or a calculator will help you go faster. Start by covering all questions and answers with a sheet of paper, except for question 1. Study the first question and write your answer in the blank provided. If you are to profit from the use of a program, it is essential that you write down your answer before proceeding further. Now slide the paper shield down the page to uncover answer 1 and question 2. If your written response corresponds to answer 1 in the righthand column, you are ready for question 2. If your answer is incorrect, restudy question 1 until you really understand why the given answer is correct. If you disagree with the answer given, make a note and plan to return to the question later.

Continue down the page by repeating this process of studying the question, writing your response, and comparing your answer with the one provided. You will be pleased with the results of your self-discipline if you resist the inclination to peek at the answers before writing in your own. Obviously, it is of little value to just copy the answers. You will profit and learn from a program only if you have a genuine desire to do so. If you feel that a question is too difficult or ambiguous, discuss it with your instructor.

PROGRAMMED SEQUENCE 1

CONVERSION OPERATIONS

Simple conversion operations are frequently employed in solving chemistry problems. You were probably first introduced to these operations in highschool algebra. This sequence is only a brief review to help you test your recall of how equations can be manipulated and rearranged to solve a problem in the desired units. PROGRAMMED SEQUENCE 2 will provide instruction and drill on using conversion with units and dimensions.

Suppose we are given the equation $x = y$. What operations can we perform on this equation without destroying the equality? For example, if $x = y$ is it also true that $cx = cy$? By assigning letter "c" a value, any value—let's choose z—it can be seen that $cx = cy$ is the same as $zx = zy$. Therefore, it is also true that $cx = cy$, and we have not changed the relationship $x = y$ by multiplying both sides (x and y) by the same value c. Using this procedure—that is, assigning a value to the unknown letters—you should be able to work through problems 1, 2, and 3 in this sequence and prove to yourself that the indicated relationships are either true or false. Problems 4, 5, 6, and 7 demonstrate how to preserve the equality while rearranging an equation to solve for a particular unknown.

1. We do not destroy the equality if we multiply or divide both sides by the same quantity (provided it is not zero). Thus, if $x = y$, which of the following are also true?

$ax = ay \qquad \frac{x}{2} = \frac{y}{2} \qquad \frac{bx}{3} = \frac{by}{3} \qquad -x = -y$

2. We do not destroy the equality if we add or subtract the same quantity to both sides. If $x = y$, which of the following are also true?

1. All are true.

$x + 2 = y + 2 \qquad x - b = y - b \qquad x - y = y - y$

3. If we start with a more complicated equation, $e/f = g/h$, which of the following equations are also true? (If an equation is not true, draw a diagonal through the equals sign.)

2. All are true.

$\frac{ea}{f} = \frac{ga}{h} \qquad \frac{e-a}{f} = \frac{g-a}{h} \qquad \frac{e}{f} - a = \frac{g}{h} - a$

4. A common operation, given the equation $e/f = g/h$, is to solve, in turn, for e, f, g, or h. To solve for e you would multiply both sides by ____, and $e =$ ______.

3. $\frac{ea}{f} = \frac{ga}{h}$

$\frac{e}{f} - a = \frac{g}{h} - a$

but

$\frac{e-a}{f} \neq \frac{g-a}{h}$

5. To solve for g, you would multiply both sides by ____, and $g =$ ______.

4. f, $e = \frac{fg}{h}$

6. To solve for f, several operations are needed. First, to remove f from the denominator, you would multiply by f, obtaining the equation $e = fg/h$. Then, to isolate f, you would multiply by ____ and divide by ____, and $f =$ ______.

7. Solving for h is done in similar fashion: first multiplying by h, then multiplying by ____, dividing by ____, and $h =$ ______.

Most of the numbers used in chemistry problems are not just pure (dimensionless) numbers, but are inseparably bound to units such as km, s, cm^3, etc. The same cancellation operations apply to units as to the numbers. These operations on units are reviewed in Programmed Sequence 2.

5. h, $g = \frac{he}{f}$

6. h, g, $f = \frac{he}{g}$

7. f, e, $h = \frac{fg}{e}$

PROGRAMMED SEQUENCE 2

UNITS AND DIMENSIONS

Instructions for programmed sequences are on page 301.

1. A scientific measurement is expressed by a number followed by a dimensional unit. The *number* shows the magnitude of the measurement, and the *unit* describes the physical relationship. For example, in the measurement 56 cm, the *number* giving the magnitude of measurement is _____ and the unit of length is _____. The expression 96 meters has __________ as the unit of length. The minute is a unit of __________, and the quart is a __________ of volume.	
2. Units are a necessary part of a complete measurement. The expression *24 g* conveys a special thought which the number *24* alone does not. The unit is as important in a measurement as the numerical portion. Kilograms, miles, years, and liters are examples of __________. In the expression *65 watts* the unit is __________.	**1.** 56, cm, meters, time, unit
3. In some cases units are simple, like centimeters (cm), liters (l), or gallons (gal). In other cases, we may use composite units like miles per hour (mi/h), grams per milliliter (g/ml), or pounds per square inch (lb/in^2 or psi). The unit meters (m) per foot would be written as ______, grams per cubic centimeter as ______, and calories (cal) per gram as ______.	**2.** units, watts
4. If distance in feet is divided by time in seconds the result is given in units of ______. If distance, miles, is divided by time in hours the result is given in units of ______. Density is defined as *mass* divided by *volume*. Thus, an object of 20 g mass with a volume of 10 ml would have a __________ of 2.0 ______.	**3.** m/ft, g/cm^3, cal/g
5. If we multiply 3 ft by 6 lb, we get 18 for the numerical portion of the answer and ft $\times$ lb for the __________. 3 ft $\times$ 6 lb = 18 ft/lb, but 3 ft $\div$ 6 lb = $\frac{1}{2}$ ft/lb. In the first case, the units are __________. In the second case, the units are __________.	**4.** ft/s, mi/h, density, g/ml
6. If we multiply grams by centimeters, we get the units ______. If we divide grams by centimeters, we get the units ______. If we divide centimeters by grams, we get the units ______.	**5.** units, ft per lb, ft/lb
7. If we multiply a volume in liters by time in minutes, we get units of ______. If we divide a volume in liters by time in minutes, we get units of ______.	**6.** g cm, g/cm, cm/g
8. If we add one volume in cubic centimeters to another volume in cubic centimeters, we get units of ______.	**7.** l per min l/min
9. In any mathematical operation, units are treated in the same manner as are numerical terms. Measured quantities can be added or subtracted if they have the same units. Thus, 2 apples + 3 apples = 5 apples because the units __________ are the same. But 2 apples + 3 peaches cannot be added because the units apples and peaches are not the ______.	**8.** cm^3 (or cc)
10. Similarly, 2 cm + 3 cm = 5 cm, but 2 cm + 3 ft cannot be added because the ______ cm and ft are not the ______. Also, ft cannot be added to ft/s because these units are not the ______.	**9.** apples, same

11. Any number of units may be combined by multiplication or division. In such cases both the number and the units follow algebraic laws of cancellation, root extraction, and the like. To illustrate:

$$\frac{a}{\not b}\times \not b = a \quad \text{is like} \quad \frac{cm}{\not s}\times \not s = cm$$

$$\frac{a}{b}\times a = \frac{a^2}{b} \quad \text{is like} \quad \frac{cm}{s}\times cm = \frac{cm^2}{s}$$

If we multiply g/m by m, the result of the multiplication must have units ____. g/ml multiplied by g will result in the units ______.

10. units, same, same

12. In making calculations, units often cancel out to leave an answer without any units at all. For example, if one density is divided by another density, the answer is a number without units:

$$\frac{3\ \not{g}/\not{ml}}{1\ g/ml} = 3 \text{ (no units remain)}$$

The answer, 3, is called a *dimensionless* number. 10 ft/s ÷ 5 ft/s gives the dimensionless number 2 for an answer. Similarly, (60 mi/h) ÷ (2 mi/h) = ______.

11. g, g^2/ml

13. Dimensionless numbers will result if all units cancel out in multiplication or division. ft/s multiplied by s/ft would give a dimensionless number, as follows:

(3 ~~ft/s~~) × (2 ~~s/ft~~) = 6 (no units)

(mi/h) × (h) has the units ______ and (mi/h) × (mi) has the units ______, but (mi/h) × (h/mi) is a ____________ number.

12. 30 (no units)

14. Logic enables us to quickly conclude that a boy who runs 9 miles in 1 hour can run 18 miles in 2 hours. Applying units, (9 mi/~~h~~) × (2 ~~h~~) = 18 mi. Use units to solve these simple relationships:

(ml/min) × (min) gives units of ______.
(ft/s) × (s/min) gives units of _________.
(yd) × (ft/yd) × (in/ft) gives units of ______.

13. mi, mi^2/h, dimensionless

15. In algebra, dividing by a fraction is the same as inverting the fraction and multiplying by it. For example, $8 \div \frac{1}{4}$ is the same as 8 × 4. Grams ÷ g/ml is the same as (g) × (ml/g). Miles ÷ mi/gal is the same as (mi) × (gal/mi), giving the units ______.

14. ml, ft/min, in

16. cm ÷ cm/in. is the same as (cm) × (in/cm), giving the units ______. 5.00 in × 2.54 cm/in = ______. 10.0 in ÷ (in/2.54 cm) = ______.

15. gal

17. It is simple to convert quantities expressed in one set of units to another set of units, like feet to meters. Such conversions are conveniently performed by using *conversion factors,* which are numerical ratios of units equal to 1. 12 in = 1 ft, so (12 in/ft) is a conversion factor. 3 ft = 1 yd, so (3 ft/yd) is a conversion factor. 24 h/day, 454 g/lb, and 5280 ft/mi are ________________.

16. in, 12.7 cm, 25.4 cm

17. conversion factors

18. Conversion factors are useful because they are equal to 1, and multiplication or division by the number 1 does not change the value of the original quantity. The conversion factor 60 s/min is equal to ___. The conversion factor 12 in/ft is also equal to 1.

$$\frac{12\text{ in}}{\text{ft}} = \frac{1\ \cancel{\text{ft}}}{1\ \cancel{\text{ft}}} = 1$$

If we were converting 12 ft to yards, we would use the conversion factor (yd/3 ft):

$$\frac{\text{yd}}{3\ \cancel{\text{ft}}} \times 12\ \cancel{\text{ft}} = 4\text{ yd}$$

18. 1

19. Conversion factors may be obtained from information given in a specific problem. The statement "3 apples cost 10 cents" has the conversion factors (3 apples/10¢) or its reciprocal, (10¢/3 apples). The problem states that 3 apples and 10¢ are one and the same, just as 12 inches is the same as 1 foot. In this problem, 3 apples/10¢ is a ________ factor equal to 1. 12 in/ft is a ________ factor applicable to *any* problem.

19. conversion, conversion

20. It is easy to convert quantities from one set of units to another but it is not always obvious when to multiply and when to divide in such conversions. This difficulty is avoided if we cancel and multiply units as though they were numbers. If we convert 87 cm to meters, we reason that the answer must be *smaller* than 87 because a cm is ______ than a meter. The conversion factor is (100 cm/m) or the inverse, (m/100 cm). But the answer units must be in ______, therefore a proper set-up is:

(87 cm) × ________ = _____ m.

20. smaller, meters, m/100 cm, 0.87

21. Units are very helpful in problem solving. Suppose we want to calculate the number of liters in a quart. The answer units are ________, therefore a proper set-up is:

$(\cancel{\text{qt}}) \times (\text{liters}/\cancel{\text{qt}}) = \text{liters}.$

How many miles can you travel on 5 gallons of gasoline if your car averages 15 mi/gal of gasoline? The answer units are miles, thus:

$(15\text{ mi}/\cancel{\text{gal}}) \times (5\ \cancel{\text{gal}}) =$ ______.

21. liters, 75 mi

22. If 1 lb of apples costs 25¢, how much will 7 lb cost? Let us assume that the answer units are required in cents (it could be dollars). The set-up of the dimensions is:

$$\frac{\text{cost}}{\cancel{\text{weight}}} \times \cancel{\text{weight}} = \text{cost}.$$

The set-up using units given is:

$$\frac{25\text{ cents}}{\cancel{\text{pound}}} \times 7\ \cancel{\text{pounds}} = 175\text{ cents}.$$

If the units were *carelessly* applied we would get *meaningless* units in the answer:

$$\frac{\text{pound}}{25\ \text{cents}} \times 7\ \text{pounds} = 7/25\ \_\_\_\_\_\_\_\_\_.$$

23. A chemical sells for three dollars per gram. We want four grams. Suppose we desire cost units in dollars. How shall we set up the problem? If we try (g/3 \$) × (4 g), the units will be ______. The method that will give dollars as answer units is: (3 \$/g) × (4 g) = ______. How many grams of chemical can we get for \$2.50? The answer units are ____, so the proper set-up including figures and units is (g/3 \$) × 2.5 \$ = _______.

22. lb^2/¢

24. The speed of a bird is 588 ft/s. How many seconds will it take the bird to fly 98 feet? The set-up including figures and units is __________ × ______. How many miles can the bird fly in 300 seconds? The answer units are ________. We must use a conversion factor to get miles because only seconds and feet are given in the problem. Tables show that there are 5280 ft/mi. The problem can now be easily solved:

(588 ft/s) × (300 s) × ________________ = __________.

23. g^2/\$
\$12,
g,
0.83 g

25. The unit ft^2 means ft × ft. The unit cm^3 means cm × cm × cm. What is the volume of a cube 2 cm on each edge? The calculation is 2 cm × 2 cm × 2 cm. Which *one* of the following is *not* a correct answer to the calculation?

2 cm^3 (2cm)^3 2^3cm^3 8 cm^3

24. s/588 ft,
98 ft,
miles,
mi/5280 ft
33.4 mi

26. Although (2cm)^3 and 2^3cm^3 represent the same quantity, 2^3cm^3 is preferred if subsequent cancellation of units is involved. For instance, which of the following is the correct conversion of the volume of the cube from cm^3 to mm^3?

a. $(2\,\cancel{cm})^{\cancel{3}}\dfrac{10^3mm^3}{1\,\cancel{cm^3}} = 2 \times 10^3mm^3$

b. $2^3\cancel{cm^3}\dfrac{10^3mm^3}{1\,\cancel{cm^3}} = 2^3 \times 10^3mm^3 = 8 \times 10^3mm^3$

25. 2 cm^3
is *not*
correct

27. Let us calculate the volume in cm^3 of a box 0.60 m long, 10 cm wide, and 50 mm deep. The answer will be in cm^3 units, thus all units must be converted to cm first, then multiplied. (0.60 $\cancel{m}$) × (100 cm/$\cancel{m}$) × (10.0 cm) × (50 $\cancel{mm}$) × (cm/10 $\cancel{mm}$) = ________ cm^3. In liters, the answer would be ____ liters.

26. b is
correct

28. Now test your knowledge of the use of units by trying some conversions. Show the complete set-up so that you can cancel units. For example, suppose we want to change 288 in^2 to ft^2.

(288 $\cancel{in^2}$) × (ft^2/144 $\cancel{in^2}$) = 2 ft^2, or
(288 $\cancel{in^2}$) × (ft/12 $\cancel{in}$) × (ft/12 $\cancel{in}$) = 2 ft^2.

Change 12 m^2 to cm^2. ________________
Change 64.00 in to mm. ________________
Change 200 mm^2 to cm^2. ________________

27. 3,000,
3

29. Make the following conversions:

0.020 g/ml to g/liter __________

28.00 g/cm to g/m __________

65.0 gal/h to gal/min __________

7.000 \$/g to \$/kg __________

30. 2 liters of gas have a mass of 4 grams. What is the mass of 500 ml of the gas?

We can work the problem stepwise, like this:

a. The density of the gas in g/l is ________.

b. Changing density to g/ml __________.

c. Weight of 500 ml is density in g/ml × 500 ml = ______.

31. A stove can burn 840 kg of wood/21 min. How much can it burn in 4 h?

a. The rate of burning in kg/min is __________.

b. The rate of burning in kg/h is __________.

c. The amount burning in 4 h is __________.

32. A particle moves 800 mi in 2 h. How many ft will it move in 10 min?

a. Speed in mi/h __________.

b. Convert speed to mi/min __________.

c. Convert speed to ft/min __________.

d. Determine distance moved in 10 min __________.

Note that the entire problem could be combined in one step:

$$\frac{800\ \cancel{mi}}{2\ \cancel{h}} \times \frac{\cancel{h}}{60\ \cancel{min}} \times \frac{5280\ \text{ft}}{\cancel{mi}} \times 10\ \cancel{min} = \underline{\hspace{3cm}}\ \text{ft.}$$

28. 120,000 cm², 1626 mm, 2.00 cm²

29. 20 g/l, 2800 g/m, 1.08 gal/min, 7000 \$/kg

30. 2 g/l, 0.002 g/ml, 1 g

31. 40 kg/min, 2400 kg/h, 9600 kg

32. 400 mi/h, 6.67 mi/min, 35,200 ft/min, 352,000 ft, 352,000

PROGRAMMED SEQUENCE 3

SCIENTIFIC MEASUREMENT AND SIGNIFICANT FIGURES

Instructions for programmed sequences are on page 301.

1. Reference to counted things like dollars or cars are exact numbers. 50 dollars means *exactly* 50 dollars, not *approximately* 50 dollars. 7 cars means ________ 7 cars, not approximately 7 cars. 3 apples means *exactly* 3 apples, no more, no less.

2. A *measured* value is never known exactly no matter how carefully made. Each measuring device has limitations that fix its accuracy. Therefore, any observation has some uncertainty associated with it. We may *count* exactly 20 beakers but a *measurement* of 20 g means *about* 20 g. We can have ________ 3 pencils, each being ________ 8 inches in length.

1. exactly

3. A scientific statement is careless if it says less than is known and misleading if it conveys more than is known. If we measure a length of 7.23 cm we mean that the length is *about* 7.23 cm. The true value is closer to 7.23 cm than to 7.22 or 7.24 cm, but it is clear that even the value 7.23 cm is only an approximation. We say that the last digit, 3, in 7.23 is *uncertain*.

2. exactly, about

4. By convention, we assume that a measured result is reliable within 1 unit in the last digit. Thus a measured value of 7.23 cm implies that any value between 7.22 and 7.24 cm is acceptable and that the true value lies somewhere between these limits. By convention, then, a measurement of 56.7 g implies that any value between ________ g and ________ g is acceptable and that the true value is somewhere between these limits.

5. Since measured results are not exact, we must use numbers to properly indicate the reliability of the result. It is convenient to incorporate the accepted range of values into one expression. If by 56.7 g we mean any value above 56.6 g but below 56.8 g, we can write 56.7 g and include its highest and lowest values by writing it as 56.7 ± 0.1 g. The 0.1 represents the uncertainty of the measurement. If we add the uncertainty to the measured value the highest acceptable value of ________ g is obtained; if the uncertainty is subtracted from 56.7 g, the lowest value of ________ g is obtained. The true value lies between the highest and lowest values.

4. 56.6, 56.8

6. A measured value of 62.54 g with an uncertainty of ± 0.01 g would be written as ________ $\pm$ ________ g. The true value of this measurement would be between a low value of 62.53 g and a high value of ________ g.

5. 56.8, 56.6

7. If the measurement 32.2 ml is reliable to ± 0.1 ml, the true value should be between ________ and ________ ml. If the volume were measured in a flask having an uncertainty of ± 0.01 ml, the reading may be 32.17 ml. This information may be written in a single expression as ________ $\pm$ ________ ml and shows a greater reliability than the cruder measurement 32.2 ± 0.1 ml.

6. 62.54, 0.01, 62.55

8. The usual method for indicating the uncertainty in a number is to indicate the uncertainty by the number of figures shown. The last figure shown is the one in which there is some uncertainty. The digits that are certain *and one more* are called *significant figures*. The number 4.56 im-

7. 32.1, 32.3, 32.17, 0.01

plies that the 4 and 5 are completely reliable but that there is some uncertainty in the last figure, 6. We say that 4.56 contains 3 significant figures. Thus, 2.89 contains ______ significant figures and 2.897 contains ______ significant figures.

9. Decimal points are used to show magnitude only, so the position of the decimal is not related to the number of significant figures. The numbers 0.258 and 2.58 contain 3 significant figures. The numbers 0.00258 and 258 also contain ______ significant figures. The position of the decimal point has no effect on the number of ____________________. The numbers 0.001234, 0.01234, 0.1234, 1.234, 12.34, and 123.4 each contain ______ significant figures.

8. 3, 4

10. Significant figures, then, express a measured result such that only the last digit is uncertain or in doubt. A number having 4 significant figures has 4 digits in the number. The first 3 digits are known exactly, the 4th digit is in doubt. The measurement 12.34 cm contains 4 significant figures. The 1, 2, and 3 are known exactly but the last digit, 4, is in doubt. The measurement 98654.8 mm contains ______ significant figures, ______ of which are known exactly. Only the digit ______ is uncertain or in doubt.

9. 3, significant figures, 4

11. A volume recorded as 28.2 ml contains ______ significant figures. If this volume were written in other units, such as liters, it would still contain ______ significant figures. The numbers 28.2 and 0.0282 have the same number of significant figures. The decimal point position has no bearing on the number of ____________________ contained in a number.

10. 6, 5, 8

12. Zeros appearing as first figures of a number merely indicate magnitude and are not significant. However, if there is a number to the left of the decimal, all zeros to the right of that number *are* significant. To illustrate, 0.00286 contains 3 significant figures but 1.00286 contains 6. 0.021 contains ______ significant figures but 2.021 contains 4. 0.00021 contains ______ significant figures but 5.00021 contains ______ significant figures.

11. 3, 3, significant figures

13. There is a useful rule for the determination of significant figures: "Lefthand zeros don't count, righthand zeros do." For example, 0.026 has 2 significant figures, but 0.0260 has 3. (The righthand zero is not needed for magnitude; thus it must have been measured.)

12. 2, 2, 6

14. There is one situation where you will not be able to determine the number of significant figures by looking at the number alone. A number like 921 has ______ significant figures, but the number 9210 may have either 3 or 4 significant figures. If the zero is used for magnitude only, the number of significant figures is 3. If the zero was actually measured, the number of significant figures is ______.

15. A number like 96,500 may contain either 3, or 4, or 5 significant figures. If the zeros are the results of measurement, 96,500 contains 5 significant figures. One sure way to convey information regarding the proper number of significant figures is to express the measured value in *exponential notation*. Thus 9.65×10^4 shows 96,500 to 3 significant figures. 9.650×10^4 shows it to ______ significant figures, and 9.6500×10^4 shows the number to 5 significant figures.

14. 3, 4

16. The number 900 is written exponentially as 9×10^2, 9.0×10^2, or 9.00×10^2. Thus the number of significant figures in 900 may be ______, or ______, or ______. If we write 900 as 9.0×10^2, we are showing that it was measured to the nearest tenth of a unit and the number of significant figures is 2. If we write 900 as 9.00×10^2, the measurement was to the nearest ________th of a unit and the number of significant figures is ______.

15. 4

17. The number 2.637 has 4 ________________________. Suppose that we wish to write this value to 3 significant figures. This involves *rounding off* by dropping the 7. Since 2.637 is closer to 2.64 than it is to 2.63, the rounded-off value will be ________. Likewise, 5.8224 rounded off to 4 significant figures will be ________.

16. 1, 2, 3, 100, 3

18. A rounded-off number is that which is closest to the original number. The number 46.7 rounded off to 2 significant figures is 47. The number 46.85 to 2 significant figures is also ______. When a 5 is dropped in the last digit remaining, the last figure always becomes an even number. For example, 46.5 to 2 significant figures is 46. 45.5 to 2 significant figures is also 46. In both cases, we rounded off to the nearest even digit.

17. significant figures, 2.64, 5.822

19. 55.5 rounded off to 2 significant figures is 56. 56.6 rounded off to 2 significant figures is 57. 98.2 to 2 significant figures is ______. 98.56 to 2 significant figures is ______. 734.5 to 3 significant figures is ________. 7869.5 to 4 significant figures is __________.

18. 47

20. The results of scientific measurements are often combined. A change in temperature, for example, is the result of two measurements, not just one. It is a *derived quantity.* Likewise, the density of a substance is a derived quantity expressed by dividing mass by volume. Derived quantities cannot be more accurate than the least accurate measurement used in the calculation. For example, the volume of a rectangular block 3.04 cm × 2.5 cm × 1.54 cm is not 11.7040 cm^3 (the product of the 3 measurements). Such an answer gives a false impression of the accuracy of the measurements.

19. 98, 99, 734, 7870

21. When using measured data in operations of multiplication or division, the results must not have more significant figures than the factor with the fewest significant figures. In the above problem (3.04 cm × 2.5 cm × 1.54 cm) the factor 2.5 has the fewest number of significant figures, 2, so that the answer, 11.7040, must be rounded off to 2 significant figures. The answer is 12 cm^3. Similarly, 225 cm/27 s = 8.3333 . . . cm/s. But the 27 s has only ____ significant figures so the answer is ______.

22. If we multiply 56 ft × 345 s, the answer will contain ____ significant figures. 1352.32 ÷ 101 would give an answer with ____ significant figures. Consider the operation 8.6 × 7.24 = 62.264. The number 8.6 contains ____ significant figures and the 7.24 contains ____ significant figures. The answer will have the smaller number of significant figures of either of these, namely ____ significant figures, so the answer is 62.

21. 2, 8.3

23. Consider 2.21 × 0.3. The answer should have ____ significant figure(s). Hence, the answer to this multiplication is ______. Consider the division 97.52 ÷ 2.54 = 38.4. The number with the smaller number of significant figures will determine the answer. The 2.54 has 3 significant figures and therefore the answer, 38.4, also has ____ significant figures.

22. 2, 3, 2, 3, 2

24. Consider $22.000 \div 4 = 5.50$. The 22.000 has ____ significant figures, and the 4 has ____ significant figure(s), so the answer must have ____ significant figure(s). When 5.50 is rounded off to 1 significant figure, the answer is ____.

23. 1, 0.7, 3

25. Now see whether you understand the application of the rules for using significant figures in calculations involving multiplication or division.

a. $2.06 \times 4.223 =$ ________
b. $82.4 \times 0.098 =$ ________
c. $97.87 \div 2.54 =$ ________

24. 5, 1, 1, 6

26. In addition or subtraction of measurements, the least accurate measurement determines the accuracy of the sum or difference. For example, in the addition of 230.4 m + 51.36 m = 281.76 m, the sum should be reported as 281.8 m. The 51.36 is correctly measured to the nearest hundredth, but the 230.4 is only reliable to the nearest tenth. Therefore, the figures after tenths in the sum are misleading, and the answer can be given only to the nearest tenth as ________ m.

25. 8.70, 8.1, 38.9

27. In addition or subtraction, the rule is to retain as many digits to the right of the decimal point as are included in the number with the fewest digits to the right of the decimal. For example, consider $2.03 + 142.0501$. The 2.03 has 2 digits to the right of the decimal and the 142.0501 has 4. The sum cannot have more than 2 digits to the right of the decimal point. Consider $7.803 + 2 = 9.803$. What should the sum be after properly rounding off? The 7.803 has 3 digits to the right of the decimal, and the 2 has none. Thus the 9.803 must be rounded off to an answer of ______, containing ______ digits to the right of the decimal.

26. 281.8

28. Consider the sum $104.75 + 0.2865 + 72.3 + 10.215 = 187.5515$. The answer should have ____ digit(s) to the right of the decimal point because 72.3 has only 1 digit to the right of the decimal point. The answer, when properly rounded off, is ________.

27. 10, no

29. Note that we are not concerned with significant figures in addition or subtraction, just the number of digits to the right of the decimal point. Consider the difference $23.05 - 22.5321 = 0.52$. Because the number 23.05 has 2 digits to the right of the decimal point, the answer must also have ____ digits to the right of the decimal point, so that the correct answer is ________.

28. 1, 187.6

30. The answer in multiplication or division depends on the factor with the least number of significant figures. For example, 3.04 has ____ significant figures, and 2.0 has ____ significant figures. The product of 3.04×2.0 will have ____ significant figures.

29. 2, 0.52

31. But, in addition or subtraction, the answer depends not on the number of significant figures but on the least number of digits to the right of the decimal. If we add $3.06 + 2.0$, the answer will have only ____ digit(s) to the right of the decimal point. The numerical answer to the sum of $3.06 + 2.0$ is ______.

30. 3, 2, 2

32. How many significant figures are contained in

a. 39.6 ____
b. 82.250 ____
c. 0.0020 ____
d. 2.45×10^2 ____
e. 0.6×10^{-6} ____

31. 1, 5.1

33. Perform the following operations with proper application of the rules of significant figures and rounding off.

a. $26.0 \times 8.21 \times 976.1 =$ ____________

b. $(1.84 + 1.24 + 23.1) \times 3.1 =$ ______

c. $34.8 \div (3.1 \times 10^{-4}) =$ ____________

32. 3, 5, 2, 3, 1

33. 2.08×10^5, 81, 1.1×10^5

PROGRAMMED SEQUENCE 4

THE MOLE CONCEPT

Instructions for programmed sequences are on page 301.

1. Several words in our language are used to represent numbers of things. For example:

The word *pair* represents the *number* 2.
The word *dozen* represents the *number* ______.
The word *score* represents the *number* ______.
The word *gross* represents the ____________ 144.

2. In chemistry, the word *mole* represents the *number* 6.02×10^{23}. A mole of things is 6.02×10^{23} things. Just as a dozen nails means ______ nails, a mole of nails means 6.02×10^{23} nails. A mole of atoms is ________ ________ atoms; a mole of molecules is ______________ molecules.

1. 12, 20, number

3. We must specify what we are counting. A mole of automobiles requires 1 mole of engines, but rides on ____ moles of wheels. A mole of tricycles requires ____ moles of wheels. A mole of anything is 6.02×10^{23} of those things. The number 6.02×10^{23} is called *Avogadro's number.*

2. 12, 6.02×10^{23}, 6.02×10^{23}

4. Avogadro's number of automobiles, 6.02×10^{23} automobiles, would require ______________ engines to move them on ______________ wheels.

3. 4, 3

5. Suppose that a shirt has two pockets and six buttons. A mole of shirts requires ____ moles of pockets and ____ moles of buttons. A mole of buttons is ______________ buttons.

4. 6.02×10^{23}, 24.1×10^{23}

6. A mole of silver atoms contains ______________ silver atoms. If we had 12.04×10^{23} silver atoms, we would have ____ moles of silver atoms.

5. 2, 6, 6.02×10^{23}

7. A mole of water molecules H_2O, contains ____ moles of hydrogen atoms H, and ____ mole of oxygen atoms, O. A mole of ammonia molecules, NH_3, contains ____ moles of hydrogen atoms, H, and ____ mole of nitrogen atoms, N.

6. 6.02×10^{23}, 2

8. Three moles of ethane molecules, C_2H_6, contain ______ moles of hydrogen atoms and ______ moles of carbon atoms. The total number of moles of atoms in a 4 mole sample of ethane is ______.

7. 2, 1, 3, 1

9. The words gross (144), dozen (12), or mole (6.02×10^{23}), can be used to represent not only a *number* but also a *mass*. A gross of dimes weighs 349.83 grams and a gross of quarters weighs 898.20 grams. Thus we can say that a gross of dimes, 144 dimes, is 349.82 grams of dimes, and that a gross of quarters is 898.20 grams.

8. 18, 6, 32

10. In chemistry, there is frequent use of the mole in representing both a number and a mass. We speak of a mole of chlorine molecules, Cl_2, and mean 6.02×10^{23} *molecules* and 71.0 *grams.* A mole of water molecules is 6.02×10^{23} *molecules* and 18.0 *grams.* A mole of HCl molecules is ______________ *molecules* and ______ *grams.*

11. Atomic masses are not *absolute* masses of atoms but *relative* masses in atomic mass units (amu). 1 atom of isotope ^{12}C has 12 amu. If an element is one-third the mass of carbon, its atomic mass is listed as ____ amu. An element of atomic mass of 24 amu is approximately ____ times the mass of ^{12}C.

10. 6.02×10^{23}, 36.5

12. 12.01 amu is the value listed for the atomic mass of carbon. This means that a mole of carbon atoms, 6.02×10^{23} carbon atoms, has a mass of 12.01 grams. Avogadro's number (6.02×10^{23}) of helium atoms has a mass of 4.00 grams, therefore the mass of He must be (*greater, less*) than that of C.

11. 4, 2

13. The atomic mass of hydrogen is 1.008 amu, and the atomic mass of oxygen is 16.00 amu. There are ________________ oxygen atoms in a mole of oxygen atoms. A mole of oxygen atoms is ________ g.

12. less

14. The molecular mass is the sum of the atomic masses of a compound and is given in amu. Since the atomic mass of hydrogen is 1.008 amu and the atomic mass of oxygen is 16.00 amu, the molecular mass of H_2O (to 3 significant figures) is ______ amu.

13. 6.02×10^{23}, 16.00

15. The molecular mass of water is 18.0 amu. Thus Avogadro's number of water molecules has a mass of ______ grams, and one ________ of water is 18.0 __________.

14. 18.0

16. The atomic mass is given in atomic mass units (amu). The *gram* atomic mass is the mass of grams in Avogadro's number of atoms. The terms *mole of atoms* and *gram atomic mass* mean the same thing. The atomic mass of N is 14.0 amu. The gram atomic mass of N is 14.0 ________.

15. 18.0, mole, grams

17. The atomic mass of F is ______ amu. A gram atomic mass of F is ______ grams.

16. grams

18. The molecular mass is given in ______. The *gram* molecular mass is the mass in grams of Avogadro's number of molecules. Thus the molecular mass of methane, CH_4, is 16.0 ______, and the gram molecular mass of methane is __________.

17. 19.0, 19.0

19. The molecular mass of hydrogen sulfide, H_2S, is ________ amu. The gram molecular mass of H_2S is ____________.

18. amu, amu, 16.0 g

20. The terms *mole of atoms* and *gram atomic mass* mean the same thing and are used interchangeably. A mole of hydrogen atoms is Avogadro's number of hydrogen atoms and has a mass of 1.01 g. A gram atomic mass of hydrogen has a mass of 1.01 g too, so a mole of hydrogen atoms is the same thing as a gram atomic mass of hydrogen.

19. 34.1, 34.1 g

21. For any element, the mass of a mole of atoms is equal to the atomic mass expressed in grams. A mole of S atoms is 32.1 g. A mole of Fe atoms is __________ g. 12.01 g is the mass of a mole of C atoms, or ____________ ____ C atoms.

22. A gram atomic mass means a mole of __________. A gram atomic mass of H is 1.01 ____. A mole of H atoms is 1.01 ____. A gram atomic mass of O is ________ g. A mole of O atoms is ________ g.

21. 55.8, 6.02×10^{23}

23. A mole of helium atoms is ________ g because a mole of helium atoms has a mass of ________ g. There are ________________ helium atoms in a mole.

22. atoms, g, g, 16.0, 16.0

24. A mole of water molecules is Avogadro's number of water molecules. It has a mass of 18.0 g. One gram molecular mass of water has a mass of 18.0 g. The terms *mole of molecules* and *gram molecular mass* mean the same thing and are used interchangeably, just as mole of atoms and gram atomic mass are used interchangeably.

23. 4.00, 4.00, 6.02×10^{23}

25. A mole of water molecules contains ____ moles of hydrogen atoms and ____ mole of oxygen atoms.

26. A mole of ammonia, NH_3, is 17.0 ____. A mole of sulfur dioxide, SO_2, is ________ g.

25. 2, 1

27. The molecular mass of H_2SO_4 is 98.1 ________. A mole of H_2SO_4 is ________ g.

26. g, 64.1

28. A 294.3 g sample of H_2SO_4 contains ____ moles of H_2SO_4. There are ____ moles of hydrogen atoms in the 294.3 g sample. There are ________________ sulfur atoms in the 294.3 g sample.

27. amu, 98.1

29. Oxygen, O_2, is a diatomic molecule. A mole of oxygen molecules contains ________________ oxygen molecules or ________________ oxygen atoms. A mole of oxygen molecules is ________ g. 2 moles of oxygen atoms, O, are ________ g.

28. 3, 6, 18.06×10^{23}

30. A mole of ethyl alcohol, C_2H_5OH, contains 6 moles of hydrogen atoms in addition to ____ moles of carbon atoms and ____ mole of oxygen atoms. Three moles of ethyl alcohol contain ______ moles of hydrogen atoms. The total number of all atoms in 3 moles of ethyl alcohol is ______ moles.

29. 6.02×10^{23}, 12.04×10^{23}, 32.0, 32.0

31. A mole of carbon dioxide, CO_2, contains a total of ____ moles of atoms; there are ____ moles of oxygen atoms but only ____ mole of carbon atoms. If 1 mole of CO_2 contains 2 moles of oxygen atoms, then 6 moles of CO_2 would contain six times that number of oxygen atoms, or ______ moles of oxygen atoms.

30. 2, 1, 18, 27

32. A 17.0 g sample of ammonia contains ____ moles of hydrogen atoms and ____ mole of nitrogen atoms. A 17.0 g sample of ammonia is a mole of ammonia.

31. 3, 2, 1, 12

32. 3, 1

PROGRAMMED SEQUENCE 5

EMPIRICAL FORMULAS AND PERCENT COMPOSITION

Instructions for programmed sequences are on page 301.

1. "Percent" means a particular kind of decimal whose denominator is 100. Percent is exactly the same as *hundredths*. 5% = 5/100. 20% = 20/100.

60% = ________

2% = ________

2. The word *percentage* is used in chemistry to denote *percent composition by mass*. Percentage is the ratio of the *part* (or portion) to the whole (or total), multiplied by 100. To calculate the percentage of anything, one must know the total amount and the portion of the total amount whose percentage is sought.

1. 60/100, 2/100

3. Percentage (or percent composition) is obtained by dividing the *part* by the *whole* and multiplying by 100:

$$\% = \frac{\text{part (or portion)}}{\text{whole (or total)}} \times 100$$

4. Suppose that you want to find the percentage of apples in a basket of fruit containing 250 apples and 750 peaches.

The *part* is 250 apples.
The *whole* is ________ fruit.

The set-up and units for the calculation is:

$$\%\ \text{apples} = \frac{250\ \text{apples (part)}}{1000\ \text{fruit (whole)}} \times 100 = \_\_\_\_$$

5. A basket of fruit contains 120 apples and oranges. If 90 of the fruit are apples, what is the percentage of apples in the basket?

4. 1000, 25%

The *part* is ______ apples.
The *whole* is ________ fruit.

The set-up and units for this calculation is

% apples = ________________ × ________ = ______

6. An alloy contains 100 g of copper, 300 g of nickel, and 600 g of zinc. The percentage zinc is ______. We are simply determining what contribution each *part* makes to the *total* mass.

5. 90, 120, $\frac{90\ \text{apples}}{120\ \text{fruit}}$, 100, 75%

7. Suppose that we wish to calculate the percentage of oxygen in water. It is convenient to let a mole of water represent the *whole* and a mole of oxygen atoms represent the *part*. 1 mole of water is 18.0 g. 1 mole of water contains ____ mole of oxygen atoms, having a mass of ________ g.

6. 60%

8. The *part* is ________ g of oxygen. The *whole* is ________ g of water.

$$\% \text{ oxygen} = \frac{\text{g of oxygen (part)}}{\text{g of water (whole)}} \times 100$$

% oxygen = ________________ × ________ = ________

Always be sure to show set-up and units.

7. 1, 16.0

9. Calculate the percentage oxygen in carbon dioxide. 1 mole of CO_2 is ________ g. 1 mole of CO_2 contains ____ moles of oxygen atoms, having a mass of ________ g.

8. 16.0, 18.0, $\frac{16.0 \text{ g O}}{18.0 \text{ g } H_2O}$, 100, 88.9%

10. The *part* is ________ g of oxygen. The *whole* is ________ g of carbon dioxide.

$$\% \text{ oxygen} = \frac{\text{g of oxygen (part)}}{\text{g of carbon dioxide (whole)}} \times 100$$

% oxygen = ________________ × ______ = __________

9. 44.0, 2, 32.0

11. It is frequently desired to calculate the percentage of *water of hydration* in a hydrated salt like $CuSO_4 \cdot 5H_2O$ or $BaCl_2 \cdot 2H_2O$. In these cases, the *part* is the water of hydration and the *whole* is the total weight of hydrated salt.

$$\% \text{ water} = \frac{\text{g of water (part)}}{\text{g of hydrated salt (whole)}} \times 100$$

10. 32.0, 44.0, $\frac{32.0 \text{ g O}}{44.0 \text{ g } CO_2}$, 100, 72.7%

12. Calculate the percentage water of hydration in $BaCl_2 \cdot 2H_2O$. Taking 1 mole of $BaCl_2 \cdot 2H_2O$ as the *whole,* each mole of $BaCl_2 \cdot 2H_2O$ contains ____ moles of water.

13. The *whole,* 1 mole of $BaCl_2 \cdot 2H_2O$, is 244.2 g. The *part,* 2 moles of H_2O, is ________ g.

$$\% \text{ water} = \frac{\text{g of water (part)}}{\text{g of } BaCl_2 \cdot 2H_2O \text{ (whole)}} \times 100$$

% water = ____________ × ________ = __________.

12. 2

14. The formula CH_4 shows that there are ____ atoms of hydrogen to each atom of carbon. The formula C_3H_8O shows that there are ____ carbon atoms, to 8 ________ atoms, to 1 oxygen atom.

13. 36.0, $\frac{36.0 \text{ g}}{244 \text{ g}}$, 100, 14.8%

15. The formula C_3H_8O may also be interpreted as relating 3 *moles* of C, to ____ moles of H, to ____ mole of O.

14. 4, 3, H

16. One mole of the molecular compound ethene (C_2H_4) contains 2 moles of carbon atoms and 4 moles of hydrogen atoms. The ratio of moles of carbon atoms to hydrogen atoms is ____ to ____. (Remember to reduce subscripts to the smallest whole numbers.)

15. 8, 1

16. 1, 2

17. 1 mole of the molecular compound cyclobutane, C_4H_3, contains ____ moles of carbon atoms and 8 ________ of hydrogen atoms. The ratio of moles of carbon atoms to hydrogen atoms is ____ to ____.

17. 4, moles, 1, 2

18. When reduced to smallest whole numbers, the ratio of moles of carbon atoms to hydrogen atoms for the compounds C_2H_4 and C_4H_8 is the same, namely ____ to ____. We say that the *simplest formula, or empirical formula* for these compounds is CH_2.

18. 1, 2

19. CH_2 is called the *simplest formula* or *empirical formula* for any compound having a 1:2 ratio of carbon to hydrogen. The actual formula will have some multiple of this 1:2 ratio and cannot be determined without knowing the molecular mass.

20. The *empirical formula,* or *simplest formula,* of a compound shows the simplest relationship of atoms in a compound as obtained from analytical data. The *actual formula* (called molecular formula or true formula) shows not only the ratio between the numbers of atoms in a compound but also the total number of each atom in the molecule or formula unit.

21. Compounds may have the same empirical formula and different true formulas. Thus, the empirical formula may or may not be the same as the true formula. For example, the empirical formula for benzene is CH, but its true formula is C_6H_6. The empirical formula for hydrogen peroxide is ______, but its true formula is H_2O_2. For water, the __________ formula and __________ formula are the same.

21. HO, empirical, true (or molecular)

22. Empirical formulas of compounds are calculated from experimental data. Analysis shows that a compound contains 5.0 moles of hydrogen and 2.5 moles of sulfur. The empirical formula for the compound would be written as ________. A compound containing 0.3 moles of carbon and 0.6 moles of hydrogen would have an empirical formula of ________.

22. H_2S, CH_2 (or C_1H_2)

23. But analytical data are not presented as a ratio of moles of atoms, it is usually presented in *mass* relations or *percentage* composition. It is simple to convert mass into moles of atoms through the use of the conversion factor relating moles and the atomic mass in grams: grams × mole/gram = mole. For example, 0.24 g of carbon is 0.24 g × mole/12.0 g = ________.

23. 0.020 mole

24. A 2-gram sample of a compound contains 0.946 g of copper and 1.054 g of chlorine. Calculate its empirical formula.

moles Cu = 0.946 g × mole/63.5 g = 0.0149
moles Cl = 1.054 g × mole/35.5 g = __________

24. 0.0297

25. To get *whole* numbers for subscripts, divide each number of moles by the smaller of the numbers. The smaller number of moles is __________ mole of Cu.

0.0149 mole Cu/0.0149 mole Cu = 1.00
0.0297 mole Cl/0.0149 mole Cu = 1.99 mole Cl/mole Cu

The empirical formula is ________.

25. 0.0149, $CuCl_2$

26. An organic compound contains 0.450 g of carbon, 0.400 g of oxygen, and 0.0500 g of hydrogen. Calculate its empirical formula.

moles of C = 0.450 g × mole/12.0 g = 0.0375

Similarly, there are __________ moles of oxygen atoms and __________ moles of hydrogen atoms in the compound.

27. Dividing the decimal values of moles by the smallest number of moles: | **26.** 0.0250, 0.0500

0.0375 mole C/0.0250 mole O = 1.50 mole C/mole O
0.0500 mole H/0.0250 mole O = ________ mole H/mole O

Because we use smallest *whole* numbers as subscripts, the empirical formula is __________.

28. The composition of a compound may be given in percentage by mass rather than in grams. In such cases, assume that you are working with 100 g of compound; the percentage will represent the grams of each component in 100 g total. | **27.** 2.00, $C_3H_4O_2$

29. Calculate the empirical formula of a compound having the composition 60.0% carbon, 13.4% hydrogen, and 26.6% oxygen. A 100 g sample of this compound would contain 60.0 g of C, ________ g of H, and ________ g of O.

30. The number of moles of each atom may be obtained by using the relationship between moles and mass. For moles of carbon atoms: | **29.** 13.4, 26.6

moles C = g of C × mole of C/g of C
= 60.0 g × mole/12.0 g = 5.00

31. The number of moles of hydrogen atoms is:

moles H = 13.4 g × mole/1.01 g = ________.

The number of moles of oxygen atoms is:

moles O = 26.6 g × mole/16.0 g = ________.

32. Of the three values obtained, the number of moles of oxygen atoms is the smallest, namely ________ moles. Dividing the values obtained by the smallest number of moles we get: 5.00 mole C/1.66 mole O = 3.01 moles C/mole O, 13.3 mole H/1.66 mole O = ________ mole H/mole O, from which the empirical formula is __________. | **31.** 13.3, 1.66

33. Calculate the empirical formula of a compound that contains 62.0% carbon, 10.5% hydrogen, and 27.5% oxygen. Again, assuming a 100 g sample, the amount of carbon is ________ g; of hydrogen is ________ g; and of oxygen is ________ g. | **32.** 1.66, 8.01, C_3H_8O

34. The number of *moles* of atoms will be: | **33.** 62.0, 10.5, 27.5

moles C = 62.0 g × mole/12.0 g = ________,
moles H = 10.5 g × mole/1.01 g = ________,
moles O = __________ × ______________ = ________.

35. To get a whole number of moles of atoms we must divide the moles of atoms by the smallest number, namely, 1.72 moles. Therefore, moles of carbon will be ________, moles of oxygen will be ________, and moles of hydrogen will be 6.05. The empirical formula is __________. | **34.** 5.17, 10.4, 27.5 g, mole/16.0 g, 1.72

36. Calculate the empirical formula of a compound that contains 52.0% zinc, 9.6% carbon, and 38.4% oxygen. The empirical formula is ______.

35. 3.00, 1.00, C_3H_6O

36. $ZnCO_3$

PROGRAMMED SEQUENCE 6

NEGATIVE NUMBERS

Negative numbers are important in that part of chemistry concerned with oxidation numbers and ionic charges. You can review operations with negative and positive numbers most simply by performing the following calculations.

1. Addition

$(+3) + (+2) =$ ____

$(+3) + (-3) =$ ____

$(-3) + (-3) =$ ____

$(-3) + (+2) =$ ____

$(-3) + (+4) =$ ____

2. Subtraction

$(+3) - (+3) =$ ____

$(+3) - (+2) =$ ____

$(+3) - (+4) =$ ____

$(-3) - (-3) =$ ____

$(-3) - (-4) =$ ____

$(-3) - (-2) =$ ____

3. Multiplication

$(+4) \times (+2) =$ ____

$(-4) \times (-2) =$ ____

$(+4) \times (-2) =$ ____

$(+4) \times 0 \quad =$ ____

4. Division

$(+4)/(+2) =$ ____

$(-4)/(-2) =$ ____

$(+4)/(-2) =$ ____

$(+4)/0 \quad =$ ____

1. +5, 0, −6, −1, +1

2. 0, +1, −1, 0, +1, −1

3. +8, +8, −8, 0

4. +2, +2, −2, has no meaning

PROGRAMMED SEQUENCE 7

STOICHIOMETRY

Instructions for programmed sequences are on page 301.

1. A chemical equation relates experimental facts. It states that substances react in the ratio of their coefficients in the balanced equation. It also says that the combined mass of the reactants is exactly equal to the combined mass of the products.

2. The equation for the electrolysis of water is $2H_2O(l) \rightarrow 2H_2(g) + O_2(g)$. The equation relates that two molecules of water, when electrolyzed, will produce ____ molecules of hydrogen and ____ molecule of oxygen. Furthermore, the mass of the reactant water is exactly the same as the combined masses of the products ____ and ____.

3. It is convenient to let the *symbol* of an *element* or the *formula* of a *compound* represent the number of grams indicated by the atomic or molecular (or formula) mass. We say that a *mole* of atoms, 6.02×10^{23} atoms, is numerically equal to the atomic mass in grams, and that a *mole* of molecules is numerically equal to the molecular mass in grams.

2. 2, 1, H_2, O_2

4. In terms of moles, then, the above equation for the electrolysis of water states that when 1 mole of water is completely electrolyzed, ____ mole of H_2 and ____ mole of O_2 are produced.

5. Methane burns in oxygen in accordance with the equation $CH_4(g) + 2O_2(g) \rightarrow CO_2(g) + 2H_2O(l)$. The equation states that ____ mole of CH_4 and ____ moles of O_2 react to form ____ mole of CO_2 and ____ moles of water.

4. 1, $\frac{1}{2}$

6. When propane burns in oxygen the equation is $C_3H_8(g) + 5O_2(g) \rightarrow 3CO_2(g) + 4H_2O(l)$. For each mole of propane burned, ____ moles of O_2 are consumed, and ____ moles of water are produced.

5. 1, 2, 1, 2

7. If we desired to burn 3 moles of propane, ____ moles of oxygen would be required, and ____ moles of CO_2 and ____ moles of H_2O would be produced.

6. 5, 4

8. To burn 5 moles of propane would require ____ moles of oxygen.

7. 15, 9, 12

9. To produce 0.60 mole of hydrogen in the reaction $Zn(s) + 2HCl(aq) \rightarrow ZnCl_2(aq) + H_2(g)$, ______ moles of HCl would be needed.

8. 25

10. Calculations from chemical equations (called stoichiometric calculations) require three simple steps:

1. Determination of the number of moles of the *given* substance.
2. Use of the chemical equation to relate the number of moles of *given* substance to moles of *desired* substance.
3. Conversion of the moles of desired substance into whatever units are required in the answer.

9. 1.2

11. How many moles of hydrogen could be produced in the reaction of Zn with HCl if we started with 0.100 g of Zn and an *excess* of HCl? Using the steps listed in frame 10,

Step #1

$$\text{moles of Zn required} = \frac{0.100 \text{ g}}{65.4 \text{ g/mole}} = \_\_\_\_\_\_\_\_\_\_$$

12. *Step #2*
The chemical equation in frame 9 shows that for every mole of Zn used, ____ mole of hydrogen is produced. Thus __________ mole of hydrogen is produced.

11. 0.00153

13. This problem could be worked conveniently by combining Steps #1 and #2 as follows:

$$\text{moles } H_2 = 0.100 \text{ g Zn} \times \frac{\text{mole Zn}}{65.4 \text{ g Zn}} \times \frac{\text{mole } H_2}{\text{mole Zn}} = 0.00153$$

12. 1, 0.00153

14. Oxygen is prepared by heating potassium chlorate to give the equation $2KClO_3(s) \rightarrow 2KCl(s) + 3O_2(g)$. How many grams of oxygen may be made from 4.38 g of $KClO_3$?

Step #1

$$\text{moles of } KClO_3 = \frac{4.38 \text{ g}}{123 \text{ g/mole}} = \_\_\_\_\_\_\_\_\_\_$$

15. The equation shows that 2 moles of $KClO_3$ produce ____ moles of O_2. Thus,

14. 0.0356

Step #2

$$\text{moles } O_2 = 0.0356 \text{ mole } KClO_3 \times \frac{3 \text{ moles } O_2}{2 \text{ moles } KClO_3} = \_\_\_\_\_\_\_\_\_\_$$

16. *Step #3*
$\text{g } O_2 = 0.0534 \text{ mole } O_2 \times 32.0 \text{ g/mole } O_2 = \_\_\_\_\_\_$

15. 3, 0.0534

17. The 3 steps for solving this problem could be combined as follows:

16. 1.71

$$\text{g } O_2 = 4.38 \text{ g } KClO_3 \times \frac{\text{mole } KClO_3}{123 \text{ g } KClO_3} \times \frac{3 \text{ moles } O_2}{2 \text{ moles } KClO_3} \times \frac{32.0 \text{ g } O_2}{\text{mole } O_2} = 1.71 \text{ g } O_2$$

18. How many grams of $KClO_3$ are needed to produce 12.8 g of O_2?

Step #1

$$\text{moles } O_2 = \frac{12.8 \text{ g}}{32.0 \text{ g/mole}} = \_\_\_\_\_\_\_\_$$

19. *Step #2*

18. 0.400

$$\text{moles } KClO_3 = 0.400 \text{ mole } O_2 \times \frac{2 \text{ moles } KClO_3}{3 \text{ moles } O_2} = \_\_\_\_\_\_\_\_$$

20. *Step #3*
$\text{g } KClO_3 = 0.267 \text{ mole } KClO_3 \times 123 \text{ g/mole} = \_\_\_\_\_\_$

19. 0.267

21. Work the same problem in frame 18 combining the 3 steps into a single expression as was done in frame 17.

20. 32.8

21. $\frac{12.8 \text{ g } O_2}{32.0 \text{ g } O_2\text{/mole } O_2} \times \frac{2 \text{ mole } KClO_3}{3 \text{ moles } O_2} \times \frac{123 \text{ g } KClO_3}{\text{mole } KClO_3} = 32.8 \text{ g } KClO_3$

22. When magnesium is added to hydrochloric acid, the reaction is $Mg(s) + 2HCl(aq) \rightarrow MgCl_2(aq) + H_2(g)$. If we add exactly 1 mole of Mg to exactly 2 moles of HCl, exactly ____ mole of H_2 is produced. Neither the Mg nor the HCl is in excess.

23. If 1 mole of Mg were added to only $\frac{1}{2}$ mole of HCl, the Mg would be in excess, and only ______ mole of hydrogen would be produced. The HCl in this case is called the *limiting reactant.* ______ mole of Mg would remain unconsumed at the end of the reaction.

22. 1

24. When sodium metal is allowed to react with chlorine gas to form sodium chloride, the reaction is $2Na(s) + Cl_2(g) \rightarrow 2NaCl(s)$. Exactly ____ moles of Na will react with exactly ____ mole of Cl_2 to form exactly ____ moles of NaCl.

23. 0.25, 0.75

25. If 2 moles of Na were added to 3 moles of Cl_2 in the above reaction, the amount of NaCl formed would be ____ moles, and ____ moles of Cl_2 would remain unreacted. Any amount of Cl_2 greater than that required for the reaction would remain unreacted.

24. 2, 1, 2

26. If 0.100 mole of Na is added to 0.100 mole of Cl_2, the substance present in excess is ________. ________ mole of NaCl will be produced, and ________ g of Cl_2 will remain unreacted.

25. 2, 2

27. Consider the reaction $4NH_3(g) + 5O_2(g) \rightarrow 6H_2O(l) + 4NO(g)$. If 4 moles of NH_3 are added to 5 moles of O_2, ____ moles of NO will be produced. If 4 moles of NH_3 are added to 6 moles of O_2 the ____ is in excess. The 4 moles of NH_3 will all be consumed to produce ____ moles of NO.

26. Cl_2, 0.100, 1.78

28. If 5 moles of NH_3 were added to 5 moles of O_2, the ______ is in excess. At the end of the reaction, ____ mole of ______ would remain unreacted.

27. 4, O_2, 4

29. If we add 34.0 g of NH_3 to 5 moles of O_2, how many moles of NO would be produced? First, convert g of NH_3 to moles of NH_3. 34.0 g $NH_3 \times$ (mole NH_3/17.0 g NH_3) = ________ mole NH_3.

28. NH_3, 1, NH_3

30. The equation shows that 4 moles of NH_3 are required to produce 4 moles of NO, thus 2 moles of NH_3 added to 5 moles of O_2 would produce ____ moles of NO. There would be ______ moles of ____ remaining unreacted.

29. 2.00

31. Consider the reaction $2Al(s) + 3H_2SO_4(aq) \rightarrow Al_2(SO_4)_3(aq) + 3H_2(g)$. How many grams of H_2 can be produced from 3 moles of Al added to 196 g of H_2SO_4? First convert grams of H_2SO_4 to moles. 196 g $H_2SO_4 \times$ mole H_2SO_4/98.1 g H_2SO_4 = ________________.

30. 2, 2.5, O_2

32. Referring to the above equation, we see that 3 moles of H_2SO_4 would produce 3 moles of H_2, so 2 moles of H_2SO_4 would produce ____ moles of H_2 or ____ g of H_2.

31. 2.00 moles H_2SO_4

32. 2, 4

PROGRAMMED SEQUENCE 8

MOLARITY

Instructions for programmed sequences are on page 301.

1. A solution is defined as a homogeneous mixture. A solution commonly used in chemistry is that in which a solid is dissolved in a liquid. The solid is called the *solute*, and the liquid the *solvent*. The solute is dissolved in the solvent. When solid NaCl is mixed with water to form a solution, the ________ is the solute, and ____________ is the solvent.

2. Solids, liquids, or gases can act as solutes or solvents in solutions. Two types of solutions are particularly important to us. One is the solution made by dissolving a solid solute in a liquid ________________; the other is the result of dissolving a liquid solute in a liquid ________________. In the latter case, we call the liquid present in larger quantity the solvent, and that present in smaller quantity the solute.

1. NaCl, water

3. If 20 ml of liquid glycerin are mixed with 1000 ml water, the glycerin, present in smaller quantity, is called the ________________, and the water is called the ________________. The smaller amount of liquid, the solute, is considered to be dissolved in the larger amount, the solvent.

2. solvent, solvent

4. When solid $CuSO_4$ is mixed with water, the solute is the ____________, and the solvent is the ____________. If 500 ml of water is mixed with 50 ml of liquid alcohol, the solute is the ________________, and the solvent is the ____________.

3. solute, solvent

5. In a homogeneous mixture the relative amount of solute and solvent, or solute and solution, is referred to as the *concentration*. Concentration of solutions may be expressed in many ways. Some of those are:

a. molarity
b. percent by mass
c. percent by volume

Of these, molarity is most often used, and our discussion will be confined to it alone.

4. $CuSO_4$, water, alcohol, water

6. The molarity of a solution is the *number of moles of solute per liter of solution*. A liter of solution containing 1 mole of solute is called a 1 *molar* solution (abbreviated, 1 *M* solution). One mole of sucrose in a liter of solution is called a 1 ____________ sucrose solution; 2 moles of NaCl per liter of solution is a ____ molar NaCl solution.

7. A 1 molar solution contains 1 mole of solute in a liter of *solution* (solution, not solvent). A 2 molar solution (abbreviated 2 *M*) contains 2 moles of ________________ per liter of solution. A 0.50 *M* solution would contain ________ moles of solute per liter of ________________.

6. molar (or *M*), 2

8. Suppose that a solution were made by mixing alcohol (molecular mass = 30 amu) with sufficient water to make a liter of 3 *M* solution. A 3 *M* solution contains ____ moles of alcohol per liter of ________________. The number of *grams* of alcohol per liter of solution would be ______.

7. solute, 0.50, solution

9. One liter of 1 *M* sulfuric acid (molecular mass = 98 amu) would contain ____ mole, or ______ g of H_2SO_4. Five hundred ml of this solution would contain only half as much solute, namely ______ g of H_2SO_4. Five

8. 3, solution, 90

hundred ml of 2 *M* sulfuric acid solution contains ______ g, or ____ mole of H_2SO_4.

10. To make 1 liter of 2 *M* sulfuric acid would require 2 moles, or ________ g of H_2SO_4. Five hundred ml of this 2 *M* solution contains ____ mole, or ______ g of solute.

9. 1, 98, 49, 98, 1

11. If we used 250 ml of the 2 *M* solution in some chemical process, we would be using $\frac{1}{2}$ mole of H_2SO_4. This calculation would be set up as follows:

10. 196, 1, 98

2 *M* sulfuric acid solution contains 2 moles of H_2SO_4 per 1000 ml of solution, so

$$\frac{2 \text{ mole } H_2SO_4}{1000 \not{\text{ml solution}}} \times 250 \text{ ml solution} = \tfrac{1}{2} \text{ mole } H_2SO_4$$

12. What is the molarity of a solution made by adding 49 g of H_2SO_4 to sufficient water to make 2000 ml of solution? Remember that molarity units are moles/liter. The problem is one of changing grams to moles and milliliters to liters:

$$\frac{49 \not{\text{g } H_2SO_4}}{2000 \not{\text{ml}} \text{ soln}} \times \frac{1000 \not{\text{ml}}}{\text{liter}} \times \frac{\text{mole } H_2SO_4}{98 \not{\text{g } H_2SO_4}} = \text{\_\_\_\_\_\_\_\_ } M$$

13. The molecular mass of NaOH is 40 amu. A solution is made which contains 40 g of NaOH in 500 ml of solution. What is the molarity of the solution? Molarity is the number of moles of NaOH present in a liter of solution. Since 40 g of NaOH is ____ mole, the concentration of solute is ____ mole of NaOH per 500 ml solution, or ____ moles of NaOH per 1000 ml solution. The solution concentration is ____ molar.

12. 0.25

14. There is no such thing as a molecule of ionic compound. We are using the term *mole* to denote the gram formula mass of an ionic compound or gram molecular mass of a molecular covalent compound.

13. 1, 1, 2, 2

15. If we wanted to make *less* than a liter of solution, say 250 ml of 0.20 *M* NaOH solution, how many grams of solute would we need? A *liter* of 0.20 *M* solution would require ________ mole of NaOH, or ______ g of NaOH. Thus 250 ml of solution will require only ______ g of NaOH. The following is a set-up for this problem with the proper units:

$$\frac{0.20 \not{\text{mole NaOH}}}{1000 \not{\text{ml soln}}} \times 250 \not{\text{ml soln}} \times \frac{40 \text{ g NaOH}}{\not{\text{mole NaOH}}} = 2.0 \text{ g NaOH.}$$

16. A 0.80 *M* NaCl solution contains ________ mole of NaCl in a liter of solution, or ________ mole of NaCl in 250 ml of solution. If we wanted only 0.20 mole of NaCl, we would use ________ ml of the solution.

15. 0.20, 8.0, 2.0

17. Suppose that we diluted 500 ml of 2 *M* NaCl by adding enough water to make a liter of solution. What is the concentration of the diluted solution? The answer is obvious, but let's reason it through so that we can apply the principle to *any* dilution problem. A 2 *M* solution contains ____ moles of NaCl per 1000 ml of solution, or ____ mole of NaCl per 500 ml of solution. Dilution changes the solution volume, but when we dilute, the number of moles of solute remains constant. Thus, we still have ____ mole of NaCl in twice the volume (1000 ml) and the concentration is ____ *M*.

16. 0.80, 0.20, 250

17. 2, 1, 1, 1

18. Diluting a solution does not change the number of moles of solute; only the volume of solution changes. Since the moles before dilution = moles after dilution, we can set up a very useful relationship for working dilution problems: moles solute *before* dilution = moles solute *after* dilution;

$$(\text{moles}/\cancel{\text{liter}}) \times \cancel{\text{liter}} = (\text{moles}/\cancel{\text{liter}}) \times \cancel{\text{liter}}.$$

19. The above relationship is briefly written:

$M_1V_1 = M_2V_2$ where M = molarity and V = volume.

Apply this relationship to the following dilutions:

a. 300 ml of a 6 M solution must be diluted to ______ ml to make it 3 M.

b. If 250 ml of 8 M solution is diluted to 2000 ml, the new molarity is ____.

19. 600, 1

20. A student mixes 300 ml of 0.20 M NaCl with 200 ml of 0.70 M NaCl. What is the concentration of the resulting solution? The total volume of solution will be ______ ml. The number of moles of NaCl in 300 ml of 0.20 M solution is ________. The number of moles NaCl in 200 ml of 0.70 M solution is ______. The total number of moles of NaCl is ______. The set-up for calculation of molarity is:

$$(\_\_\_\_\_\_ \text{ mole}/500\ \cancel{\text{ml}}) \times (1000\ \cancel{\text{ml}}/\text{liter}) = \_\_\_\_\_\_.$$

20. 500, 0.060, 0.14, 0.20 0.20, 0.40

21. The *total* volume will be 300 ml + 200 ml = 500 ml. If we can determine the total moles of NaCl in the 500 ml, we have the set-up for calculating molarity.

a. Moles of NaCl in 300 ml of 0.20 M NaCl:
$(0.20 \text{ moles}/1000\ \cancel{\text{ml}}) \times 300\ \cancel{\text{ml}} = 0.06$ moles

b. Moles of NaCl in 200 ml of 0.70 M NaCl:
$(0.70 \text{ moles}/1000\ \cancel{\text{ml}}) \times 200\ \cancel{\text{ml}} = 0.14$ moles

c. Molarity = moles/liter:
$0.20 \text{ mole}/500\ \cancel{\text{ml}} \times 1000\ \cancel{\text{ml}}/\text{liter} = 0.40$ mole/liter

22. A short cut in working this problem is to reason that each solution will contribute its *volume fraction* toward the total molarity. In short,

$$\left(\frac{300\ \cancel{\text{ml}}}{500\ \cancel{\text{ml}}} \times 0.20\ M\right) + \left(\frac{200\ \cancel{\text{ml}}}{500\ \cancel{\text{ml}}} \times 0.70\ M\right) = 0.40\ M$$

23. If 500 ml of 2 M NaOH is added to 1500 ml of 2 M NaOH, the total number of *moles* of solute in the 2000 ml of solution is _____. The concentration will be ______. Since NaOH is completely dissociated in water to form Na^+ and OH^- ions, the molarity of each of these in the solution will be equal, namely _____.

23. 4.0, 2.0 M, 2.0

24. $BaCl_2$ is a soluble salt which completely dissociates in water to form the ions Ba^{2+} and Cl^-. Since there are 2 Cl^- ions to each Ba^{2+} ion, a 1 M $BaCl_2$ solution is said to be 1 M in Ba^{2+} and 2 M in Cl^-. In 500 ml of 1 M $BaCl_2$, there is _____ mole of Ba^{2+}, and ____ mole of Cl^-. In 250 ml of 0.20 M $BaCl_2$, there is ________ mole of Ba^{2+} and ________ mole of Cl^-.

24. 0.5, 1, 0.050, 0.10

25. If we mix 500 ml of 2 M NaCl with 500 ml of 2 M $BaCl_2$ the resulting solution is 1 M with respect to Na^+. What would the molarity of the Cl^- ion be?

a. In 500 ml of 2 *M* NaCl the number of moles of Cl^- is ____.

b. In 500 ml of 2 *M* $BaCl_2$ the number of moles of Cl^- is ____.

c. So the total moles of Cl^- is ____ and the molarity is ____ moles of Cl^-/1000 ml or ____.

25. 1, 2, 3, 3, 3 *M*

PROGRAMMED SEQUENCE 9

NET IONIC EQUATIONS

Instructions for using programmed sequences are on page 301. Because of the size of some of the equations, answers are given below the frame rather than in the right margin.

1. Metallic zinc reacts with a water solution of copper(II) sulfate to produce solid copper and a solution of zinc sulfate. Using the symbols (*s*) for solid and (*aq*) for aqueous (water) solutions, the equation for this reaction is:

$Zn(s) + CuSO_4(aq) \rightarrow Cu(s) + ZnSO_4(aq)$

This type of equation is called a *molecular* equation. It shows no distinction between ionic or nonionic species.

2. But $CuSO_4$ and $ZnSO_4$ are soluble salts that exist as ions in solution. Writing the same equation to show the ions gives

$Zn(s) + Cu^{2+} + SO_4^{2-} \rightarrow Cu(s) + Zn^{2+} + SO_4^{2-}$

This equation, written to show ions in solution, is called a *total ionic* equation.* Notice that the sulfate ions appear exactly the same in number and form on both sides of the equation.

3. The only species remaining unchanged in the reaction is the sulfate ion. Ions remaining the same in form after a reaction as before are called *spectator* ions. The SO_4^{2-} ions are ________________ ions. Spectator ions are not shown in ionic equation because they really have no part in a reaction. If the SO_4^{2-} ions are removed from the right and left side of the equation, we have:

$Zn(s) + Cu^{2+} \rightarrow Cu(s) + Zn^{2+}$

This equation, with spectator ions omitted, is called the *net ionic* equation, or simply the *ionic* equation.

---

spectator

---

4. The reaction between hydrogen gas, H_2, and solid iodine, I_2, is written:

$H_2(g) + I_2(s) \rightarrow 2HI(g)$

Notice that hydrogen and hydrogen iodide are properly identified as gases and that iodine is shown as a solid. None of these substances is ionic. There cannot be an ionic equation for nonionic species.

* For simplicity, the (*aq*) notation has been omitted from the designation for ions in aqueous solution.

5. In the total ionic equation,

$Ag^+ + NO_3^- + Na^+ + Cl^- \rightarrow AgCl(s) + Na^+ + NO_3^-$

the spectator ions are ________ and ________. In the total ionic equation,

$K^+ + OH^- + H_3O^+ + Cl^- \rightarrow K^+ + Cl^- + 2HOH,$

the spectator ions are ________ and ________.

NO_3^-, Na^+
K^+, Cl^-

---

6. The net ionic equation for the reaction between solutions of silver nitrate and sodium chloride is:

________ + ________ → ________(*s*)

The net ionic equation for the reaction between solutions of potassium hydroxide and hydrochloric acid is:

________ + ________ → $2H_2O(l)$

Notice that (*l*) is used to indicate liquids.

Ag^+, Cl^-, AgCl
OH^-, H_3O^+

---

7. When zinc is added to an aqueous solution of lead(II) nitrate the precipitation of lead is observed, and an ionic solution remains. The *molecular* equation for this reaction is:

____

$Zn(s) + Pb(NO_3)_2(aq) \rightarrow Pb(s) + Zn(NO_3)_2(aq)$

---

8. When the spectator ion, ____________, is omitted from the total ionic equation, the *net ionic* equation becomes:

____

NO_3^-
$Zn(s) + Pb^{2+} \rightarrow Pb(s) + Zn^{2+}$

---

9. Sodium metal reacts with water to form hydrogen gas and sodium hydroxide remains in solution. Using the symbols (*s*) for solid, (*l*) for liquid, (*g*) for gas, and (*aq*) for salts dissolved in aqueous solution, the equation is:

$2Na(s) + 2H_2O(l) \rightarrow 2NaOH(aq) + H_2(g)$

This type of equation is called a ________________ equation.

---

molecular

---

10. Of the four species in the above molecular equation, only NaOH is ionic. The same equation in *total ionic* form is:

$2Na(s) + 2H_2O(l) \rightarrow 2Na^+ + 2OH^- + H_2(g)$

Notice that 2NaOH(*aq*) is written ionically as $2Na^+ + 2OH^-$.

11. There are no spectator ions in the above equation. Therefore, the *net ionic* equation is:

____.

$2Na(s) + 2H_2O(l) \rightarrow 2Na^+ + 2OH^- + H_2(g)$

---

12. Metallic barium will react with water to form hydrogen gas and an ionic solution of barium hydroxide. Barium hydroxide is not very soluble but is considered a strong base and is shown as an ion:

$\text{Water} + Ba(OH)_2 \rightarrow Ba^{2+} + 2OH^-$

The net ionic equation for the reaction between Ba(*s*) and water is:

____.

$Ba(s) + 2H_2O(l) \rightarrow Ba^{2+} + 2OH^- + H_2(g)$

---

13. Copper replaces silver from a solution of silver nitrate to form a solution of copper(II) nitrate. The molecular equation is:

____.

The spectator ion is ____________.

The net ionic equation is:

____.

$Cu(s) + 2AgNO_3(aq) \rightarrow 2Ag(s) + Cu(NO_3)_2(aq)$
NO_3^-
$Cu(s) + 2Ag^+ \rightarrow 2Ag(s) + Cu^{2+}$

---

14. Hydrogen gas may be made from the reaction of hydrochloric acid with zinc. The molecular equation is:

____.

The spectator ion(s) are ____________.

The net ionic equation is:

____.

$Zn(s) + 2HCl(aq) \rightarrow H_2(g) + ZnCl_2(aq)$
Cl^-
$Zn(s) + 2H_3O^+ \rightarrow H_2(g) + Zn^{2+} + 2H_2O$

15. Liquid bromine (Br_2) will displace the iodide ion (I^-) from an aqueous solution of an iodide salt, like KI. The molecular equation for the reaction between liquid bromine and KI(*aq*) is:

____.

$Br_2(l) + 2KI(aq) \rightarrow I_2(s) + 2KBr(aq)$

16. The total ionic equation for this reaction is:

____.

$Br_2(l) + 2K^+ + 2I^- \rightarrow I_2(s) + 2K^+ + 2Br^-$

17. The spectator ion(s) are the ____________ ion(s). Therefore, the net ionic equation is:

____.

K^+
$Br_2(l) + 2I^- \rightarrow I_2(s) + 2Br^-$

18. Aqueous solutions of barium chloride and silver nitrate are mixed to produce solid silver chloride and a solution containing barium and nitrate ions. The molecular equation for the reaction is:

$BaCl_2(aq) + 2AgNO_3(aq) \rightarrow 2AgCl(s) + Ba(NO_3)_2(aq)$

$BaCl_2$ dissociates into three ions per formula unit, one barium ion and two chloride ions.

19. The total ionic equation for this reaction is:

____.

$Ba^{2+} + 2Cl^- + 2Ag^+ + 2NO_3^- \rightarrow 2AgCl(s) + Ba^{2+} + 2NO_3^-$

20. The spectator ions are ____________ and ____________, so the net ionic equation is:

____.

This net ionic equation represents the net reaction for *any* solution containing Ag^+ and Cl^-.

Ba^{2+}, NO_3^-
$Ag^+ + Cl^- \rightarrow AgCl(s)$

21. Suppose we mix water solutions of aluminum sulfate and barium nitrate to form a precipitate of barium sulfate. Write the molecular equation for the reaction.

____.

$Al_2(SO_4)_3(aq) + 3Ba(NO_3)_2(aq) \rightarrow 3BaSO_4(s) + 2Al(NO_3)_3(aq)$

---

22. Before writing the *total ionic* equation, look at the dissociation of the salts $Ba(NO_3)_2$ and $Al_2(SO_4)_3$ in water:

$Ba(NO_3)_2$ + water $\rightarrow Ba^{2+} + 2NO_3^-$

$Al_2(SO_4)_3$ + water $\rightarrow 2Al^{3+} + 3SO_4^{2-}$

Now, the *total ionic* equation is:

____.

$2Al^{3+} + 3SO_4^{2-} + 3Ba^{2+} + 6NO_3^- \rightarrow 3BaSO_4(s) + 2Al^{3+} + 6NO_3^-$

---

23. To get the net ionic equation the spectator ions, __________ and __________, must be eliminated. This leaves a net ionic equation of:

____.

Al^{3+}, NO_3^-
$SO_4^{2-} + Ba^{2+} \rightarrow BaSO_4(s)$

---

appendix B

answers to problems in problem sets A

The answers given below apply only to Set A in each chapter.

CHAPTER 2

1. a. 0.50 m, 5.0×10^{-4} km
   b. 8.7×10^4 ml, 8.7×10^4 cc
   c. 9.0×10^{-2} kg, 9.0×10^4 mg
   d. 7.63×10^{-4} km, 763 mm
   e. 0.349 liter, 349 cc
   f. 3.90×10^{-2} g, 3.90×10^{-5} kg
2. 10.9 m
3. 0.990 liter
4. 6.0×10^3 liters
5. 6.0×10^6 g
6.

μg	*mg*	*g*	*kg*	*Mg*
830	0.830	8.30×10^{-4}	8.30×10^{-7}	8.30×10^{-10}
1.50×10^3	1.50	1.50×10^{-3}	1.50×10^{-6}	1.50×10^{-9}
6.32×10^6	6.32×10^3	6.32	6.32×10^{-3}	6.32×10^{-6}
9.33×10^9	9.33×10^6	9.33×10^3	9.33	9.33×10^{-3}
3.87×10^{14}	3.87×10^{11}	3.87×10^8	3.87×10^5	3.87×10^2

7. a. 1.08×10^6 s
   b. 4.32×10^5 s
   c. 3.15×10^{11} s
8. 68.1 kg, 178 cm
9. a. 107 m
   b. 2.19×10^3 cm
   c. 6.21×10^2 miles
   d. 33.9 in
   e. 3.77 liters
   f. 22.6 liters
   g. 0.265 qt
   h. 95.4 qt
   i. 3.18×10^3 g
   j. 11.0 lb
10. 755 g
11. 33.9 ¢/gal
12. $23.60
13. 1.50 g/ml
14. 167 ml

15. 375 g
16. 389 g
17. 4.18
18. 0.790 g/ml
19. 1.28 g/liter
20. a. 2.45×10^5
    b. 6.54×10^{-8}
    c. 9.00×10^6
    d. 7.89×10^5
21. a. 14.60
    b. 121
    c. 2.7×10^3
    d. 0.68
22. a. 9.64×10^7
    b. 9.0×10^{11}
    c. 3.0×10^{-10}
    d. 1.4×10^{-39}
23. 500 cal
24. a. 22°C
    b. 572°F
    c. −6.7°C
    d. −148°F
    e. −51°C
    f. −40°F

CHAPTER 3

1. Mass of N is 14.0 times that of H.
2. The Berzelius values are: Fe, 53.253; Ca, 40.193; Na, 45.669.
   The accepted values are: Fe, 55.847; Ca, 40.08; Na, 22.9898.
3. Ag atoms are 1.698 times as heavy as a Cu atom.
4. Yes, because of the existence of isotopes.
5. The values reported in the atomic mass scale represent the average atomic mass of the isotopes as they exist in nature.
6. 56.5% is the 107 isotope.
7. 9.48×10^{23} atoms
8. 1.09×10^{-22} g
9. 7.00 g
10. H atoms have a mass of 1.67×10^{-24} g; Lr atoms have a mass of 4.27×10^{-22}. Lr atoms are 256 times heavier than H atoms.
11. 5.38 moles Fe/300 g
12. 30.0 g C/2.5 moles
13. 6.02×10^{23} amu/g
14. 1.51×10^{-23} ml

CHAPTER 4

1. 75.2% Ag
2. The percentage of Na in both samples is 39.3%.

3. a. 3.48 g Fe
   b. 2.32 g Fe
   c. 1.50/1.00, or 3/2
4. 0.270 g O
5. 39.7% O
6. 5.91×10^{10} ft/min
7. a. 32.0 amu
   b. 108 amu
   c. 136 amu
   d. 60.0 amu
   e. 123 amu
8. a. 32.0 g
   b. 108 g
   c. 136 g
   d. 60.0 g
   e. 123 g
9. 1.81×10^{24} molecules, 3.61×10^{24} atoms
10. a. 342 g
    b. 98.1 g
    c. 74.1 g
    d. 114 g
    e. 101 g
11. a. 2.78 moles
    b. 0.621 mole
    c. 0.625 mole
    d. 0.833 mole
    e. 0.237 mole
12. a. 660 g
    b. 42.5 g
    c. 20.1 g
13. About 5.9×10^3 units
14.

H_2O	18.0 amu	18.0 g	5.56 moles	5.56 moles
H_2O_2	34.0 amu	34.0 g	2.94 moles	5.88 moles
CH_3OH	32.0 amu	32.0 g	3.12 moles	3.12 moles
CH_3CO_2H	60.0 amu	60.0 g	1.67 moles	3.33 moles

15.

H_2O	2.78×10^{-2}	1.67×10^{22}	5.56×10^{-2}	3.34×10^{22}
H_2O_2	1.47×10^{-2}	8.85×10^{21}	2.94×10^{-2}	1.77×10^{22}
CH_3OH	1.56×10^{-2}	9.39×10^{21}	6.24×10^{-2}	3.76×10^{22}
CH_3CO_2H	8.33×10^{-3}	5.01×10^{21}	3.33×10^{-2}	2.00×10^{22}

16. 257 g/mole S_8
17. 4.24×10^{24} molecules/500 g, 8.48×10^{24} atoms/500 g Cl_2

CHAPTER 5

1. CuO
2. Fe_2O_3, Fe_3O_4, FeO
3. U_3O_8
4. $Na_2S_2O_8$
5. 67.1% Zn, 32.9% S
6. 22.6 g S
7. 20.9 g As
8. 6.67% H
9. 64.1% Pb, 29.7% C, 6.2% H
10. 31.6 g Mn

11. a. $2NaOH(aq) + H_2S(g) \rightarrow Na_2S(aq) + 2H_2O$
    b. $4Fe(s) + 3O_2(g) \rightarrow 2Fe_2O_3(s)$
    c. $4FeS_2(s) + 11O_2(g) \rightarrow 2Fe_2O_3(s) + 8SO_2(g)$
    d. $4NH_3(g) + 7O_2(g) \rightarrow 4NO_2(g) + 6H_2O(g)$
    e. $2C_4H_{10}(g) + 13O_2(g) \rightarrow 8CO_2(g) + 10H_2O(g)$

12. a. $2Al(s) + Cr_2O_3(s) \rightarrow Al_2O_3(s) + 2Cr(s)$ This is a replacement or substitution reaction.
    b. $2NO(g) + O_2(g) \rightarrow 2NO_2(g)$ This is a combination reaction.
    c. $K_2CO_3(aq) + Sr(NO_3)_2(aq) \rightarrow 2KNO_3(aq) + SrCO_3(s)$
    This appears to be a double displacement reaction, but writing it in ionic form makes it a combination reaction. The topic of ionic equations is discussed in Chapter 15.
    d. $C_7H_{14}(g) \rightarrow C_7H_8(g) + 3H_2(g)$
    This is a decomposition reaction.
    e. $AgNO_3(aq) + NaI(aq) \rightarrow AgI(s) + NaNO_3(aq)$
    This is another example of a combination reaction that looks like double displacement. Ionic equations are discussed in Chapter 15.

13. 50.5 g. The law of conservation of mass-energy says that the mass of products must equal the mass of the reactants if the reactants are totally consumed.

CHAPTER 6

1. FeO: $Fe = +2$, $O = -2$
   H_2S: $H = +1$, $S = -2$
   Sb_2S_3: $Sb = +3$, $S = -2$
   $KMnO_4$: $K = +1$, $Mn = +7$, $O = -2$
   $Ca(NO_3)_2$: $Ca = +2$, $N = +5$, $O = -2$

2. The least common multiple is 6. The number 3 must be multiplied by 2, and the number 2 by 3. The correct formula is M_3A_2.

3. AX A_2Y A_3Z
   BX_2 BY B_3Z_2
   CX_3 C_2Y_3 CZ

4.

	Periodic group	Oxidation number	Oxide formula
Se	VIA	+6	SeO_3
Cu	IB	+1	Cu_2O
Cr	VIB	+6	CrO_3
Li	IA	+1	Li_2O
Ce	IIIB	+3	Ce_2O_3
Zr	IVB	+4	ZrO_2

5. Rubidium iodide, RbI; calcium iodide, CaI_2; tin(IV) iodide, SnI_4; antimony(V) iodide, SbI_5.

6. Aluminum sulfide, Al_2S_3; aluminum chloride, $AlCl_3$; aluminum sulfate, $Al_2(SO_4)_3$; aluminum nitrate, $Al(NO_3)_3$; aluminum phosphate, $AlPO_4$
   Iron(II) sulfide, FeS; iron(II) chloride, $FeCl_2$; iron(II) sulfate, $FeSO_4$; iron(II) nitrate, $Fe(NO_3)_2$; iron(II) phosphate, $Fe_3(PO_4)_2$
   Mercury(II) sulfide, HgS; mercury(II) chloride, $HgCl_2$; mercury(II) sulfate, $HgSO_4$; mercury(II) nitrate, $Hg(NO_3)_2$; mercury(II) phosphate, $Hg_3(PO_4)_3$

Sodium sulfide, Na_2S; sodium chloride, NaCl; sodium sulfate, Na_2SO_4; sodium nitrate, $NaNO_3$; sodium phosphate, Na_3PO_4.

7. a. magnesium oxide
   b. mercury(II) oxide or mercuric oxide
   c. hydrogen bromide
   d. carbon tetrafluoride
   e. potassium hydride
   f. zinc sulfide
   g. carbon disulfide
   h. oxygen difluoride
8. a. beryllium chromate
   b. zinc phosphate
   c. magnesium chlorate
   d. potassium sulfate
   e. sodium carbonate
   f. calcium nitrate
9. a. ammonium bromide
   b. acetic acid
   c. tin(II) cyanide or stannous cyanide
   d. potassium hydroxide
   e. iron(II) hydroxide or ferrous hydroxide
   f. aluminum hydroxide
10. a. zinc sulfite
    b. potassium hypochlorite
    c. sodium nitrite
    d. aluminum perchlorate
    e. calcium chlorite
11. a. HCl = hydrogen chloride, HCl(*aq*) = hydrochloric acid
    b. H_2S = hydrogen sulfide, $H_2S(aq)$ = hydrosulfuric acid
    c. HF = hydrogen fluoride, HF(*aq*) = hydrofluoric acid
    d. H_2SO_4 = hydrogen sulfate, $H_2SO_4(aq)$ = sulfuric acid
12. a. hypochlorous acid
    b. sulfurous acid
    c. nitrous acid
    d. perchloric acid
13. a. $CoCl_2$
    b. $Fe_2(SO_4)_3$
    c. $HgBr_2$
    d. $Zn_3(AsO_4)_2$
    e. $PbCrO_4$
    f. $Cu_3(PO_4)_2$
    g. Cu_2O
    h. PbO_2
14. a. NaOH
    b. $(NH_4)_2CO_3$
    c. SnC_2O_4
    d. $KMnO_4$
    e. LiCN
    f. $Zn(CH_3CO_2)_2$
15. a. $Ca(ClO)_2$
    b. $MnSO_3$

c. $KClO_2$
d. $Ba(NO_2)_2$
e. $Fe(ClO_4)_3$
f. $H_2SO_4 \cdot 2H_2O$

CHAPTER 7

1. 60.0 liters
2. 2.29 10^3 torr or 3.01 atm
3. 333°K
4. 222 kiloliters
5. 472 ml
6. 869 ml
7. 211 atm
8. 28.2 liters
9. 0.79 atm
10. The gas in container A has a volume of 8.83 liters at STP; container B has a volume of gas equal to 9.83 liters at STP. Container B has the greater number of molecules.
11. 780 torr
12. 101 liters
13. 0.0202 mole
14. 153 g/mole
15. 3.17 g/liter
16. 12.7 liters
17. 8.75 moles

CHAPTER 8

1. Compare the height of rise of liquids in similar capillary tubes.
2. 83–84°C
3. 34–35°C
4. Mercury, acetic acid, water, benzene, carbon tetrachloride. Carbon tetrachloride has the lowest and mercury the highest boiling point.
5. −246°C, the normal boiling point.
6. When water (liquid) and ice are in equilibrium at atmospheric pressure, the temperature must be 0°C; hence the water below the ice is at 0°C.
7. 500 cal
8. 2,000 cal
9. 18,000 cal
10. 7,280 cal

CHAPTER 9

1. 10.2 moles, 373 g
2. 3.76 moles Fe, 2.82 moles O_2
3. 1.83 kg

4. 2.16 kg
5. 0.0746 mole
6. 92.8 g
7. 3.59 kg O_2, 3.30 kg CO_2
8. 2.44 moles
9. $4FeS_2(s) + 11O_2(g) \rightarrow 2Fe_2O_3(s) + 8SO_2(g)$
   $2SO_2(g) + O_2(g) \rightarrow 2SO_3(g)$
   $SO_3(g) + H_2SO_4(l) \rightarrow H_2S_2O_7(l)$
   $H_2S_2O_7(l) + H_2O \rightarrow 2H_2SO_4(l)$
10. 147 kg
11. a. 3.33 moles NO
    b. 0.0180 kg water
    c. 101 g Cu
    d. 7.09×10^{-3} mole HNO_3
12. $H_2SO_4(aq) + 2NaOH(aq) \rightarrow Na_2SO_4(aq) + 2H_2O$
    2.50 moles H_2SO_4
13. 145 g
14. 79.2%
15. a. 241 liters, b. 0.446 mole, c. 11.2 g
16. 29.4 liters
17. $N_2(g) + 3H_2(g) \rightarrow 2NH_3(g) + 22$ kcal
18. 2.35×10^5 kcal
19. 3.46×10^4 kcal

CHAPTER 10

1. Atomic number is 54, mass number is 131, number of protons is 54.
2.

Nucleus	${}^1_1H$	${}^2_1H$	${}^{12}_6C$	${}^{14}_6C$	${}^{235}_{92}U$	${}^{238}_{92}U$
Protons	1	1	6	6	92	92
Neutrons	0	1	6	8	143	146
Mass number	1	2	12	14	235	238

3. ${}^1_1H$, 1 electron; ${}^2_1H$, 1 electron; ${}^{12}_6C$, 6 electrons; ${}^{14}_6C$, 6 electrons; ${}^{235}_{92}U$, 92 electrons; ${}^{238}_{92}U$, 92 electrons.
4.

Protons	1	2	17	17	18	19
Neutrons	2	2	18	20	22	20
Mass number	3	4	35	37	40	39
Notation	${}^3_1H$ or ${}^3_1T$	${}^4_2He$	${}^{35}_{17}Cl$	${}^{37}_{17}Cl$	${}^{40}_{18}Ar$	${}^{39}_{19}K$

5. Nine molecules are possible: ${}^{12}C^{18}O$; ${}^{12}C^{17}O$; ${}^{12}C^{18}O$; ${}^{13}C^{16}O$; ${}^{13}C^{17}O$; ${}^{13}C^{18}O$; ${}^{14}C^{16}O$; ${}^{14}C^{17}O$; ${}^{14}C^{18}O$.
6. The argon ion, Ar^+
7. 8.47×10^3

CHAPTER 11

1. Gamma radiation has wavelengths in the 10^{-11} to 10^{-14} m range. These correspond to photon energies of 10^{-15} to 10^{-12} cal/photon, a very high energy when compared to

the energy of x rays, which are in the 10^{-18} to 10^{-15} cal/photon range. Gamma radiation is more dangerous because of its high energy.

2. Light of short wavelength.

3. Bohr was able to explain the origin of spectral lines by assuming that electrons could occupy only certain positions around the nucleus. Since an electron's energy in an atom depends on its distance from the nucleus, assigning a distinct average distance also assigns a distinct average energy. Bohr assumed that electrons could have only certain energy values set by their locations. When they move from one location to another they do so only by giving up or taking on a discrete amount of energy—a quantum of energy. This is the basic postulate of the quantum theory: The energy of the electrons is quantized.

4. Greater energy is liberated in the 4 to 2 jump.

5. The Pauli exclusion principle is a statement about the behavior of nature; no exceptions have been found. It is not a theory, for it does not explain why electrons cannot have similar sets of quantum numbers when they are in the same atom.

6. Br: $1s^22s^22p^63s^23p^64s^23d^{10}4p^5$
Hg: $1s^22s^22p^63s^23p^64s^23d^{10}4p^65s^24d^{10}5p^66s^24f^{14}5d^{10}$
Xe: $1s^22s^22p^63s^23p^64s^23d^{10}4p^65s^24d^{10}5p^6$
Ba: $1s^22s^22p^63s^23p^64s^23d^{10}4p^65s^24d^{10}5p^66s^2$
Note: these assignments are listed in the order in which the energy levels are filled and do not necessarily represent the correct orders of increasing energy in the atoms.

7. The assignments for the electrons in the first three energy levels are as follows:

n Principal	l Azimuthal	m_l Magnetic	m_s Spin	Electron notation
1	0 (*s* subshell)	0	$+\frac{1}{2}$	1*s*
1	0	0	$-\frac{1}{2}$	
2	0 (*s* subshell)	0	$+\frac{1}{2}$	2*s*
2	0	0	$-\frac{1}{2}$	
2	1 (*p* subshell)	−1	$+\frac{1}{2}$	2*p*
2	1	−1	$-\frac{1}{2}$	
2	1	0	$+\frac{1}{2}$	
2	1	0	$-\frac{1}{2}$	
2	1	+1	$+\frac{1}{2}$	
2	1	+1	$-\frac{1}{2}$	
3	0 (*s* subshell)	0	$+\frac{1}{2}$	3*s*
3	0	0	$-\frac{1}{2}$	
3	1 (*p* subshell)	−1	$+\frac{1}{2}$	3*p*
3	1	−1	$-\frac{1}{2}$	
3	1	0	$+\frac{1}{2}$	
3	1	0	$-\frac{1}{2}$	
3	1	+1	$+\frac{1}{2}$	
3	1	+1	$-\frac{1}{2}$	

3	2 (*d* subshell)	−2	$+\frac{1}{2}$	
3	2	−2	$-\frac{1}{2}$	
3	2	−1	$+\frac{1}{2}$	
3	2	−1	$-\frac{1}{2}$	
3	2	0	$+\frac{1}{2}$	3*d*
3	2	0	$-\frac{1}{2}$	
3	2	+1	$+\frac{1}{2}$	
3	2	+1	$-\frac{1}{2}$	
3	2	+2	$+\frac{1}{2}$	
3	2	+2	$-\frac{1}{2}$	

8. There can be a maximum of 10 *d* and 14 *f* electrons in an atom.
9. Element 57, lanthanum.
10. The electron assignments for Mn are: $1s^22s^22p^63s^23p^64s^23d^5$. Hund's rule says that the five 3d electrons cannot pair until each of the five available d orbitals has at least one electron each. We can therefore designate the ($5d^5$ electrons as $5d^15d^15d^15d^15d^1$)

CHAPTER 12

1.

Be	beryllium	4	IIA	2
B	boron	5	IIIA	3
C	carbon	6	IVA	4
N	nitrogen	7	VA	5
O	oxygen	8	VIA	6
F	fluorine	9	VIIA	7
Ne	neon	10	(noble gas)	8
Na	sodium	11	IA	1
Mg	magnesium	12	IIA	2
Al	aluminum	13	IIIA	3
Si	silicon	14	IVA	4
P	phosphorus	15	VA	5
S	sulfur	16	VIA	6
Cl	chlorine	17	VIIA	7
Ar	argon	18	(noble gas)	8
K	potassium	19	IA	1
Ca	calcium	20	IIA	2
Se	selenium	34	VIA	6
Te	tellurium	52	VIA	6
Po	polonium	84	VIA	6

2. Alkali metals: lithium, sodium, potassium
halogens: fluorine, chlorine
noble gases: helium, neon, argon
3. Elements number 15, phosphorus, and 51, antimony, resemble arsenic in properties.
4. Be, Ca, and Ba are in group IIA. S and Se are in group VIA. Ca, K, Zn, and Se are in the fourth period.
5. They all have an ns^2 outer orbital.
6. Se and Po

7. Francium, atomic number 87, would be expected to have properties quite similar to those of cesium, element number 55. Its properties are those of an alkali metal.

8. a. Rubidium + water → rubidium hydroxide + hydrogen
   $2Rb(s) + 2H_2O \rightarrow 2RbOH(aq) + H_2(g)$
   b. Sodium + sulfur → sodium sulfide
   $2Na(s) + S(s) \rightarrow Na_2S(s)$
   c. Lithium + hydrogen → lithium hydride
   $2Li(s) + H_2(g) \rightarrow 2LiH(s)$
   d. Chlorine + cesium → cesium chloride
   $Cl_2(g) + Cs(s) \rightarrow 2CsCl(s)$
   e. Sulfuric acid + calcium fluoride → calcium sulfate + hydrogen fluoride
   $H_2SO_4(aq) + CaF_2(s) \rightarrow CaSO_4(s) + 2HF(g)$
   f. no reaction
   g. Potassium iodate + potassium hydrogen sulfite → iodine + potassium sulfate + water + potassium hydrogen sulfate.
   $2KIO_3(aq) + 5KHSO_3(aq) \rightarrow I_2(s) + 2K_2SO_4(aq) + H_2O + 2KHSO_4(aq)$
   h. Fluorine + potassium bromide.
   $F_2(g) + 2KBr(aq) \rightarrow Br_2(l) + 2KF(aq)$
   i. Arsenic + bromine → arsenic(V) bromide
   $4As(s) + 5Br_2(l) \rightarrow 2As_2Br_5(s)$
   j. Fluorine + water → hydrofluoric acid + oxygen
   $2F_2(g) + 2H_2O \rightarrow 4HF(aq) + O_2(g)$
   k. Bromine + platinum → no reaction
   l. Zinc + chlorine → zinc chloride
   $Zn(s) + Cl_2(g) \rightarrow ZnCl_2(s)$

CHAPTER 13

1. $Ca \rightarrow Ca^{2+} + 2e^-$; $F + e^- \rightarrow F^-$
2. An element with one unpaired electron should be suitable; copper is one possibility.
3. 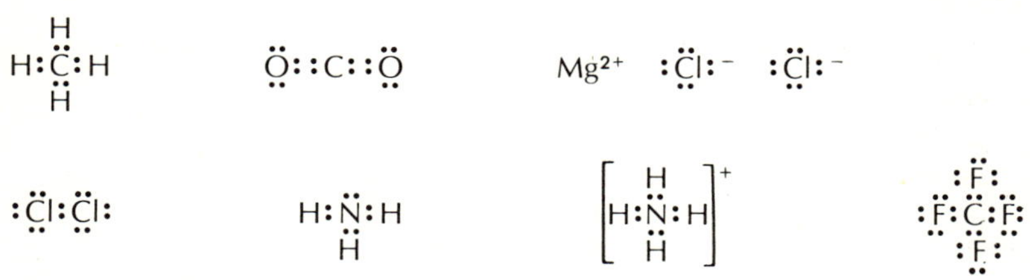

4. Refer to Figure 13–2. The bond is approximately 25% ionic.
5. The order of increasing ionic character is HI, CaS, KBr, NaCl, MgF_2.
6. The two common resonance forms are:

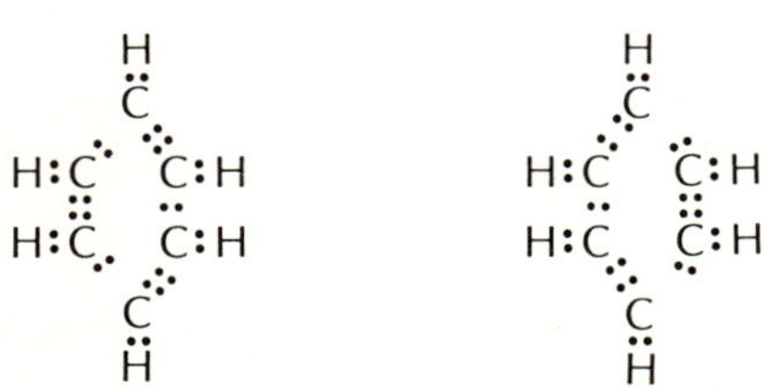

7. 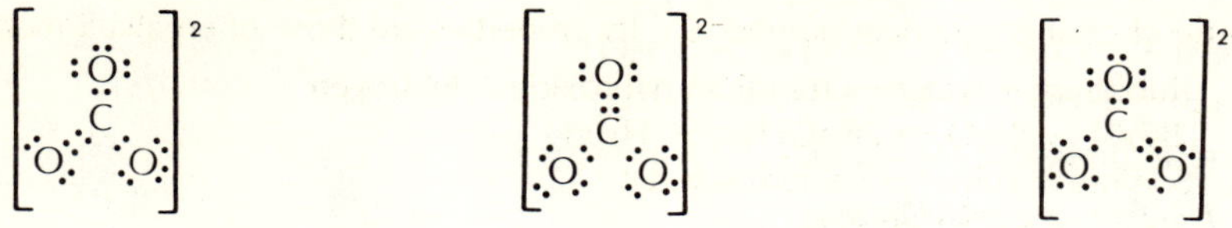

8. The water molecule is polar because it is not linear. The CCl_4 molecule has polar bonds, but the molecule is not polar due to its symmetrical structure. Each Cl atom is at the corner of a regular tetrahedral arrangement.

CHAPTER 14

1. The polar molecules of water dissolve NaCl by solvation of its ions, but dissolve sucrose by solvation of its molecules.
2. Resemble NaCl: KNO_3, $MgSO_4$. Resemble sucrose: glycerol, ethyl alcohol.
3. At 20°C, 35 g/100 g water. At 80°C, 53 g/100 g water. About 34% precipitates.
4. a. 3.00 moles, 224 g; b. 3.00 moles, 175 g; c. 3.00 moles, 175 g;
   d. 0.250 mole, 14.6 g.
5. a. 3.00 moles, 318 g; b. 3.00 moles, 696 g; c. 3.00 moles, 858 g.
6. a. 0.588 mole, 43.9 g; b. 0.0588 mole, 3.44 g; c. 0.242 mole, 26.9;
   d. 5.51×10^{-4} mole, 0.0749 g.
7. a. 0.117 *M*; b. 0.870 *M*; c. 0.300 *M*; d. 0.329 *M*.
8. a. 15.6 ml; b. 19.9 ml; c. 59.8 ml; d. 1.20 liters.
9. a. 0.400 *M*; b. 1.80 moles; c. 1.80 moles.
10. 104 ml.
11. a. 75.0 ml, 50 ml; b. 2.92 liters, 2.08 liters; c. 0.0638 *M*, 75 ml; d. 63.8 ml, 38.8 ml.
12. a. 0.333 *M*; b. 1.00 *M*; c. 1.50 *M*.
13. a. 1.00 *M*; b. 2.00 *M*.
14. a. 0.500 *M*; b. 0.500 *M*; c. 1.50 *M*.
15. 0.510 liter.
16. 3.36 liters (STP).
17. −2.79°C
18. 180 g/mole

CHAPTER 15

1. The electrolytes are calcium chloride and zinc nitrate. The nonelectrolytes are sugar, benzene, and urea.
2. 0.4 *M*
3. −11.16°
4. a. $2H_3O^+ + Zn(s) \rightarrow Zn^{2+} + H_2(g) + 2H_2O$
   b. $H_3O^+ + OH^- \rightarrow 2H_2O$
   c. $H_3O^+ + OH^- \rightarrow 2H_2O$
5. Weak electrolytes are not completely dissociated into ions in solution, strong electrolytes are.

6. $Zn(s) + 2H_3O^+ \rightarrow Zn^{2+} + H_2(g) + 2H_2O$
$Fe(s) + 3H_3O^+ \rightarrow Fe^{3+} + H_2(g) + 3H_2O$
$Ba(s) + 2H_3O^+ \rightarrow Ba^{2+} + H_2(g) + 2H_2O$
7. Use an indicator like litmus.
8. 10^{-6}, the solution is basic.
9. Basic
10. a. $H_3O^+ + OH^- \rightarrow H_2O + H_2O$
acid base acid base

b. $H_2O + CO_3^{2-} \rightarrow HCO_3^- + OH^-$
acid base acid base

c. $OH^- + NH_4^+ \rightarrow NH_3 + H_2O$
base acid base acid
11. NH_3, H_2O, CO_2, H_3O^+, H_2SO_4
12. The acetic acid molecules are not completely dissociated into ions. Acetic acid is a weak electrolyte.
13. The conjugate base of a weak acid is a strong base.
14. 4
15. 2.52
16. pOH = 4.22, pH = 9.78
17. You would expect the acidities to be the same.
18. 64 ml
19. 6
20. 0.351 *M*
21. $2H_3O^+ + CO_3^{2-} \rightarrow CO_2(g) + 3H_2O$
22. 0.0987 *M*

CHAPTER 16

1. a. $Al \rightarrow Al^{3+} + 3e^-$ (oxidation)
b. $Al^{3+} + 3e^- \rightarrow Al$ (reduction)
c. $O_2 + 4e^- \rightarrow 2O^{2-}$ (reduction)
d. $Fe^{2+} \rightarrow Fe^{3+} + e^-$ (oxidation)
2. a. $N_2 + 6e^- \rightarrow 2N^{3-}$ (reduction)
b. $2H^- \rightarrow H_2 + 2e^-$ (oxidation)
c. $S_8 + 16e^- \rightarrow 8S^{2-}$ (reduction)
d. $Cl_2 + 2e^- \rightarrow 2Cl^-$ (reduction)
3. a. $MnO_4^- + 8H_3O^+ + 5e^- \rightarrow Mn^{2+} + 12H_2O$
b. $NO_3^- + 2H_3O^+ + e^- \rightarrow NO_2 + 3H_2O$
c. $NO_3^- + 4H_3O^+ + 3e^- \rightarrow NO + 6H_2O$
d. $Cr_2O_7^{2-} + 14H_3O^+ + 6e^- \rightarrow 2Cr^{3+} + 21H_2O$
4. a. $H_2S + 2H_2O \rightarrow S + 2H_3O^+ + 2e^-$
b. $H_2O_2 + 2H_2O \rightarrow O_2 + 2H_3O^+ + 2e^-$
c. $H_2C_2O_4 + 2H_2O \rightarrow 2CO_2 + 2H_3O^+ + 2e^-$
d. $SO_3^{2-} + 3H_2O \rightarrow SO_4^{2-} + 2H_3O^+ + 2e^-$
5. a. $2MnO_4^- + 6H_3O^+ + 5H_2C_2O_4 \rightarrow 2Mn^{2+} + 10CO_2 + 14H_2O$

b. $H_2S + 2NO_3^- + 2H_3O^+ \rightarrow S + 2NO_2 + 4H_2O$
c. $Cr_2O_7^{2-} + 3H_2O_2 + 8H_3O^+ \rightarrow 2Cr^{3+} + 3O_2 + 15H_2O$

6. a. $2MnO_4^- + 5Fe^{2+} + 8H_3O^+ \rightarrow 2Mn^{2+} + 5Fe^{3+} + 12H_2O$
b. $2Fe^{2+} + H_2O_2 + 2H_3O^+ \rightarrow 2Fe^{3+} + 4H_2O$
c. $2S_2O_3^{2-} + I_2 \rightarrow S_4O_6^{2-} + 2I^-$
d. $Cr_2O_7^{2-} + 14H_3O^+ + 6Cl^- \rightarrow 2Cr^{3+} + 3Cl_2 + 21H_2O$

7. a. $2Cr^{3+} + 3Cl_2 + 16OH^- \rightarrow 2CrO_4^{2-} + 6Cl^- + 8H_2O$
b. $2Cr(OH)_4^- + 3H_2O_2 + 2OH^- \rightarrow 2CrO_4^{2-} + 8H_2O$

8. $3Cu + 8HNO_3 \rightarrow 3Cu(NO_3)_2 + 2NO + 4H_2O$

9. a. $H_2O_2 + 2Fe(NO_3)_2 + 2HNO_3 \rightarrow 2H_2O + 2Fe(NO_3)_3$
H_2O_2 is the oxidizing agent, and $Fe(NO_3)_2$ the reducing agent.
b. $3Cl_2 + 6KOH \rightarrow KClO_3 + 5KCl + 3H_2O$
Cl_2 is both the oxidizing and the reducing agent.
c. $2Fe(NO_3)_3 + 2KI \rightarrow 2Fe(NO_3)_2 + I_2 + 2KNO_3$
KI is the reducing agent, and $Fe(NO_3)_3$ the oxidizing agent.

10. For silver, 1 faraday; for copper, 2 faradays; for iron(III), 3 faradays.

11. a. 28,946 coulombs
b. 0.10364 faraday

12. 0.556 faraday

13. 143 g Cu

14. 201 g Ag

15. Oxidation occurs at the hydrogen electrode, which is the anode. The copper electrode is the cathode.

16. Only reaction (a) will be spontaneous in the forward direction.

17. a. $Zn(s) + 2Ag^+ \rightleftharpoons Zn^{2+} + Ag(s)$ $\quad +1.56$ volts
b. $Fe(s) + 2Au^+ \rightleftharpoons Fe^{2+} + 2Au(s)$ $\quad +2.12$ volts
c. $Ni(s) + Pb^{2+} \rightleftharpoons Ni^2 + Pb(s)$ $\quad +0.12$ volt

18. If the lead sulfate drops to the bottom of the battery, it is no longer available for the regeneration of the Pb and PbO_2 electrodes.

CHAPTER 17

1. a. Observe volume change.
b. Observe the disappearance of the purple color of the permanganate ion.

2. Oily rags react with the oxygen of the air. If this reaction takes place in an insulated, poorly vented spot, the heat of the slow combustion does not dissipate. This heat raises the temperature of the oily rags and causes them to oxidize more rapidly. This, in turn, creates more heat. Soon the kindling temperature is reached, and the rags burst into flame.

3.

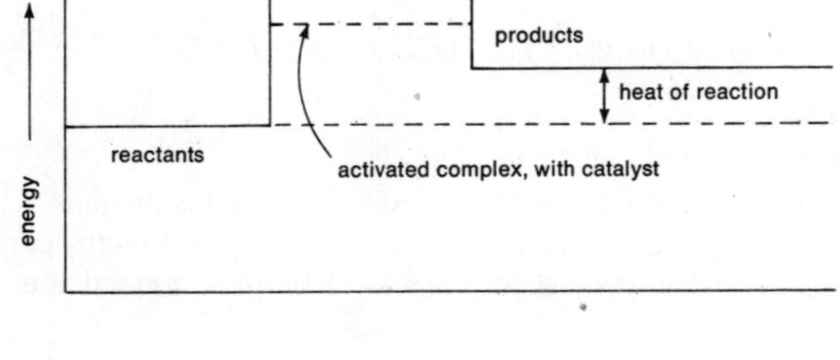

4. Reactants are continuously producing products, and products are continuously producing reactants. The rate at which the forward reaction is proceeding equals the rate of the reverse reaction. It is a dynamic, not a static equilibrium.

5. rate of forward reaction $= k_f A^a B^b$
   rate of reverse reaction $= k_r C^c D^d$
   At equilibrium the rates are equal, hence
   $k_f A^a B^b = k_r C^c D^d$, $k_f/k_r = C^c D^d / A^a B^b = K =$ equilibrium constant

6. a. $K = \dfrac{[PCl_3][Cl_2]}{[PCl_5]}$
   b. $K = \dfrac{[NH_4^+][OH^-]}{[NH_4OH(aq)]}$
   c. $K = \dfrac{[Hg]^2[O_2]}{[HgO]^2}$
   d. $K = \dfrac{[O_3]^2}{[O_2]^3}$
   e. $K = \dfrac{[NO]^4[H_2O]^6}{[NH_3]^4[O_2]^5}$

7. The presence of a catalyst will speed up the attainment of equilibrium for all the reactions but will not alter the position of the equilibrium.
   a. Exothermic reaction; an increase in temperature will favor the reverse reaction. An increase in pressure will shift the equilibrium to favor the forward reaction since there is a volume change from 3 to 2.
   b. Same as (a).
   c. Endothermic reaction; an increase in temperature favors the forward reaction. Pressure has no effect.
   d. Exothermic reaction; an increase in temperature favors the reverse reaction. Pressure has no effect.

   Increasing the concentration of a reactant shifts the equilibrium to favor the forward reaction.

8. The body uses catalysts (enzymes) to carry out the reactions.
9. $\frac{1}{5}$ or 0.2
10. 1.47×10^{-5} mole/liter
11. 5.8×10^{-7} mole/liter
12. Barium sulfate precipitates.
13. a. 5.5×10^{-11} mole/liter
    b. 4.9×10^{-3} mole/liter
    c. 1.6×10^{-2} mole/liter

CHAPTER 18

1. Aldehyde, alcohol, ketone, acid, halide.
2. Numerous answers are possible. The following are examples:
   $CH_3CH_2CH_2CH_2CH_2CO_2H$, $CH_3CH_2CH_2CH_2CH_2CH_2OH$,
   $CH_3CH_2CH_2CH_2CH_2CH_2NH_2$, $CH_3CH_2CH_2C(=O)—O—CH_2CH_3$,
   $CH_3CH_2CH_2C(=O)—NH—CH_2CH_3$, $CH_3CH_2CH_2CH_2CH_2C(H)=O$,
   $CH_3CH_2CH_2CH_2CH_2CH_2I$
3. a. 1-bromobutane (butyl bromide)
   b. n-octane (n- stands for normal)
   c. 2-iodopropane (isopropyl iodide)
   d. 3,3-dimethylpentane
   e. 2,2-diethyl-4-methylpentane
   f. 1-bromo-2-iodoethane

4. a. 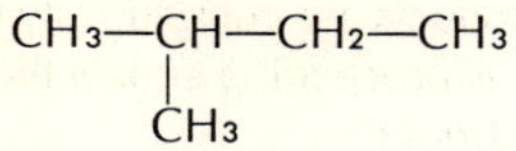

b. $CH_3—CCl_2—CH_2—CHCl—CH_3$

c. 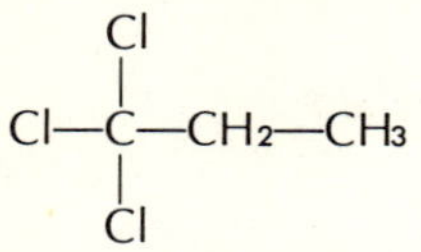

d. $CH_3—CHBr—CH_2—C(CH_3)_2—CH_2—CH(C_2H_5)—CH_2—CH_2—CH_3$

e. 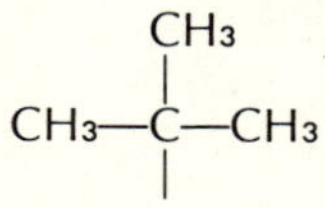

5. a. 1-pentene
   b. 5-chloro-1-pentyne
6. a. $CaCO_3 \rightarrow CaO + CO_2$
   $3C + CaO \rightarrow CaC_2 + CO$
   $CaC_2 + H_2O \rightarrow HC{\equiv}CH + Ca(OH)_2$
   b. $CH_2{=}CH—CH_3 + Br_2 \rightarrow CH_2Br—CHBr—CH_3$
   c. $CH_2{=}CH_2 + H_2SO_4 \rightarrow CH_3—CH_2—O—SO_2—OH$
   d. $CH{\equiv}C—CH_3 + 2HCl \rightarrow CHCl_2—CH_2—CH_3$

7. 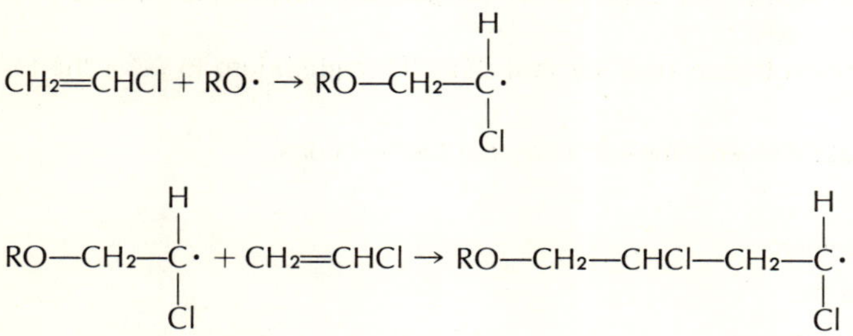

$nCH_2{=}CHCl \rightarrow {+}CH_2—CHCl{+}_n$

8. 1786
9. The —OH groups make it an alcohol, and the R—O—R linkages make it an ether.
10. 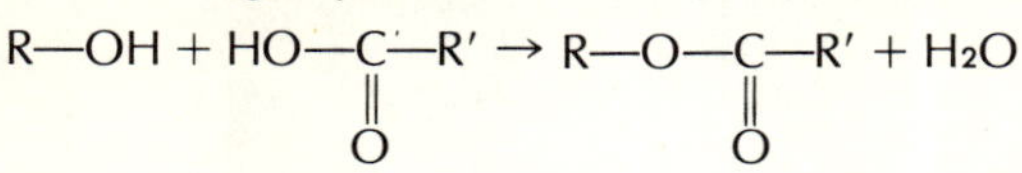
11. 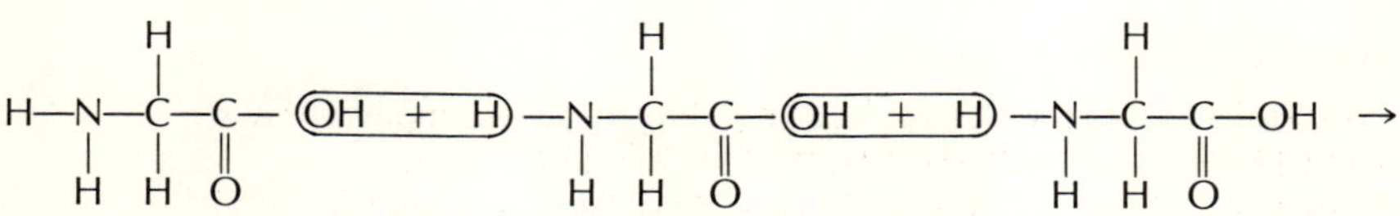

$H_2N—CH_2—CONH—CH_2—CONH—CH_2—CO_2H + 2H_2O$

12. Both are polyamides.

index*

Absolute temperature scale, see Kelvin temperature
Absolute zero temperature, 91, **99**
Absorbent cotton, 293, **297**
Accuracy of measurements, 13–15
Acetate ion, 77, 205
Acetic acid, 77, 224–224, 228, 234
Acetylenes, 298
  see also alkynes
Acid–base reactions, 229, 234–236
Acids, **79,** 203, **239,** 293
  amino, 295–296
  Arrhenius definition, 222–227, **239**
  Brønsted-Lowry definition, 229–230, **239**
  conjugate, 229–230, **239**
  functional group, 282, 297
  nomenclature, 73–74
  solutions, 227–229
  strengths of, 230
  see also pH, reactions
Actinides, 160, 170, **179**
Activated complex, 267, **278**
Activation energy, 267, **278**
Addition polymerization, 291–292, **297**
Addition reaction, 289, **297**
Air, 29, 175
Alcohol, 46, 282–283, 290–294
  see also ethanol
Aldehydes, 282, 292
Aliphatic hydrocarbons, 283, **297**
Alkali metals, 175–177, **179**
Alkanes, 283–284, 285–287, **297**
Alkenes, 287, **297**
Alkyl groups, 286, **297**
Alkynes, 287, **298**
Alloys, 200, **215**
Alpha particles, 142, 144, **146**
Amalgams, 200, **215**
Americium, 170
Amides, 282, 297, 298
Amines, 282, 295, 297, 298
Amino acids, 295–297, **298**
Aminoacetic acid, 295
Ammonia, 62, 203, 264, 273
  household, 76
Ammonium, ion, 76–77
  salts, 205
Ammonium hydroxide, 228, 235
Amorphous solids, 105
Ampere, 254, **259**
Amu (atomic mass unit), **41**
Analytical chemistry, 4
Anode, 252, **259**
Antimony, 70
Aqua ammonia, *see* ammonium hydroxide
Argon, 175
Aromatic hydrocarbons, 283, 288, 297, **298**
Arrhenius, Svante, 222
Arrhenius theory, 222–227, **239**
Arsenic, 69
Astatine, 177
Aston, Francis William, 33
-ate compounds, 74–77
Atm (atmosphere), 86
Atmosphere (atm), 86, **99**
Atmospheric pollution, 291
  see also smog
Atomic mass, 35–38, **41,** inside back cover
  determination of, 33–34
  and empirical formulas, 58
  atomic mass scale, 33–34
  atomic mass unit (amu), 33, **41**
Atomic number, 144–145, **146,** inside back cover
Atomic orbitals, *see* orbitals
Atomic spectra, 151–156
Atomic weight, *see* atomic mass
Atoms, 29, 298
  classification of, 39–41
  electronic structure of, 151–164
  relative mass of, 31–37
  size of, 38–39

* **Boldface numbers** refer to pages that contain glossary definitions of the terms.

structure of, 138–146
see *also* periodic table
Avogadro, Amedeo, 37
Avogadro's law, 94, 97, **99**
Avogadro's number, 37–38, **41**
Azimuthal quantum number, **164**
see *also* orbital quantum number

Balance, 20–22
Balancing equations, 61–63
Barometer, 85–86, **99**
Bartlett, Neil, 175
Bases, 76, **79, 239**
Arrhenius definition, 222–227, **229**
Brønsted-Lowry definition, 229–230, **239**
solutions, 227–229
strengths, 230
Battery, 259
storage, 258–259, **260**
see *also* voltaic cells
Becquerel, Antoine-Henri, 138
Benzene, 283, 288–289, 298
Berkelium, 170
Berzelius, Jöns Jakob, 31–33
Beta particles, 142
see *also* electrons
Binary compound, 68–74, **79**
see *also* nomenclature
Biochemistry, 4, 295
Bisulfite, 298
Bjerrum, Neils, 229
Bohr, Niels, 151, 153
Bohr atomic theory, 151–156, 162, 183
Boiling point, 111–113, **116,** 172–174, 214, inside back cover
Bonds, 185, **197**
carbon-carbon, 187, 297, 298
covalent, 185, 190–192, **197,** 297
double, 189, **197,** 297
electrons and, 158, 185
electrovalent, 184, **197**
hybrid, 187–188
hydrocarbon, 288, 299
hydrogen, 194, **197**
ionic, 184, **197**
metallic, 195–196
molecular, 195–196
multiple, 189, 297
peptide, 297, **298**
polar, 192–193, **197**
single, 297
triple, 189, 197
Boron, 69
Boyle, Robert, 87
Boyle's law, 87, 93, **99**
Branched-chain structures, 285, **298**
Brass, 200
Bromides, 205
Bromine, 40, 114, 177–178, 191, 289
see *also* halogens
Brønsted, J. N., 229
Brønsted-Lowry definitions, 229–230, **239**
Buret, 236, **239**
Butane, 284
Butyl group, 286

Calcium, 70
Calcium carbonate, 52
Calcium hydroxide, 228
Calculations from equations, *see* formulas, stoichiometry
Californium, 170–171
Calorie, 18, **22**
Capillary attraction, 106–107, **116**
Carbohydrates, 292
Carbon, 35, 69, 188, 195–196
see *also* organic chemistry
Carbon–carbon bonds, 187, 297, 298
Carbon dioxide, 46, 48, 72, 83, 196, 200, 203, 206
Carbon-hydrogen bond, 187
Carbon monoxide, 48, 72
Carbon tetrachloride, 186, 196
Carbonates, 205
Carbonic acid, 235
Carborundum, 195–196
Carboxylic acids, 298
see *also* amino acids
Catalyst, 122, **134,** 268, 273, 278, 295, **298**
Cathode, 252, **259**
Cathode ray, **146**
see *also* x-rays
Caustic soda, *see* sodium hydroxide
Cellophane, 293, **298**
Cells, voltaic, 251–252, 260
see *also* half-cell
Cellulose, 292–293, **298**
Celsius temperature, 19–20, **22,** 91–92
Centigrade temperature, *see* Celcius temperature
Centimeter, 6–8
Cesium, 169, 172, 175, 176, 190–191
see *also* alkali metals
Chadwick, James, 145–146
Charge, electric, 139–142, 259–260
Charles, Jacques, 89

Charles's law, 89–93, **99**
Chemical change, 27–28, **41**
Chemical equations, 45–46, **53**
  redox, 246–251
  types, 63–64
  writing and balancing, 61–63
  see also formulas
Chemical equilibrium, 269–274, **278**
Chemical property, 27, **41**
  see also properties
Chemistry, 2–3, 4
  branches, 3–4
Chlorates, 205
Chloride ion, 184, 205, 223, 235, 236, 248
Chlorine, 35, 72, 77, 160, 177–178, 183–184, 191
  see also halogens
Chloroform, 46
Chromium, 161–162
Colligative property, 213–214, **215,** 223–224
Combination reaction, 63, **64**
Combustion, 268, 290–291
Common ion effect, 275–276, **278**
Compounds, 29, **41**
  binary, 69–74
  naming of, 71–78
  ternary, 74–77
  see also formulas, nomenclature
Concentrated solutions, 206, **215**
Concentration, units of, 211–213
Conductor, **259**
  see also electrical conductivity
Conjugate acid-base pair, 229–230, 239
Conservation of mass-energy, law of, 49, **53**
Contact process, **134**
Conversion factors, 7
  metric and English, 7–11
Cooling by evaporation, 115–116
Coordinate covalence, 190, **197**
  see also bonds
Copper, 200, 222, 223, 246–247, 251–252
Copper sulfate, 78
Cosmic rays, 150
Cotton, 297, 298
Coulomb, 140, **146,** 253, **259**
Coulomb, Charles Augustin, 140
Covalent bonds, 185, 190–192, **197,** 297
Covalent solids, 195–196
Crystals, 105, **116,** 204
Curie, Marie Sklodowska, 138, 173
Curie, Pierre, 138, 173
Curium, 170–171
Cyanide ion, 76

Dacron, 293–294, **298**
Dalton, John, 89
Dalton's law, 89, **99**
de Broglie, Louis, 156
de Broglie hypothesis, 156
Debye, Peter, 225
Debye-Hückel theory, **239**
Decane, 284
Decomposition reaction, **64**
Definite composition, law of, **53**
Dehydrating agent, 124, **135**
Density, 11, **22,** inside back cover
  of selected substances, 12
  as periodic function, 173
Dextrose, *see* glucose
Diacids, 298
Dialcohols, 298
Diamines, 298
1, 6-diaminohexane, 295
Diamond, 195–196
Diatomic molecules, 40–**41**
Dibromoethane, 178, 186
Dichromate ion, 222, 248–250
Diffusion, **100,** 105–106
Dihydrate, **134**
Dilute solutions, 206, 210, **215**
Dimethyl ether, 46, 48
Dipole moment, 193, **197**
Displacement reaction, 257–258, **259**
Double bond, 189, **197,** 297
  see also bonds
Double displacement reaction, **64**
Ductility, 175, 196
Dynamic equilibrium, 109, **116**

Einstein, Albert, 49
Einsteinium, 170
Electric charge, 139–142, 259–260
Electric current, 139, 196, 221
Electrical conductivity, 196, 224–226
Electrical units, 253–254
Electrochemistry, 221–226, 251–259
Electrolysis, 175, **179**
  calculations, 253
Electrolytes, 221–227, **239**
  strong, **240**
  weak, **240**
Electromagnetic radiation, 150–151, **164**
Electromotive force (emf), 255, **259**

Electron clouds, 162–163, 164
see also orbitals
Electron levels, see energy (levels)
Electron shell, see energy (levels)
Electronegativity, 190–192, **197**
Electrons, 139–141, 144, **146,** 298
assignments, 160–162
notation, 159
and structure of atoms, 151–164
see also oxidation, periodic table, reduction
Electroscope, 138–139
Electrovalent bond, 184, **197**
see also bonds, ionic bonds
Elements, **41,** 144, **147,** inside back cover
abundance of, 29–30
actinides, 160, 170, **179**
alkali, 175–177, **179**
lanthanides, 160, 170, **179**
metals, 40, **41,** 203, 227
noble gases, 40, 41, 175, 179, 183–185
oxidation numbers, 68–78
transition, 169–170, **180**
transuranium, 170–171, **180**
Emf, 255–258, **259**
Emission spectrum, 151, **164**
Empirical formula, 48, **53,** 58–59
Endothermic reactions, 113, **116,** 132–133
Energy, **22,** 83–84
activation, 278
conservation, 49
electromagnetic, 150–151
forms, 18–19
levels, 153–154, 158–159, 164, 169–170
transfer, 113–116
English system of measurement, 9
metric conversions, 9–11
Enthalpy, **134**
-change, 132
Enzymes, 295, **298**
Epsom salts, 78
Equations, see chemical equations
Equilibrium constant, 269–277, **278**
Equilibrium vapor pressure, 109–110, 112
Essential amino acids, 296, **298**
Esters, 282, 293–294, 298
Ethane, 284, 289
Ethanoic acid, see acetic acid
Ethanol, 46, 48–50, 178, 203, 290
Ethene, 287, 289, 291–292
Ethyl alcohol, see ethanol
Ethyl group, 286, 297
Ethylene, see ethene
Ethyne, 287
Evaporation, 107–109, 115, **116,** 200
Exchange reaction, 64
Excited state, 154, **164**
Exclusion principle, see Pauli exclusion principle
Exothermic reactions, 49, **53,** 133, 267
Exponential notation, 16–18, **22**

Factors, conversion, 7–11
Fahrenheit temperature scale, 20, 91
Faraday, 253, **259**
Fermium, 170
Ferric compounds, 72
Ferrous compounds, 72
Fluorescence, 140, **147**
Fluorine, 35, 69, 172, 175, 177–178, 190–191, 194
see also halogens
Formula mass, **53**
Formulas, 45–56
and percent composition, 60–61
prediction of, 68–71
see also empirical formula, molecular formula
Francium, 175, 190
Franklin, Benjamin, 139
Free radicals, 291–292, **298**
Freezing point, 114, 214–215, 223–225
Functional group, 295, 296, **298**
Fusion, heat of, 114, 116, 173

Gallium, 161, 172–173. 174
Gamma particles, 142–143, 150–151
Gas, 24, 40–41
ideal, 95
kinetic-molecular theory, 83–84, 104
pressure-temperature relations, 84–94
solutions with liquids, 200–201, 205–206
volumes, 130–132
Gasoline, 290–291
Gay-Lussac, Joseph, 93
Gay-Lussac's law, 93, **100**
Geiger, Hans, 144
General chemistry, 4
Gerlach, Walther, 185
Germanium, 196
Ghiorso, Albert, 171
Glucose, 292, **298**
Glutamate dehydrogenase, 297
Glycerol, 283
Glycine, 295
Gold, 143–144, 200

Gram atomic mass, **53**
Granite, 28
Greek prefixes, 72
Ground state, 153, **164**
Group, **179**
Guldberg, C. M., 265
Gypsum, 78

Hahn, Otto, 171
Hahnium, 145, 170–171
Half-cell, 252, 254–258, **259**
Half-life, 170
Half-reactions, 245–246, 250–251, 256–257, **259**
Halides, **179,** 282, 289
Halogens, 177–**179,** 246, 289
  see also organic chemistry
Heat, 18–19, **22**
  of condensation, 114–115, **116**
  of freezing, 114, 214–215, 223–225
  of fusion, 114, 116, 173
  of liquefaction, 114
  of melting, 114, 172–173
  of solidification, 114, **116**
  of vaporization, 114–115, 116, 173
  see also temperature, boiling point
Heavy ion linear accelerator (HILAC), 171
Helium, 159, 175
  -ions, 142–143
  see also alpha particles, noble gases
Henry, William, 206
Henry's law, 206, **215**
Heptane, 284
Heterogeneous matter, 28, **41,** 200
Hexane, 284
Histidine, 296
Homogeneous mixtures, 28–29, **41,** 200
Homologous series, 283–284, **298**
Hückel, Erich, 221
Hückel-Debye theory, **239**
Hund's rule, **164,** 187
Hybrid bonds, 187–188
Hydrates, 78, **79,** 124
Hydrocarbons, 282–283, **298**
  aliphatic, 283, **297**
  aromatic, 283, 288, 297, **298**
  combustion, 290–291
  saturated, 283, 289, **299**
  unsaturated, 283, 287–291, 297, **299**
Hydrochloric acid, 228, 235, 272
Hydrogen, 40, 45, 68, 77, 152, 159, 176, 201
  atomic structure, 162–163
  binary compounds of, 73–74
  emission spectrum, 154–156
  ternary compounds of, 75–76
  see also organic chemistry
Hydrogen bond, 194, **197**
Hydrogen bromide, 192–193
Hydrogen chloride, 185
Hydrogen electrode, 254–255, **259**
Hydrogen peroxide, 48
Hydrolysis, 229–230, **239**
Hydronium ion, 224, 231, 236, **239**
Hydroxide ion, 76, 176, 184, 224, 232
Hypothesis, 3, **4**

-ic compounds, 72
-ide compounds, 71, 76
Ideal gas, 95
  constant, 95
  equation of state, 96, **100**
Ideal solution, **215**
Immiscibility, 204, **215**
Indicators, 228, **239**
Indium, 172–173, 174
Inert gases, *see* noble gases
Inertia, **6**
Infrared radiation, 150
Inner transition elements, 170, **179**
Inorganic chemistry, 4
Insulator, **259**
Insulin, 297
International Union of Pure and Applied Chemistry (IUPAC), 285
Iodides, 205
Iodine, 168, 177–179
  see also halogens
Ion migration, 223, **239**
Ion product of water, 274, **278**
Ionic bonds, 184, 191–192, **197**
Ionic equilibrium, 274–275
Ionic reaction, **65**
Ionic solids, 195
Ionic solutions, 201, **215**
Ionization constant, 274, **278**
Ionization potential, 171–172, **179**
Ions, **53,** 298
  oxidation number of, 70
  polyatomic, 68, **79**
  formation, 158
  see also bonds, solutions
Iron, 72, 248–250, 257
Isoleucine, 296
Isomers, 285
Isotopes, 34–36, 39, **41,** 146, **147,** 168

k, *see* equilibrium constant
k shell, *see* energy (levels)
Kekulé, Friedrich A., 288
Kekulé structure of benzine, 288–289
Kelvin temperature scale, 91–92, **100**
Ketone, 282
Kilogram, 6, **22**
Kindling temperature, 268, **278**
Kinetic energy, 84, **100**
Kinetic-molecular theory
  of crystalline solids, 104–105
  of gases, 83–84
  of liquids, 105–107, 108, 109–113
  of solids, 107–109
Kossel, Walther, 183, 185
Krypton, 175
Kurchatovium, 171
  see also rutherfordium

l shell, *see* energy (levels)
Langmuir, Irving, 185
Lanthanides, 160, 170, **179**
Laplace, Pierre Simon de, 133
Lavoisier, Antoine Laurent, 133
Law
  Avogadro's, 94, 97, **99**
  Boyle's, 87, 93, **99**
  Charles', 89–93, **99**
  conservation of mass-energy, 49, 53
  Dalton's, 89, **99**
  definite composition, 45–47, 53
  Gay-Lussac's, 93, **100**
  Henry's, 206, **215**
  mass action, 265, 278
  multiple proportions, 47–48, **53**
  periodic, 39, **41,** 168, **179**
Lawrencium, 170
Lead, 34, 258–259
Lead storage battery, 258–259
LeChatelier, Henri, 272
LeChatelier's principle, 272, 275, **278**
Length units, conversion, 7–8
Leucine, 296
Lewis, Gilbert N., 183, 185
Lewis structures, 185–187, **197**
Light, 142
  as energy, **150**
  speed of, 53
Lignin, **298**
Limiting reagent, 128, **134**
Linen, 293
Liquids, 27, **41**
  kinetic molecular theory, 105–107, 108, 109–113
  solutions, 200–201, 205–206
Liter, 6, **22**
Lithium, 159–160, 169, 175–176
  halides of, 195
Litmus, 228, **239**
Lowry, Thomas Martin, 229
Lowry-Bronsted definitions, 229–230, **239**
Lye, 46, 76
Lysine, 296

m shell, *see* energy (levels)
Macromolecular chemistry, 53, 291–292
Magnesium chloride, 187
Magnetic quantum number, 156, 158, **164**
Malleability, 175, 196
Manometer, 86–87, **100,** 213
Marsden, E., 144
Mass, 6, **22**
  conservation, 49
  conversion, 11
  determination of, 20–22
  see also atomic mass
Mass action, law of, 265, **278**
Mass-mass calculations, 127–128
Mass number, 146, **147**
  of transuranium elements, 170
Mass spectrograph, 33–34, **41**
Matter, 2, 27–29, **41,** 49
Measurement, accuracy of, 13–16
Melting point, 114, **116,** 173–175, 297, inside back cover
Mendeleev, Dmitri Ivanovich, 39, 168
Mendelevium, 168, 170, 179
Meniscus, 106–107, **116**
Mercerized cotton, 293, **298**
Mercury, 40, 106–107
  barometer, 85–86
  spectrum, 152
  thermometer, 19
Metal, 40, **41,** 203
  alkali, 175–177, 179
  ion, 227
Metallic solids, 196–197
Meter, 6, **22**
Methane, 46, 187–188, 284, 298
Methanol, 283
Methionine, 296
Methyl group, 286, 297
Metric system, 6–7
  conversions, 7–11

Meyer, Lothar, 39, 168
Millikan, Robert Andrews, 139–140
Migration of ions, 223, **239**
Miscibility, 203, **215**
Mixture, **41,** 200
  see also solutions
Molality, 212–213, **215**
Molar gas volume, 97–98, **100**
Molarity, 206–211, **215,** 231–232
Mole, 36–37, **41,** 50–52, **53**
Mole factor, **134**
  calculations, 120–122
Mole-mass calculations, 125, 126
Mole-mole calculations, 125
Molecular formulas, 48, **53**
  determination of, 59–60
Molecular mass, 49, **53**
  determination of, 33–34
  of gases, 98–99
  of proteins, 297
Molecular motion
  in gases, 84
  in liquids, 105–106
Molecular orbitals, *see* orbitals
Molecular solids, 196
Molecular weight, *see* molecular mass
Molecule, 41, 297, 298
  polar, 192–193
  size of, 53
  see also kinet c-molecular theory, polymers
Moment, dipole, 193, **197**
Monatomic gas, 40, **41**
Monomer, 291–299, 297, **298**
Moseley, Henry Gwyn-Jeffreys, 144–145
Multiple proportions, law of, 47–48, **53**

Naming of compounds, *see* nomenclature
Natural gas, **134, 298**
Natural rubber, 53
Neon, 175, 183
Neptunium, 170
Net ionic equation, 226–227, **239**
Neutral solutions, *see* pH
Neutralization reaction, **239**
Neutron, 146
Nickel, 162, 222
Nitrates, 184, 189–190, 205, 223
Nitric acid, 228, 246–247
Nitrogen, 69, 70, 189, 194, 196
  see also organic chemistry
Nobellum, 170–171
Noble gases, 40, **41,** 175, **179,** 183–184, 185
Nomenclature
  additions and exceptions, 76
  of alkanes, 285–287
  of alkenes, 287–288
  of alkynes, 287–288
  of binary compounds, 69–74
  of energy transfer processes, 113
  of hydrocarbons, 283–285
  organic, 285–288
  of oxyacids and salts, 76–78
  of ternary compounds, 74–77
Noble gas configuration, 183–185
Nonane, 284
Nonelectrolytes, 201
Nucleus, 143–144, **147**
  structure, 145
Nylon, 294–295, **298**

Octane, 284–285
Octet rule, 183–184, **197**
Ohm, 254, **259**
Olefins, *see* alkenes
Open-chain structure, 285, **298**
Orbit, **164**
Orbital quantum number, 156–157, **164**
Orbitals, 157–**164,** 168, 187–188
Organic chemistry, 4, 282–297
-ous compounds, 72
Oxalate ion, 77
Oxalic acid, 77
Oxidation, 244–252, **259,** 290
Oxidation number, 68–78, **79,** 246–247
Oxidation potential, 254–257, **259**
Oxidation state, **79**
  see also oxidation number
Oxidation-reduction reactions, 244–252
Oxides, 205
Oxidizing agent, 246, **259**
Oxyacid, 76–77, **79**
Oxygen, 45, 51, 63, 68, 77, 175, 194, 268

Palladium, 201
Paraffins, *see* alkanes
Partial pressure, 89, **100**
Pauli, Wolfgang, 158
Pauli exclusion principle, 158, **164,** 185
Pauling, Linus, 190
Pentane, 284
Peptide bond, 297, **298**
Percent composition, 60–61
Percent by mass, **215**
Percent by volume, **216**
Percent yield, 129–130, **134**

Period, 168, **179**
Periodic groups, 168
Periodic law, 39, **41,** 168, **179**
Periodic relationships, 171–174, 257
Periodic table, 39–40, 160, 168–179, **180,** inside back cover
Permanganate ions, 222, 223, 248
Peroxides, **298**
Petroleum, 298
pH, 231–233, **239**
Phenolphthalein, 228, 236, **240**
Phenylalanine, 296
Phosphate, 184
Photocell, **180**
Photon, 151, **165**
Physical change, 27–28, **41**
Physical chemistry, 4
Physical properties, 27, **42**
Planck, Max, 151
Platinum, 254
Plutonium, 170
pOH, 231–233, **240**
Polar bond, 192–193, **197**
Polar molecule, 192–193, **197**
Polonium, 138, 144, 168
Polyamides, **298**
Polyatomic ions, 68, 74–77, **79,** 184
Polyesters, 294, **298**
Polyethene (polyethylene), 291–292
Polymers, **53,** 291–295, **298**
  protein as, 297
Polyprotic acid, 235, **240**
Potassium, 169, 176
  halides of, 195
  salts, 205
Potassium carbonate, 175
Potassium hydroxide, 228
Precipitate, **65**
  pressure, 84–91, **100**
    and boiling point, 115
    and solubility, 204–206
    see also temperature
  principle quantum number, 153, 157, **165**
  propane, 284
  propanol, 283
  properties, 39, **42**
    colligative, 213–214, **215,** 223–224
    prediction of, 173–174
    of matter, 27
    of solutions, 222–226
  propyl group, 286
Protein, 295–296, **298**
  polymer structure, 297
Proton, 141–142, 144, **147**
Pure substance, 28

Qualitative analysis, 4
Quanta, 151, 153–154, **165**
Quantitative analysis, 4
Quantum numbers
  azimuthal, 156–157, **164**
  magnetic, 156, 158, **164**
  orbital, 156–157, **164**
  principal, 153, 157, **165**
  spin, 156, 158, **164**
Quartz, 195

Radar, 150
Radiation, electromagnetic, **164**
Radio waves, 150–151
Radioactivity, 138, 142–143
Radium, 138, 173
Radon, 175
Raoult, Francois, Marie, 213
Rare earth elements, *see* lathanides
Rayon, 293, **299**
Reaction mechanism, **278**
Reaction rates, 264–273, **278**
Reaction velocity, **278**
  see also reaction rates
Reactions
  acid-base, 229, 234–236
  addition, 289, **297**
  combination, 63, **64**
  decomposition, **64**
  endothermic, 113, **116,** 132–133
  exothermic, 49, **53,** 133, 267
  half-cell, 252, 254–258
  instantaneous, 223
  oxidation-reduction, 244–252
  polymerization, 291–295
  replacement, 64, **65**
  see also half reactions
Redox terminology, 245–246
  see also oxidation, reduction
Reducing agent, 246, **259**
Reduction, 244–252, **259**
Replacement reaction, 64, **65**
Resonance, 189–190
Resonance hybrid, 190, **197**
Roentgen, Wilhelm Konrad, 138
Rubber, natural, 53
Rubidium, 169, 176
Rule of Hund, **164,** 187
Rutherford, Ernest, scattering experiment, 143–144

Rutherford-Bohr theory, *see* Bohr atomic theory
Rutherfordium, 170–171
Rydberg, Johannes, 151

Salts, 77, **79,** 204–205, 227
Saturated hydrocarbons, 283, 289, **299**
Saturated solution, 202, 203, **216**
Scattering of alpha particles, 143–144
Science, 3, **4**
Scientific method, 3, **4**
"Sea of electrons," 196
Second ionization potential, 172
  see also ionization potential
Selenium, 168
Shells, **165**
  see also energy (levels)
Shrödinger, Edwin, 156–157
Sidgwick, Nevil Vincent, 190
Significant figures, 14, **22**
Silicon, 35–36, 196
Silver, 185, 223, 257
Silver bromide, 178
Silver iodide, 178
Silver nitrate, 223
Smog, 178, 235
Sodium, 151, 169, 176
  halides, 195
  ion, 183–184, 223, 234, 236
  salts, 205
  see also alkali metals
Sodium carbonate, 46, 175
Sodium chloride, 105, 195, 201–202, 223–225
Sodium hydroxide, 46, 76, 228, 234, 236
Solids, 27, **42**
  amorphous, 105
  bonds in, 194–196
  crystalline, 104–105, 204–205
  molecular motion, 107–109
  in solution, 200–201
Solubility, 203–206
Solubility product, 276–277, **278**
Solubility product constants, 276
Solute, 200, **216**
Solutions, **42,** 200–215, **216,** 221–239
Solutions of electrolytes, *see* ionic solutions
Solvation, 201–202
Solvent, 200, **216**
Specific rate constant, 265, **278**
Spectator ion, 226, **240**
Spectrograph, 151–152
Spectroscope, 151, **165**
Spectrum, *see* atomic spectra
Spin quantum number, 156, 158, **165**
Standard solution, 236, **240**
Standard temperature and pressure, **100**
Stern, Otto, 185
Stock, Albert, 72
Stock system, 72
Stoichiometry, 119–132, **134,** 210–211
Storage battery, 258–259, **260**
STP, 87, **100**
Strengths of acids and bases, 230–236
Strong electrolyte, 221, **240**
Structural formula, 48
  see also structural isomerism
Structural isomerism, 49, **54,** 285, **299**
  see also isomers
Sublimation, 107–109, **116, 180,** 200
Substitution reactions, 64, **65,** 289, **299**
Sucrose, 46, 202–203, 213–214
Sugar, 46, 50, 196, 292, 298
Sulfate ion, 74, 173, 184, 205
Sulfides, 205
Sulfur, 69
  see also organic chemistry
Sulfur dioxide, 298
Sulfur trioxide, 60–61
Sulfuric acid, 119, 228
  contact process, 122–123
  and marble, 235
  properties of, 123–124
Surface compound, 268, **278**
Surface tension, 106, **116**
Symbols, chemical, 30–31, inside back cover

Table salt, 46, 71
Tellurium, 168
Temperature, **22,** 83, 273
  absolute, 91–92, **99, 100**
  Celsius, 19–20, **22,** 91–92
  conversions, 20
  Fahrenheit, 20, **91**
  of gases, 84–94
  Kelvin, 91–92, **100**
  kinetic energy and, 83–84, 91
  measurement of, 18–20
  and reaction rates, 266–268
  of solutions, 203–205
Ternary compound, 68, 74–77, **79**
Tetraethyl lead, 178
Thallium, 172–173, 174
Theory, 3, **4**
  Arrhenius, 222–227, **239**
  atomic, 151–156, 162, 183

Bohr, 151–156, 162, 183
Debye-Hückel, **239**
kinetic molecular, 83–84, 104–113
wave-mechanics, 155–157
Thermochemical equations, 132–133, **134**
Thermometer, 19, **100**
Thomson, Joseph John, 140
Threonine, 296
Tin, 35, 146, 196
Titration, 236–239, **240**
Torr, 86, **100**
Torricelli, Evangelista, 85, 86
Transition elements, 169–170, **180**
Transuranium elements, 170–171, **180**
Triple bond, **197**
Trytophan, 296

Ultraviolet radiation, 138, 150
Unsaturated hydrocarbons, 283, 287–291, 297, **299**
Uranium, 170–171, 178

Valine, 296
Vanadium pentoxide, 122
Vapor, **100**
Vapor pressure, 109–111, **116,** 213–214
Vinegar, *see* acetic acid
Viruses, 295
Viscose, 293, **299**
Volt, 254, **260**
Voltaic cells, 251–252, **260**
see also half-cells
Volume units, conversion, 10
Volumetric analysis, 236, **240**

Waage, Peter, 265
Washing soda, 46, 78
Water
as acid, 230
density, 6, 11
mass, 6
structure and properties, 27–28, 163, 193–194, 230
surface tension, 106–107
vapor pressure, 111
Water of crystallization, 78
Wave equations, **165**
Wave-theory of atomic structure, 155–157
see also orbitals
Wavelength, **165**
Weak electrolyte, 221, **240**
Weight, 6, **22**
Williamson, Alexander W., 269

x-rays, 138, 144–145, **147,** 150
Xanthate, 293, **299**
Xenon, 175

Zinc, 200, 203, 244, 251–252, 272
Zinc sulfide, 140, 143–144

Element	Symbol	Atomic Mass[a]	Atomic No.	Melting Point, ° C.	Boiling Point, ° C.	Density, Grams/M or Specific Gravity
Actinium	Ac	227.02	89	1050	3200	10.07
Aluminum	Al	26.98154	13	660.2	2467	2.70
Americium	Am	(243)	95	>850		11.7
Antimony	Sb	121.75	51	630.5	1380	6.69
Argon	Ar	39.948	18	−189.2	−185.7	0.00178
Arsenic	As	74.9216	33			
Yellow						1.97
Gray				817	613	5.73
Black						4.73
Astatine	At	(210)	85			
Barium	Ba	137.34	56	725	1140	3.5
Berkelium	Bk	(247)	97			
Beryllium	Be	9.0122	4	1278	2970	1.848
Bismuth	Bi	208.9804	83	271.3	1560	9.747
Boron	B	10.811	5	2300	2550	2.34
Bromine	Br	79.909	35	−7.2	58.78	3.12
Cadmium	Cd	112.40	48	320.9	765	8.65
Calcium	Ca	40.08	20	842 to 848	1487	1.55
Californium	Cf	(251)	98			
Carbon	C	12.01115	6	3550	4827	
Diamond						3.15 to 3.53
Graphite						1.9 to 2.3
Cerium	Ce	140.12	58	795	3468	6.67 to 8.23
Cesium	Cs	132.9054	55	28.5	690	1.873
Chlorine	Cl	35.453	17	−100.98	−34.6	0.003214
Chromium	Cr	51.996	24	1890	2482	7.18 to 7.20
Cobalt	Co	58.9332	27	1495	2900	8.9
Copper	Cu	63.54	29	1083.0	2595	8.96
Curium	Cm	(247)	96			~7
Dysprosium	Dy	162.50	66	1407	2600	8.536
Einsteinium	Es	(254)	99			
Erbium	Er	167.26	68	1497	2900	9.051
Europium	Eu	151.96	63	826	1439	5.259
Fermium	Fm	(253)	100			
Fluorine	F	18.99840	9	−219.62	−188.14	0.001696(0° C., 1 at
Francium	Fr	(223)	87			
Gadolinium	Gd	157.25	64	1312	3000	7.895, 7.8
Gallium	Ga	69.72	31	29.78	2403	5.907
Germanium	Ge	72.59	32	937.4	2830	5.323 (25° C.)
Gold	Au	196.967	79	1063.0	2966	19.32
Hahnium	Ha		105			
Hafnium	Hf	178.49	72	2150	5400	13.29
Helium	He	4.0026	2	<−272.2[c]	−268.6	0.000177
Holmium	Ho	164.9304	67	1461	2600	8.803
Hydrogen	H	1.0079	1	−259.14	−252.5	0.00008988
Indium	In	114.82	49	156.61	2000	7.31
Iodine	I	126.9044	53	113.5	184.35	4.93
Iridium	Ir	192.2	77	2410	4527	22.42 (17° C.)
Iron	Fe	55.847	26	1535	3000	7.874
Krypton	Kr	83.80	36	−156.6	−152.30	0.003733 (0° C.)
Lanthanum	La	138.91	57	920	3469	5.98 to 6.186
Lawrencium	Lr	(257)	103			
Lead	Pb	207.19	82	327.5	1744	11.35
Lithium	Li	6.939	3	179	1317	0.534
Lutetium	Lu	174.97	71	1652	3327	9.872
Magnesium	Mg	24.312	12	651	1107	1.738
Manganese	Mn	54.9380	25	1244	2097	7.21 to 7.44
Mendelevium	Md	(256)	101			
Mercury	Hg	200.59	80	−38.87	356.58	13.546
Molybdenum	Mo	95.94	42	2610	5560	10.22

(From "The Elements/Material for a Growing World of Knowledge," *Chemistry, 39, 20–22* [October, 1966] Copyright American Chemical Society.)

ement	Symbol	Atomic Mass[a]	Atomic No.	Melting Point, ° C.	Boiling Point, ° C.	Density, Grams/Ml., or Specific Gravity[b]
odymium	Nd	144.24	60	1024	3027	6.80, 7.004
on	Ne	20.183	10	−248.67	−245.92	0.0008999
ptunium	Np	(237)	93	640		18.0 to 20.45
ckel	Ni	58.71	28	1453	2732	8.902 (25° C.)
obium	Nb	92.906	41	2468	4927	8.57
trogen	N	14.0067	7	−209.86	−195.8	0.0012506
belium	No	(254)	102			
smium	Os	190.2	76	3000	5000	22.57
ygen	O	15.9994	8	−218.4	−183.0	0.001429 (0° C.)
lladium	Pd	106.4	46	1552	2927	12.02
osphorus	P	30.97376	15			
White				44.1	280	1.82
Red						2.20
Black						2.25 to 2.69
atinum	Pt	195.09	78	1769	3827	21.45
utonium	Pu	(239)	94	639.5	3235	19.84 (25° C.)
lonium	Po	(210)	84	254	962	9.32
tassium	K	39.098	19	63.65	774	0.862
aseodymium	Pr	140.907	59	935	3127	6.782, 6.64
omethium	Pm	(147)	61	1035	2730	
otactinium	Pa	(231)	91	1230		15.37
dium	Ra	(226)	88	700	<1737	5
don	Rn	(222)	86	−71	−61.8	0.00973
enium	Re	186.2	75	3180	5627	21.02
odium	Rh	102.905	45	1966	3727	12.41
bidium	Rb	85.47	37	38.89	688	1.532
thenium	Ru	101.07	44	2250	3900	12.41
therfordium	Rh		104			
marium	Sm	150.35	62	1072	1900	7,536, 7.40
andium	Sc	44.956	21	1539	2727	2.992
lenium	Se	78.96	34			
Gray				217	684.9	4.79
Vitreous						4.28
icon	Si	28.086	14	1410	2355	2.33 (25° C.)
ver	Ag	107.870	47	960.8	2212	10.50
dium	Na	22.98977	11	97.81	892	0.971
ontium	Sr	87.62	38	769	1384	2.54
lfur	S	32.064	16		444.6	
Rhombic				112.8		2.07
Monoclinic				119.0		1.957
ntalum	Ta	180.948	73	2996	5425	16.6
chnetium	Tc	(99)	43	2200		11.50
llurium	Te	127.60	52	449.5	989.8	6.24
rbium	Tb	158.924	65	1356	2800	8.272
allium	Tl	204.37	81	303.5	1457	11.85
orium	Th	232.038	90	1700	4000	11.66
ulium	Tm	168.934	69	1545	1727	9.332
	Sn	118.69	50	231.89	2270	
Gray						5.75
White						7.31
anium	Ti	47.90	22	1675	3260	4.54
ngsten	W	183.85	74	3410	5927	19.3
anium	U	238.03	92	1132.3	3818	18.95
nadium	V	50.942	23	1890	3000	6.11 (18.7° C.)
non	Xe	131.30	54	−111.9	−107.1	0.005887
erbium	Yb	173.04	70	824	1427	6.977, 6.54
rium	Y	88.905	39	1495	2927	4.45
c	Zn	65.38	30	419.4	907	7.133 (25° C.)
conium	Zr	91.22	40	1852	3578	6.53

asses in parentheses are those of the most stable or most common isotopes.
20° C. and standard pressure unless otherwise noted.
26 atm.

For detailed list
of the elements,
turn back one page.